THE MECHANICS' HANDBOOK

A CONVENIENT REFERENCE BOOK

FOR ALL PERSONS INTERESTED IN

Mechanical Engineering, Steam Engineering, Electrical Engineering, Railroad Engineering, Hydraulic Engineering, Bridge Engineering, Etc.

BY

INTERNATIONAL CORRESPONDENCE SCHOOLS

SCRANTON, PA.

7th Edition, 389th Thousand, 27th Impression

SCRANTON, PA.

INTERNATIONAL TEXTBOOK COMPANY

Press of
International Textbook Company
Scranton, Pa.

28635

PREFACE.

The first edition (2,000 copies) of the handbook of which this is the outcome was issued in October, 1893, in the form of a notebook containing 74 printed pages, with about the same number of blank pages for memoranda, under the title of Mechanics' Pocket Memoranda. The little book proved so popular that a new edition (10,000 copies) enlarged to 110 pages was issued 8 months later. In June, 1897, the blank pages were discarded, the work was entirely recast and enlarged to 318 pages, and the edition (third) consisted of 25,000 copies. Before printing the fifth edition (March, 1898), a large amount of matter relating especially to Plumbing, Heating, and Ventilation and the Building Trades was taken- out, replaced by tables of logarithms, trigonometric functions, etc., together with directions for using them, and other new matter, the result being to confine the work more particularly to the different branches of engineering and mechanics.

It has been the aim of the publishers, from the first, to present to the public a handbook of a size convenient to carry in the coat or hip pocket—a pocketbook in reality—which would contain rules, formulas, tables, etc. in most common use by engineers, together with explanations concerning them and practical examples illustrating their use. We have not endeavored to produce a condensed cyclopedia of engineering or of any branch of it, but we have striven to anticipate the daily wants of the user and to give him the information sought in the manner best suited to his needs. Our aim has been to meet the necessities not only of the engineer but of all in any manner interested in engineering, and in accomplishing this we have selected that rule, formula, or process which was, in our opinion, best adapted to the circumstances of the case, describing it fully, giving full directions how and when to use it, and not mentioning other methods (when such were available); in other words, *we* have made the selection instead of leaving the choice to the judgment of the user, which is frequently at fault. The exceedingly large sale proves that the idea was popular and has vindicated our judgment. We hope that succeeding editions will meet and merit the same approval that has been accorded those preceding.

The present (seventh) edition contains the most convenient table of powers, roots, and reciprocals of numbers yet printed. This table was arranged and computed by us and will be of great use to all having occasion to use it.

INTERNATIONAL CORRESPONDENCE SCHOOLS.

December 1, 1903.

INDEX.

A.

B.

M.

2

W.

THE MECHANICS' HANDBOOK.

USEFUL TABLES.

WEIGHTS AND MEASURES.

LINEAR MEASURE.

12	inches (in.)	= 1 foot ....ft.
3	feet	= 1 yard.....................yd.
5.5	yards	= 1 rod ....rd.
40	rods	= 1 furlong ....fur.
8	furlongs	= 1 mile ....mi.

in.	ft.	yd.	rd.	fur.	mi.
36 =	3 =	1			
198 =	16.5 =	5.5 =	1		
7,920 =	660 =	220 =	40 =	1	
63,360 =	5,280 =	1,760 =	320 =	8 =	1

SURVEYOR'S MEASURE.

7.92	inches	= 1 link ....li.
25	links	= 1 rod.....................rd.
4 rods / 100 links / 66 feet		= 1 chain ....ch.
80	chains	= 1 mile.....................mi.

1 mi. = 80 ch. = 320 rd. = 8,000 li. = 63,360 in.

SQUARE MEASURE.

144 square inches (sq. in.)...... = 1 square foot ....sq. ft.
  9 square feet.................... = 1 square yard ....sq. yd.
30¼ square yards.................... = 1 square rod ....sq. rd.
160 square rods .... = 1 acre ....A.
640 acres..................... = 1 square mile ....sq. mi.

sq. mi.	A.	sq. rd.	sq. yd.	sq. ft.	sq. in.
1	= 640	= 102,400	= 3,097,600	= 27,878,400	= 4,014,489,600

SURVEYOR'S SQUARE MEASURE.

625 square links (sq. li.) .... = 1 square rod ....sq. rd.
 16 square rods .... = 1 square chain ......sq. ch.
 10 square chains.................... = 1 acre ....A.
640 acres .... = 1 square mile ....sq. mi.
 36 square miles (6 mi. square) = 1 township ....Tp.

1 sq. mi. = 640 A. = 6,400 sq. ch. = 102,400 sq. rd.
= 64,000,000 sq. li.

The acre contains 4,840 sq. yd., or 43,560 sq. ft., and is equal to the area of a square measuring 208.71 ft. on a side.

CUBIC MEASURE.

1,728 cubic inches (cu. in.)...... = 1 cubic foot .... cu. ft.
   27 cubic feet .... = 1 cubic yard...........cu. yd.
  128 cubic feet .... = 1 cord...................cd.
 24¾ cubic feet .... = 1 perch ....P.

1 cu. yd. = 27 cu. ft. = 46,656 cu. in.

MEASURE OF ANGLES OR ARCS.

 60 seconds (″) .... = 1 minute ....′
 60 minutes .... = 1 degree ....°
 90 degrees.................... = 1 rt. angle or quadrant □
360 degrees.................... = 1 circle ....cir.

1 cir. = 360° = 21,600′ = 1,296,000″

AVOIRDUPOIS WEIGHT.

437.5 grains (gr.) .... = 1 ounce ....oz.
16 ounces .... = 1 pound ....lb.
100 pounds .... = 1 hundredweight ......cwt.
20 cwt., or 2,000 lb................. = 1 ton..............................T.
    1 T. = 20 cwt. = 2,000 lb. = 32,000 oz. = 14,000,000 gr.
The avoirdupois pound contains 7,000 grains.

LONG TON TABLE.

16 ounces .... = 1 pound ....lb.
112 pounds................................. = 1 hundredweight ......cwt.
20 cwt., or 2,240 lb. .... = 1 ton................................T.

TROY WEIGHT.

24 grains (gr.) .... = 1 pennyweight ....pwt.
.20 pennyweights........................ = 1 ounce ....oz.
12 ounces................................. = 1 pound ....lb.
    1 lb. = 12 oz. = 240 pwt. = 5,760 gr.

DRY MEASURE.

2 pints (pt.) .... = 1 quart ....qt.
8 quarts .... = 1 peck ....pk.
4 pecks................................... = 1 bushel ....bu.
    1 bu. = 4 pk. = 32 qt.= 64 pt.

The U. S. struck bushel contains 2,150.42 cu. in. = 1.2444 cu. ft. By law, its dimensions are those of a cylinder 18¼ in. in diameter and 8 in. deep. The heaped bushel is equal to 1¼ struck bushels, the cone being 6 in. high. The dry gallon contains 268.8 cu. in., being ⅛ of a struck bushel.

For approximations, the bushel may be taken at 1¼ cu. ft.; or a cubic foot may be considered ⅘ of a bushel.

The British bushel contains 2,218.19 cu. in. = 1.2837 cu. ft. = 1.032 U. S. bushels.

LIQUID MEASURE.

4 gills (gi.)	= 1 pint	pt.
2 pints	= 1 quart	qt.
4 quarts	= 1 gallon	gal.
31½ gallons	= 1 barrel	bbl.
2 barrels, or 63 gallons	= 1 hogshead	hhd.

1 hhd. = 2 bbl. = 63 gal. = 252 qt. = 504 pt. = 2,016 gi.

The U. S. gallon contains 231 cu. in. = .134 cu. ft., nearly; or 1 cu. ft. contains 7.481 gal. The following cylinders contain the given measures very closely:

	Diam.	*Height.*		*Diam.*	*Height.*
Gill	1¾ in.	3 in.	Gallon	7 in.	6 in.
Pint	3½ in.	3 in.	8 gallons	14 in.	12 in.
Quart	3½ in.	6 in.	10 gallons	14 in.	15 in.

When water is at its maximum density, 1 cu. ft. weighs 62.425 lb. and 1 gallon weighs 8.345 lb.

For approximations, 1 cu. ft. of water is considered equal to 7½ gal., and 1 gal. as weighing 8⅓ lb.

The British imperial gallon, both liquid and dry, contains 277.274 cu. in. = .16046 cu. ft., and is equivalent to the volume of 10 lb. of pure water at 62° F. To reduce British to U. S. liquid gallons, multiply by 1.2. Conversely, to convert U. S. into British liquid gallons, divide by 1.2; or, increase the number of gallons ⅙.

MISCELLANEOUS TABLE.

12 articles	=	1 dozen.	20 quires	=	1 ream.
12 dozen	=	1 gross.	1 league	=	3 miles.
12 gross	=	1 great gross.	1 fathom	=	6 feet.
2 articles	=	1 pair.	1 hand	=	4 inches.
20 articles	=	1 score.	1 palm	=	3 inches.
24 sheets	=	1 quire.	1 span	=	9 inches.

1 sea mile (U. S.) = 6,080 ft. = 1⅛ statute miles (roughly).

1 meter = 3 feet 3⅜ inches (nearly).

THE METRIC SYSTEM.

The metric system is based on the meter, which, according to the U. S. Coast and Geodetic Survey Report of 1884, is equal to 39.370432 inches. The value commonly used is 39.37 inches, and is authorized by the U. S. government. The meter is defined as one ten-millionth the distance from the pole to the equator, measured on a meridian passing near Paris.

There are three principal units—the meter, the liter (pronounced lee-ter), and the gram, the units of length, capacity, and weight, respectively. Multiples of these units are obtained by prefixing to the names of the principal units the Greek words deca (10), hecto (100), and kilo (1,000); the submultiples, or divisions, are obtained by prefixing the Latin words deci ($\frac{1}{10}$), centi ($\frac{1}{100}$), and milli ($\frac{1}{1000}$). These prefixes form the key to the entire system. In the following tables, the abbreviations of the principal units of these submultiples begin with a small letter, while those of the multiples begin with a capital letter; they should always be written as here printed.

MEASURES OF LENGTH.

10 millimeters (mm.)	= 1 centimeter	cm.
10 centimeters	= 1 decimeter	dm.
10 decimeters	= 1 meter	m.
10 meters	= 1 decameter	Dm.
10 decameters	= 1 hectometer	Hm.
10 hectometers	= 1 kilometer	Km.

MEASURES OF SURFACE (NOT LAND).

100 square millimeters (mm².)	= 1 square centimeter	cm².
100 square centimeters	= 1 square decimeter	dm².
100 square decimeters	= 1 square meter	m².

MEASURES OF VOLUME.

1,000 cubic millimeters (mm³.)	= 1 cubic centimeter	cm³.
1,000 cubic centimeters	= 1 cubic decimeter	dm³.
1,000 cubic decimeters	= 1 cubic meter	m³.

MEASURES OF CAPACITY.

10 milliliters (ml.)	= 1 centiliter	cl.
10 centiliters	= 1 deciliter	dl.
10 deciliters	= 1 liter	l.
10 liters	= 1 decaliter	Dl.
10 decaliters	= 1 hectoliters	Hl.
10 hectoliters	= 1 kiloliters	Kl.

NOTE.—The liter is equal to the volume that is occupied by 1 cubic decimeter.

MEASURES OF WEIGHT.

10 milligrams (mg.)	= 1 centigram	cg.
10 centigrams	= 1 decigram	dg.
10 decigrams	= 1 gram	g.
10 grams	= 1 decagram	Dg.
10 decagrams	= 1 hectogram	Hg.
10 hectograms	= 1 kilogram	Kg.
1,000 kilograms	= 1 ton	T.

NOTE.—The gram is the weight of 1 cubic centimeter of pure distilled water at a temperature of 39.2° F.; the kilogram is the weight of 1 liter of water; the ton is the weight of 1 cubic meter of water.

TEMPERING OF STEEL.

The following colors may be made use of in tempering steel-cutting tools:

	Corresponding Temperature F.
Lancets	Pale yellow .... 430°
Razors	Straw yellow .... 450°
All kinds of wood-cutting tools	Darker straw yellow ...... 470°
Screw taps	Yellow .... 490°
Chipping chisels, hatchets, and saws	Brown yellow .... 500°
All kinds of percussive tools	Brown (slightly tinged purple) .... 520°
	Light purple .... 530°
Springs	Clear black .... 570°
	Dark blue .... 600°

CONVERSION TABLES.

By means of the tables on pages 8 and 9, metric measures can be converted into English, and *vice versa*, by simple addition. All the figures of the values given are not required, four or five digits being all that are commonly used; it is only in very exact calculations that all the digits are necessary. Using table, proceed as follows: Change 6,471.8 feet into meters. Any number, as 6,471.8, may be regarded as 6,000 + 400 + 70 + 1 + .8; also, 6,000 = 1,000 × 6; 400 = 100 × 4, etc. Hence, looking in the left-hand column of the upper table, page 8, for figure 6 (the first figure of the given number), we find opposite it in the third column, which is headed "Feet to Meters," the number 1.8287838. Now, using but five digits and increasing the fifth digit by 1 (since the next is greater than 5), we get 1.8288. In other words, 6 feet = 1.8288 meters; hence, 6,000 feet = 1,000 × 1.8288 = 1,828.8, simply moving the decimal point three places to the right. Likewise, 400 feet = 121.92 meters; 70 feet = 21.336 meters; 1 foot = .3048 meter, and .8 foot = .2438 meter. Adding as shown above, we get 1,972.6046 meters.

$$\begin{array}{r} 1,828.8 \\ 121.92 \\ 21.336 \\ .3048 \\ .2438 \\ \hline 1,972.6046 \end{array}$$

Again, convert 19.635 kilos into pounds. The work should be perfectly clear from the explanation given above. The result is 43.2875 pounds.

$$\begin{array}{r} 22.046 \\ 19.8416 \\ 1.3228 \\ .0661 \\ .0110 \\ \hline 43.2875 \end{array}$$

The only difficulty in applying these tables lies in locating the decimal point; it may always be found thus: If the figure considered lies to the left of the decimal point, count each figure in order, beginning with units (but calling unit's place zero), until the desired figure is reached, then move the decimal point to the *right* as many places as the figure being considered is to the left of the unit figure. Thus, in the first case above, 6 lies three places to the left of 1, which is in unit's place; hence, the decimal point is moved three places to the *right*. By exchanging the words "right" and "left," the statement will also apply to decimals. Thus, in the second case above, the 5 lies three places to the *right* of unit's place; hence, the decimal point in the number taken from the table is moved three places to the *left*.

CONVERSION TABLE—ENGLISH MEASURES INTO METRIC.

English.	Metric. Inches to Meters.	Metric. Feet to Meters.	Metric. Pounds to Kilos.	Metric. Gallons to Liters.
1	.0253998	.3047973	.4535925	8.7853122
2	.0507996	.6095946	.9071850	7.5706244
8	.0761993	.9143919	1.3607775	11.3559366
4	.1015991	1.2191892	1.8143700	15.1412488
5	.1269989	1.5239865	2.2679625	18.9265610
6	.1523987	1.8287838	2.7215550	22.7118732
7	.1777984	2.1335811	3.1751475	26.4971854
8	.2031982	2.4383784	3.6287400	30.2824976
9	.2285980	2.7431757	4.0823325	34.0678098
10	.2539978	8.0479730	4.5359250	37.8531220

CONVERSION TABLE—ENGLISH MEASURES INTO METRIC.

English.	Metric. Square Inches to Square Meters.	Metric. Square Feet to Square Meters.	Metric. Cubic Feet to Cubic Meters.	Metric. Pounds per Square Inch to Kilo per Square Meter.
1	.000645150	.092901394	.028316094	703.08241
2	.001290300	.185802788	.056632188	1,406.16482
3	.001935450	.278704182	.084948282	2,109.24723
4	.002580600	.371605576	.113264376	2,812.32964
5	.003225750	.464506970	.141580470	3,515.41205
6	.003870900	.557408364	.169896564	4,218.49446
7	.004516050	.650309758	.198212658	4,921.57687
8	.005161200	.743211152	.226528752	5,624.65928
9	.005806350	.836112546	.254844846	6,327.74169
10	.006451500	.929013940	.283160940	7,030.82410

CONVERSION TABLE—METRIC MEASURES INTO ENGLISH.

Metric.	English. Meters to Inches.	English. Meters to Feet.	English. Kilos to Pounds.	English. Liters to Gallons.
1	89.370432	3.2808693	2.2046223	.2641790
2	78.740864	6.5617386	4.4092447	.5283580
8	118.111296	9.8426079	6.6138670	.7925371
4	157.481728	13.1234772	8.8184894	1.0567161
5	196.852160	16.4043465	11.0231117	1.3208951
6	236.222592	19.6852158	13.2277340	1.5850741
7	275.593024	22.9660851	15.4323564	1.8492531
8	314.963456	26.2469544	17.6369787	2.1134322
9	354.333888	29.5278237	19.8416011	2.3776112
10	393.704320	32.8086930	22.0462234	2.6417902

CONVERSION TABLE—METRIC MEASURES INTO ENGLISH.

Metric.	English. Square Meters to Square Inches.	English. Square Meters to Square Feet.	English. Cubic Meters to Cubic Feet.	English. Kilos per Square Meter to Pounds per Square Inch.
1	1,550.03092	10.7641034	35.3156163	.001422310
2	3,100.06184	21.5282068	70.6312326	.002844620
3	4,650.09276	32.2923102	105.9468489	.004266930
4	6,200.12368	43.0564136	141.2624652	.005689240
5	7,750.15460	53.8205170	176.5780815	.007111550
6	9,300.18552	64.5846204	211.8936978	.008533860
7	10,850.21644	75.3487238	247.2093141	.009956170
8	12,400.24736	86.1128272	282.5249304	.011378480
9	13,950.27828	96.8769306	317.8405467	.012800790
10	15,500.30920	107.6410340	353.1561630	.014223100

SPECIFIC GRAVITY.

The specific gravity of a body is the ratio between its weight and the weight of a like volume of distilled water at a temperature of 39.2° F. For gases, air is taken as the unit. One cubic foot of water at 39.2° F. weighs 62.425 pounds.

Name of Substance.	Specific Gravity.	Weight per Cu. In. Pounds.
METALS.		
Platinum, rolled	22.009	.819
Platinum, wire	21.042	.760
Platinum, hammered	20.337	.735
Gold, hammered	19.361	.699
Gold, pure cast	19.258	.696
Gold, 22 carats fine	17.486	.682
Mercury, solid at − 40° F.	15.632	.565
Mercury, at + 32° F.	13.619	.492
Mercury, at 60° F.	13.580	.491
Mercury, at 212° F.	13.375	.483
Lead, pure	11.330	.409
Lead, hammered	11.388	.411
Silver, hammered	10.511	.380
Silver, pure	10.474	.378
Bismuth	9.746	.352
Copper, wire and rolled	8.878	.321
Copper, pure	8.788	.317
Bronze, gun metal	8.500	.307
Brass, common	8.500	.307
Steel, cast steel	7.919	.286
Steel, common soft	7.833	.283
Steel, hardened and tempered	7.818	.282
Iron, pure	7.768	.281
Iron, wrought and rolled	7.780	.281
Iron, hammered	7.789	.281
Iron, cast	7.207	.260
Tin, from Böhmen	7.312	.264
Tin, English	7.201	.263
Zinc, rolled	7.101	.260
Antimony	6.712	.242
Aluminum	2.660	.096
STONES AND EARTHS.		
Emery	4.000	.145
Limestone	2.700	.098
Asbestos, starry	3.073	.111

TABLE—(*Continued*).

Name of Substance.	Specific Gravity.	Weight per Cu. In. Pounds.
Glass, flint	3.500	.1260
Glass, white	2.900	.1050
Glass, bottle	2.732	.0987
Glass, green	2.642	.0954
Marble, Parian	2.838	.1025
Marble, African	2.708	.0978
Marble, Egyptian	2.668	.0964
Mica	2.800	.1012
Chalk	2.784	.1006
Coral, red	2.700	.0975
Granite, Susquehanna	2.704	.0977
Granite, Quincy	2.652	.0958
Granite, Patapsco	2.640	.0954
Granite, Scotch	2.625	.0948
Marble, white Italian	2.708	.0978
Marble, common	2.686	.0970
Talc, block	2.900	.0105
Quartz	2.660	.0961
Slate	2.800	.1012
Pearl, oriental	2.650	.0957
Shale	2.600	.0939
Flint, white	2.594	.0937
Flint, black	2.582	.0933
Stone, common	2.520	.0910
Stone, Bristol	2.510	.0907
Stone, mill	2.484	.0897
Stone, paving	2.416	.0873
Gypsum, opaque	2.168	.0783
Grindstone	2.143	.0774
Salt, common	2.130	.0769
Saltpeter	2.090	.0755
Sulphur, native	2.033	.0734
Common soil	1.984	.0717
Rotten stone	1.981	.0716
Clay	1.900	.0686
Brick	2.000	.0723
Niter	1.900	.0686
Plaster Paris {	1.872	.0676
	2.473	.0893
Ivory	1.822	.0659
Sand	2.650	.0957
Phosphorus	1.770	.0639
Borax	1.714	.0619
Coal, anthracite {	1.640	.0592
	1.436	.0519

TABLE—(*Continued*).

Name of Substance.	Specific Gravity.	Weight per Cu. In. Pounds.
Coal, Maryland	1.355	.0490
Coal, Scotch	1.300	.0470
Coal, Newcastle	1.270	.0459
Coal, bituminous	1.350	.0488
Earth, loose	1.360	.0491
Lime, quick	1.500	.0542
Charcoal	.441	.0159
Woods (Dry).		
Alder	.800	.0289
Apple tree	.793	.0287
Ash, the trunk	.845	.0305
Bay tree	.822	.0297
Beech	.852	.0308
Box, French	.960	.0347
Box, Dutch	1.328	.0480
Box, Brazilian red	1.031	.0372
Cedar, wild	.596	.0215
Cedar, Palestine	.613	.0221
Cedar, American	.561	.0203
Cherry tree	.672	.0243
Cork	.250	.0090
Ebony, American	1.220	.0441
Elder tree	.695	.0251
Elm	.560	.0202
Filbert tree	.600	.0217
Fir, male	.550	.0199
Fir, female	.498	.0180
Hazel	.600	.0217
Lemon tree	.703	.0254
Lignum-vitæ	1.330	.0481
Linden tree	.604	.0218
Logwood	.913	.0330
Mahogany, Honduras	.560	.0202
Maple	.790	.0285
Mulberry	.897	.0324
Oak	.950	.0343
Orange tree	.705	.0255
Pear tree	.661	.0239
Poplar	.383	.0138
Poplar, white Spanish	.529	.0191
Sassafras	.482	.0174
Spruce	.500	.0181
Spruce, old	.460	.0166

TABLE—(*Continued*).

Name of Substance.	Specific Gravity.	Weight per Cu. In. Pounds.
Pine, southern	.720	.0260
Pine, white	.400	.0144
Walnut	.610	.0220
LIQUIDS.		
Acid, acetic	1.062	.0384
Acid, nitric	1.217	.0440
Acid, sulphuric	1.841	.0665
Acid, muriatic	1.200	.0434
Acid, phosphoric	1.558	.0563
Alcohol, commercial	.833	.0301
Alcohol, pure	.792	.0286
Beer, lager	1.034	.0374
Champagne	.997	.0360
Cider	1.018	.0368
Ether, sulphuric	.739	.0267
Egg	1.090	.0394
Honey	1.450	.0524
Human blood	1.054	.0381
Milk	1.032	.0373
Oil, linseed	.940	.0340
Oil, olive	.915	.0331
Oil, turpentine	.870	.0314
Oil, whale	.932	.0337
Proof spirit	.925	.0334
Vinegar	1.080	.0390
Water, distilled (62.425 lb. per cu. ft.)	1.000	.0361
Water, sea	1.030	.0372
Wine	.992	.0358
MISCELLANEOUS.		
Beeswax	.965	.0349
Butter	.942	.0340
India rubber	.933	.0337
Fat	.923	.0333
Gunpowder, loose	.900	.0325
Gunpowder, shaken	1.000	.0361
Gum arabic	1.452	.0525
Lard	.947	.0342
Spermaceti	.943	.0341
Sugar	1.605	.0580
Tallow, sheep	.924	.0334
Tallow, calf	.984	.0337
Tallow, ox	.923	.0333
Atmospheric air	.0012	

TABLE—(*Continued*).

Name of Substance.	Specific Gravity.	Weight per Cu. Ft. Grains.
GASES AND VAPORS.		
At 32° and a tension of 1 atmosphere.		
Atmospheric air	1.0000	565.11
Ammonia gas	.5894	333.1
Carbonic acid	1.5201	859.0
Carbonic oxide	.9673	546.6
Light carbureted hydrogen	.5527	312.3
Chlorine	2.4502	1,384.6
Olefiant gas	.9672	546.6
Hydrogen	.0692	39.1
Oxygen	1.1056	624.8
Sulphureted hydrogen	1.1747	663.8
Nitrogen	.9713	548.9
Vapor of alcohol	1.5890	898.0
Vapor of turpentine spirits	4.6978	2,654.8
Vapor of water	.6219	351.4
Smoke of bituminous coal	.1020	57.6
Smoke of wood	.9000	508.6
Steam at 212° F.	.4880	275.8

The weight of a cubic foot of any solid or liquid is found by multiplying its specific gravity by 62.425 lb. avoirdupois. The weight of a cubic foot of any gas at atmospheric pressure and at 32° F. is found by multiplying its specific gravity by .08073 lb. avoirdupois.

WROUGHT-IRON CHAIN CABLES.

The strength of a chain link is less than twice that of a straight bar of a sectional area equal to that of one side of the link. A weld exists at one end and a bend at the other, each requiring at least one heat, which produces a decrease in the strength. The report of the committee of the U. S. Testing Board, on tests of wrought-iron and chain cables, contains the following conclusions:

"That beyond doubt, when made of American bar iron, with cast-iron studs, the studded link is inferior in strength to the unstudded one.

" That, when proper care is exercised in the selection of material, a variation of 5% to 17% of the strongest may be expected in the resistance of cables. Without this care the variation may rise to 25%.

" That with proper material and construction the ultimate resistance of the chain may be expected to vary from 155% to 170% of that of the bar used in making the links, and show an average of about 163%.

" That the proof test of a chain cable should be about 50% of the ultimate resistance of the weakest link."

From a great number of tests of bars and unfinished cables, the committee considered that the average ultimate resistance and proof tests of chain cables made of the bars, whose diameters are given, should be such as are shown in the accompanying table.

ULTIMATE RESISTANCE AND PROOF TESTS OF CHAIN CABLES.

Diam. of Bar. Inches.	Average Resist. = 163% of Bar. Pounds.	Proof Test. Pounds.	Diam. of Bar. Inches.	Average Resist. = 163% of Bar. Pounds.	Proof Test. Pounds.
1	71,172	33,840	1 1/16	162,283	77,159
1 1/16	79,544	37,820	1 5/8	174,475	82,956
1 1/8	88,445	42,053	1 11/16	187,075	88,947
1 3/16	97,731	46,468	1 3/4	200,074	95,128
1 1/4	107,440	51,084	1 13/16	213,475	101,499
1 5/16	117,577	55,903	1 7/8	227,271	108,058
1 3/8	128,129	60,920	1 15/16	241,463	114,806
1 7/16	139,103	66,138	2	256,040	121,737
1 1/2	150,485	71,550			

TYPE METALS.

Name.	Proportions.
Smallest type	3 L, 1 A
Small type	4 L, 1 A
Medium type	5 L, 1 A
Large type	6 L, 1 A
Largest type	7 L, 1 A

In the above table, L represents the lead, and A the antimony in the alloy.

3

TABLE OF ELEMENTS.

	Symbol.	Atomic Weight.*
Aluminum	Al	27.04
Antimony (stibium)	Sb	119.96
Arsenic	As	74.9
Barium	Ba	136.9
Beryllium	Be	9.08
Bismuth	Bi	207.5
Boron	B	10.9
Bromine	Br	79.76
Cadmium	Cd	111.7
Cæsium	Cs	133.0
Calcium	Ca	39.91
Carbon	C	11.97
Cerium	Ce	141.2
Chlorine	Cl	35.37
Chromium	Cr	52.45
Cobalt	Co	58.6
Columbium	Cb	93.7
Copper (cuprum)	Cu	63.18
Didymium	D	147.0
Erbium	E	169.0
Fluorine	F	19.06
Gallium	G	69.8
Germanium	Ge	72.32
Gold (aurum)	Au	196.2
Hydrogen	H	1.0
Indium	In	113.4
Iodine	I	126.54
Iridium	Ir	196.7
Iron (ferrum)	Fe	55.88
Lanthanum	La	139.0
Lead (plumbum)	Pb	206.39
Lithium	Li	7.01
Magnesium	Mg	23.94
Mercury (hydrargyrum)	Hg	199.8
Manganese	Mn	54.8
Molybdenum	Mo	95.6
Nickel	Ni	58.6
Niobium	Nb	94.0
Nitrogen	N	14.01
Osmium	Os	198.6
Oxygen	O	15.96

*Principally from the 16th edition *Des Ingenieurs Taschen- buch*. The names of the non-metals are printed in heavy type.

TABLE—(*Continued*).

	Symbol.	Atomic Weight.
Palladium	*Pd*	106.2
Phosphorus	*P*	30.96
Platinum	*Pt*	194.43
Potassium (kalium)	*K*	39.04
Rhodium	*Rh*	104.1
Rubidium	*Rb*	85.2
Ruthenium	*Ru*	103.5
Scandium	*Sc*	44.04
Selenium	*Se*	78.00
Silicon	*Si*	28.00
Silver (argentum)	*Ag*	107.66
Sodium (natrium)	*Na*	23.0
Strontium	*Sr*	87.3
Sulphur	*S*	31.98
Tantalum	*Ta*	182.0
Tellurium	*Te*	128.0
Thallium	*Tl*	203.6
Thorium	*Th*	231.5
Tin (stannum)	*Sn*	117.35
Titanium	*Ti*	48.0
Tungsten (wolfram)	*W*	183.6
Uranium	*U*	240.0
Vanadium	*V*	51.2
Ytterbium	*Yb*	93.0
Yttrium	*Y*	172.6
Zinc	*Zn*	64.88
Zirconium	*Zr*	90.0

TABLE OF SPECIFIC HEATS.

SOLIDS.

Copper	.0951	Cast iron	.1298
Gold	.0824	Lead	.0314
Wrought iron	.1138	Platinum	.0824
Steel (soft)	.1165	Silver	.0570
Steel (hard)	.1175	Tin	.0562
Zinc	.0956	Ice	.5040
Brass	.0939	Sulphur	.2026
Glass	.1937	Charcoal	.2410

LIQUIDS.

Water	1.0000	Lead (melted)	.0402
Alcohol	.7000	Sulphur (melted)	.2340
Mercury	.0333	Tin (melted)	.0637
Benzine	.4500	Sulphuric acid	.3350
Glycerine	.5550	Oil of turpentine	.4260

GASES.

Air	.23751	Superheated steam	.4805
Oxygen	.21751	Carbonic oxide (CO)	.2479
Nitrogen	.24380	Carbonic acid (CO_2)	.2170
Hydrogen	8.40900	Olefiant gas	.4040

TEMPERATURES AND LATENT HEATS OF FUSION AND OF VAPORIZATION.

Substance.	Temperature of Fusion.	Temperature of Vaporization.	Latent Heat of Fusion.	Latent Heat of Vaporization.
Water	32°	212°	142.65	966.6
Mercury	−37.8°	662°	5.09	157
Sulphur	228.3°	824°	13.26	
Tin	446°		25.65	
Lead	626°		9.67	
Zinc	680°	1,900°	50.63	493
Alcohol	Unknown	173°		372
Oil of turpentine ...	14°	313°		124
Linseed oil		600°		
Aluminum	1,400°			
Copper	2,100°			
Cast iron	2,192°	3,300°		
Wrought iron	2,912°	5,000°		
Steel	2,520°			
Platinum	3,632°			
Iridium	4,892°			

EXAMPLE.—How many units of heat are required to melt 10 lb. of zinc from a temperature of 60° F.?

SOLUTION.—The specific heat of zinc is found from the table to be .0956. Hence, the number of heat units necessary to raise it to the melting point is $10 \times (680 - 60) \times .0956 = 592.72$. Latent heat of fusion $= 50.63$ heat units. Hence, the total number of heat units required is $592.72 + 10 \times 50.63 = 1,099.02$.

HEAT.

COEFFICIENT OF EXPANSION FOR A NUMBER OF SUBSTANCES.

Name of Substance.	Linear Expansion.	Surface Expansion.	Cubic Expansion.
Cast iron	.00000617	.00001234	.00001850
Copper	.00000955	.00001910	.00002864
Brass	.00001037	.00002074	.00003112
Silver	.00000690	.00001390	.00002070
Bar iron	.00000686	.00001372	.00002058
Steel (untempered)	.00000599	.00001198	.00001798
Steel (tempered)	.00000702	.00001404	.00002106
Zinc	.00001634	.00003268	.00004903
Tin	.00001410	.00002820	.00003229
Mercury	.00003334	.00006668	.00010010
Alcohol	.00019259	.00038518	.00057778
Gases	....	....	.00203252

EXAMPLE.—A wrought-iron bar 22 ft. long is heated from 70° to 300°. How much will it lengthen?

SOLUTION.— $22 \times (300 - 70) \times .00000686 = .0347116$ ft. $= .41654$ in.

ALLOYS.

NOTE.—A = Antimony, B = Bismuth, C = Copper, G = Gold, I = Iron, L = Lead, N = Nickel, S = Silver, T = Tin, Z = Zinc.

Name.	*Proportions.*
Brass, common yellow	2 C, 1 Z
Brass, to be rolled	32 C, 10 Z, 1.5 T
Brass castings, common	20 C, 1.25 Z, 2.5 T
Brass castings, hard	25 C, 2 Z, 4.5 T
Brass propellers	8 C, .5 Z, 1 T
Gun metal	8 C, 1 T

ALLOYS—(*Continued*).

Name.	Proportions.
Copper flanges	9 *C*, 1 *Z*, .26 *T*
Muntz's metal	6 *C*, 4 *Z*
Statuary	91.4 *C*, 5.53 *Z*, 1.7 *T*, 1.87 *L*
German silver	2 *C*, 7.9 *N*, 6.3 *Z*, 6.5 *I*
Britannia metal	50 *A*, 25 *T*, 25 *B*
Chinese silver	65.1 *C*, 19.8 *Z*, 13 *N*, 2.58 *S*, 12 *I*
Chinese white copper	20.2 *C*, 12.7 *Z*, 1.3 *T*, 15.8 *N*
Medals	100 *C*, 8 *Z*
Pinchbeck	5 *C*, 1 *Z*
Babbitt's metal	25 *T*, 2 *A*, .5 *C*
Bell metal, large	3 *C*, 1 *T*
Bell metal, small	4 *C*, 1 *T*
Chinese gongs	40.5 *C*, 9.2 *T*
Telescope mirrors	33.3 *C*, 16.7 *T*
White metal, ordinary	3.7 *C*, 3.7 *Z*, 14.2 *T*, 28.4 *A*
White metal, hard	35 *C*, 13 *Z*, 2.2 *T*
Sheeting metal	56 *C*, 45 *Z*, 12 arsenic
Metal, expands in cooling	75 *L*, 16.7 *A*, 8.3 *B*

ALLOYS FOR SOLDERS.

Name.	Proportions.	Melting Point.
Newton's fusible	8 *B*, 5 *L*, 3 *T*,	212°
Rose's fusible	2 *B*, 1 *L*, 1 *T*,	201°
A more fusible	5 *B*, 3 *L*, 2 *T*,	199°
Still more fusible	12 *T*, 25 *L*, 50 *B*, 13 cadmium,	155°
For tin solder, coarse,	1 *T*, 3 *L*,	500°
For tin solder, ordinary	2 *T*, 1 *L*,	360°
For brass, soft spelter	1 *C*, 1 *Z*,	550°
Hard, for iron	2 *C*, 1 *Z*,	700°
For steel	19 *S*, 3 *C*, 1 *Z*	
For fine brasswork	1 *S*, 8 *C*, 8 *Z*	
Pewterer's soft solder	2 *B*, 4 *L*, 3 *T*	
Pewterer's soft solder	1 *B*, 1 *L*, 2 *T*	
Gold solder	24 *G*, 2 *S*, 1 *C*	
Silver solder, hard	4 *S*, 1 *C*	
Silver solder, soft	2 *S*, 1 brass wire	
For lead	16 *T*, 33 *L*	

WEIGHT OF ROUND AND SQUARE ROLLED IRON.

From $\frac{1}{16}$ in. to 9½ in. in Diameter, and 1 ft. in Length.

Side or Diam. Inches.	Weight. Lb. per ft.		Side or Diam. Inches.	Weight. Lb. per ft.	
	Round.	Square.		Round.	Square.
$\frac{1}{16}$	.010	.013	3⅞	39.864	50.756
⅛	.041	.053	4	42.464	54.084
$\frac{3}{16}$	.093	.118	4⅛	45.174	57.517
¼	.165	.211	4¼	47.952	61.055
⅜	.373	.475	4⅜	50.815	64.700
½	.663	.845	4½	53.760	68.448
⅝	1.043	1.320	4⅝	56.788	72.305
¾	1.493	1.901	4¾	59.900	76.264
⅞	2.032	2.588	4⅞	63.094	80.333
1	2.654	3.380	5	66.350	84.480
1⅛	3.359	4.278	5⅛	69.731	88.784
1¼	4.147	5.280	5¼	73.172	93.168
1⅜	5.019	6.390	5⅜	76.700	97.657
1½	5.972	7.604	5½	80.304	102.240
1⅝	7.010	8.926	5⅝	84.001	106.953
1¾	8.128	10.352	5¾	87.776	111.756
1⅞	9.333	11.883	5⅞	91.634	116.671
2	10.616	13.520	6	95.552	121.664
2⅛	11.988	15.263	6¼	103.704	132.040
2¼	13.440	17.112	6½	112.160	142.816
2⅜	14.975	19.066	6¾	120.960	154.012
2½	16.588	21.120	7	130.048	165.632
2⅝	18.293	23.292	7¼	139.544	177.672
2¾	20.076	25.560	7½	149.328	190.136
2⅞	21.944	27.939	7¾	159.456	203.024
3	23.888	30.416	8	169.856	216.336
3⅛	25.926	33.010	8¼	180.696	230.068
3¼	28.040	35.704	8½	191.808	244.220
3⅜	30.240	38.503	8¾	203.260	258.800
3½	32.512	41.408	9	215.040	273.792
3⅝	34.886	44.418	9¼	227.152	289.220
3¾	37.332	47.534	9½	239.600	305.056

WEIGHT OF SHEET LEAD.

Thickness. Inches.	W'ght. Lb.	Thickness. Inches.	W'ght. Lb.	Thickness. Inches.	W'ght. Lb.
.017	1	.085	5	.152	9
.034	2	.101	6	.169	10
.051	3	.118	7	.186	11
.068	4	.135	8	.203	12

PROPORTIONS OF THE UNITED STATES STANDARD SCREW THREADS, NUTS, AND BOLT HEADS.

Diam. of Screw.	Threads per In.	Diam. of Core.	Width of Flat.	Inside Diam.	Outside Diam.	Diago-nal.	Height of Head.
1-4	20	.185	.0062	1-2	37-64	45-64	1-4
5-16	18	.240	.0070	19-32	11-16	27-32	19-64
3-8	16	.294	.0078	11-16	51-64	31-32	11-32
7-16	14	.344	.0089	25-32	29-32	1 7-64	25-64
1-2	13	.400	.0096	7-8	1 1-64	1 15-64	7-16
9-16	12	.454	.0104	31-32	1 1-8	1 3-8	31-64
5-8	11	.507	.0113	1 1-16	1 15-64	1 1-2	17-32
3-4	10	.620	.0125	1 1-4	1 7-16	1 3-4	5-8
7-8	9	.731	.0140	1 7-16	1 21-32	2 1-32	23-32
1	8	.837	.0156	1 5-8	1 7-8	2 19-64	13-16
1 1-8	7	.940	.0180	1 13-16	2 3-32	2 9-16	29-32
1 1-4	7	1.065	.0180	2	2 5-16	2 53-64	1
1 3-8	6	1.160	.0210	2 3-16	2 17-32	3 3-32	1 3-32
1 1-2	6	1.284	.0210	2 3-8	2 3-4	3 23-64	1 3-16
1 5-8	5 1-2	1.389	.0227	2 9-16	2 31-32	3 5-8	1 9-32
1 3-4	5	1.490	.0250	2 3-4	3 11-64	3 57-64	1 3-8
1 7-8	5	1.615	.0250	2 15-16	3 25-64	4 5-32	1 15-32
2	4 1-2	1.712	.0250	3 1-8	3 39-64	4 27-64	1 9-16
2 1-4	4 1-2	1.962	.0280	3 1-2	4 3-64	4 61-64	1 3-4
2 1-2	4	2.175	.0310	3 7-8	4 15-32	5 31-64	1 15-16
2 3-4	4	2.425	.0310	4 1-4	4 29-32	6 1-64	2 1-8
3	3 1-2	2.628	.0357	4 5-8	5 11-32	6 35-64	2 5-16
3 1-4	3 1-2	2.878	.0357	5	5 25-32	7 5-64	2 1-2
3 1-2	3 1-4	3.100	.0384	5 3-8	6 13-64	7 19-32	2 11-16
3 3-4	3	3.317	.0410	5 3-4	6 41-64	8 1-8	2 7-8
4	3	3.566	.0410	6 1-8	7 5-64	8 21-32	3 1-16
4 1-4	2 7-8	3.798	.0435	6 1-2	7 1-2	9 3-16	3 1-4
4 1-2	2 3-4	4.027	.0460	6 7-8	7 15-16	9 23-32	3 7-16
4 3-4	2 5-8	4.255	.0180	7 1-4	8 3-8	10 1-4	3 5-8
5	2 1-2	4.480	.0500	7 5-8	8 13-16	10 25-32	3 13-16
5 1-4	2 1-2	4.730	.0500	8	9 15-64	11 5-16	4
5 1-2	2 3-8	4.953	.0526	8 3-8	9 45-64	11 27-32	4 3-16
5 3-4	2 3-8	5.203	.0526	8 3-4	10 7-64	12 3-8	4 3-8
6	2 1-4	5.423	.0555	9 1-8	10 35-64	12 13-16	4 9-16

The threads have an angle of 60°, with flat tops and bottoms, and are of the following proportions:

Notation of letters. All dimensions in inches.

D = outside diameter of screw;
d = diameter of root of thread, or of hole in the nut;
p = pitch of screw;
t = number of threads per inch;
f = flat top and bottom;
o = outside diameter of hexagon nut or bolt head;

i = inside diameter of hexagon, or side of square nut or bolt head;
s = diagonal of square nut or bolt head;
h = height of rough or unfinished bolt head.

The height of finished nut or bolt head is made equal to the diameter D of the screw.

$$p = \frac{\sqrt{16\,D + 10} - 2.909}{16.64},$$

$$t = \frac{1}{p}. \qquad s = 1.414\,i.$$

$$d = D - \frac{1.299}{t}. \qquad i = \frac{3\,D}{2} + \frac{1}{8}. \qquad o = 1.155\,i. \qquad f = \frac{p}{8}.$$

WEIGHT OF CAST-IRON PIPE PER FOOT IN POUNDS.

These weights are for plain pipe. For hautboy pipe add 8 in. in length for each joint. For copper add ¼; for lead, ⅝; for welded iron, add $\frac{1}{15}$, or multiply by 1.0667.

Diameter of Bore. Inches.	Thickness of Pipe in Inches.									
	¼	⅜	½	⅝	¾	⅞	1	1⅛	1¼	1⅜
1	3.07	5.07	7.38							
1¼	3.69	6.00	8.61							
1½	4.30	6.92	9.84							
1¾	4.92	7.84	11.10							
2	5.53	8.76	12.30	16.2						
2¼	6.15	9.69	13.50	17.7						
2½	6.76	10.60	14.80	19.2	24.0					
2¾	7.37	11.50	16.00	20.8	25.9					
3	7.98	12.50	17.20	22.3	27.7	33.4				
3½	9.21	14.30	19.70	25.4	31.4	37.7				
4	10.30	16.10	22.20	28.5	35.1	42.0				
4½	11.70	18.00	24.60	31.5	38.8	46.3				
5	12.90	19.80	27.10	34.6	42.5	50.6				
5½	14.20	21.70	29.50	37.7	46.1	54.9				
6	15.40	23.50	32.00	40.8	49.8	59.2	68.9			
6½	16.60	25.40	34.50	43.8	53.5	63.5	73.8	84.4		
7	17.80	27.20	36.90	46.9	57.2	67.8	78.7	89.4		
7½	19.10	29.10	39.40	50.0	60.9	72.1	83.7	95.5	108	
8	20.30	30.90	41.80	53.1	64.6	76.4	88.6	101.0	114	127
8½	21.50	32.80	44.30	56.1	68.3	80.7	93.5	107.0	120	134
9	22.80	34.60	46.80	59.2	72.0	85.1	98.4	112.0	126	140
9½	24.00	36.40	49.20	62.3	75.7	89.3	103.0	118.0	132	147
10	25.10	38.30	51.70	65.3	79.4	93.6	108.0	123.0	138	161
11	27.60	42.00	56.60	71.5	86.7	102.0	118.0	134.0	151	168
12	30.00	45.70	61.50	77.7	94.1	111.0	128.0	145.0	163	181
13	32.50	49.40	66.40	83.8	102.0	120.0	138.0	156.0	175	195
14	35.00	53.10	71.40	89.4	109.0	128.0	148.0	168.0	188	208
15	37.40	56.70	76.30	96.1	116.0	137.0	158.0	179.0	200	222
16	39.10	60.40	81.20	102.0	124.0	145.0	167.0	190.0	212	235
17	42.30	64.10	86.10	108.0	131.0	154.0	177.0	201.0	225	249
18	44.80	67.80	91.00	115.0	139.0	163.0	187.0	212.0	237	262
19	47.30	71.50	96.00	121.0	146.0	171.0	197.0	223.0	249	276
20	49.70	75.20	101.00	127.0	153.0	180.0	207.0	234.0	261	289
22	54.60	82.60	111.00	139.0	168.0	196.0	227.0	256.0	286	316
24	59.60	89.90	121.00	152.0	183.0	214.0	246.0	278.0	311	343
26	64.50	97.30	131.00	164.0	198.0	231.0	266.0	300.0	335	370
28	69.40	105.00	140.00	176.0	212.0	249.0	286.0	323.0	360	397
30	74.20	112.00	150.00	188.0	227.0	266.0	305.0	345.0	384	424

TABLE OF STANDARD DIMENSIONS OF WROUGHT-IRON WELDED PIPES.

Nominal Diameter.	External Diameter.	Thickness.	Internal Diameter.	Internal Circumference.	External Circumference.	Length of Pipe per Sq. Ft. of Internal Surface.	Length of Pipe per Sq. Ft. of External Surface.	Internal Area.	Weight per Foot.	No. of Threads per Inch of Screw.
In.	In.	In.	In.	In.	In.	Ft.	Ft.	In.	Lb.	
⅛	.40	.068	.27	.85	1.27	14.15	9.440	.057	.24	27
¼	.54	.088	.36	1.14	1.70	10.50	7.075	.104	.42	18
⅜	.67	.091	.49	1.55	2.12	7.67	5.657	.192	.56	18
½	.84	.109	.62	1.96	2.65	6.13	4.502	.305	.84	14
¾	1.05	.113	.82	2.59	3.30	4.64	3.637	.533	1.13	14
1	1.31	.134	1.05	3.29	4.13	3.66	2.903	.863	1.67	11½
1¼	1.66	.140	1.38	4.33	5.21	2.77	2.301	1.496	2.26	11½
1½	1.90	.145	1.61	5.06	5.97	2.37	2.010	2.038	2.69	11½
2	2.37	.154	2.07	6.49	7.46	1.85	1.611	3.355	3.67	11½
2½	2.87	.204	2.47	7.75	9.03	1.55	1.328	4.783	5.77	8
3	3.50	.217	3.07	9.64	11.00	1.24	1.091	7.388	7.55	8
3½	4.00	.226	3.55	11.15	12.57	1.08	0.955	9.887	9.05	8
4	4.50	.237	4.03	12.65	14.14	.95	0.849	12.730	10.73	8
4½	5.00	.247	4.51	14.15	15.71	.85	0.765	15.939	12.49	8
5	5.56	.259	5.04	15.85	17.47	.78	0.629	19.990	14.56	8
6	6.62	.280	6.06	19.05	20.81	.63	0.577	28.889	18.77	8
7	7.62	.301	7.02	22.06	23.95	.54	0.505	38.737	23.41	8
8	8.62	.322	7.98	25.08	27.10	.48	0.444	50.039	28.35	8
9	9.69	.344	9.00	28.28	30.43	.42	0.394	63.633	34.08	8
10	10.75	.366	10.02	31.47	33.77	.38	0.355	78.838	40.64	8

FLUXES FOR SOLDERING OR WELDING.

Iron ....Borax
Tinned iron ....Resin
Copper and brass
          Sal ammoniac

Zinc............Chloride of zinc
Lead ....Tallow or resin
Lead and tin pipes
        Resin and sweet oil

Steel.—Pulverize together 1 part of sal ammoniac and 10 parts of borax and fuse until clear. When solidified, pulverize to powder.

STEAM TABLES.

Whenever the pressure of saturated steam is changed, there are other properties that change with it. These properties are the following:

1. The temperature of the steam, or, what is the same thing, the boiling point.

2. The number of B. T. U. required to raise a pound of water from 32° (freezing) to the boiling point corresponding to the given pressure. This is called the *heat of the liquid.*

3. The number of B. T. U. required to change the water at the boiling temperature into steam at the same temperature. This is called the *latent heat of vaporization*, or, simply, the *latent heat.*

4. The number of heat units required to change a pound of water at 32° to steam of the required temperature and pressure. This is called the *total heat of vaporization*, or, simply, the *total heat.*

It is plain that the *total heat* is the sum of the *heat of the liquid* and the *latent heat.* That is, total heat = heat of liquid + latent heat.

5. The *specific volume* of the steam at the given pressure; that is, the number of cubic feet occupied by a pound of steam of the given pressure.

6. The *density* of the steam; that is, the weight of 1 cubic foot of the steam at the given pressure.

All the above properties are different for different pressures. For example, if steam boils under atmospheric pressure, the temperature is 212°; the heat of the liquid is 180.531 B. T. U.; the latent heat, 966.069 B. T. U.; the total heat, 1,146.6 B. T. U. A pound of steam at this pressure occupies 26.37 cu. ft., and a cubic foot of the steam weighs about .037928 lb. When the pressure is 70 lb. per sq. in. above vacuum, the temperature is 302.774°; the heat of the liquid is 272.657 B. T. U.; the latent heat is 901.629 B. T. U.; the total heat is 1,174.286 B. T. U. A pound of the steam occupies 6.076 cu. ft., and a cubic foot of the steam weighs .164584 lb.

These properties have been determined by direct experiment for all ordinary steam pressures. They are given in the table of the properties of saturated steam, pages 29-31.

EXPLANATION OF THE TABLE.

Column 1 gives the pressures from 1 to 300 lb. These pressures are above vacuum. The steam gauges fitted on steam boilers register the pressure above the atmosphere. That is, if the steam is at atmospheric pressure, 14.7 lb. per sq. in., the gauge registers 0. Consequently, the atmospheric pressure must be added to the reading of the gauge to obtain the pressure above vacuum. In using the table, care must be taken *not* to use the gauge pressures without first adding 14.7 lb. per sq. in.

Pressures registered above vacuum are called *absolute pressures*. The pressures given in column 1 are *absolute*. Absolute pressure per square inch = gauge pressure per square inch + 14.7.

Column 2 gives the temperature of the steam when at the pressure shown in column 1.

Column 3 gives the *heat of the liquid*. It will be noticed that the values in column 3 may be obtained approximately by subtracting 32° from the temperature in column 2. If the specific heat of water were exactly 1.00, it would, of course, take exactly 212 − 32 = 180 B. T. U. to raise a pound of water from 32° to 212°. But experiment shows that the specific heat of water is slightly greater than 1.00 when the temperature of the water is above 62°, and it therefore takes 180.531 B. T. U. to raise a pound of water from 32° to 212°.

Column 4 gives the *latent heat of vaporization*, which is seen to decrease slightly as the pressure increases.

Column 5 gives the *total heat of vaporization*. The values in column 5 may be obtained by adding together the corresponding values in columns 3 and 4.

Column 6 gives the weight of a cubic foot of steam in pounds. As would be expected, the steam becomes denser as the pressure rises, and weighs more per cubic foot.

Column 7 gives the number of cubic feet occupied by 1 pound of steam at the given pressure. It will be noticed that the corresponding values of columns 6 and 7 multiplied together always produce 1. Thus, for 31.3 pounds pressure, gauge, .11088 × 9.018 = 1.000, nearly.

Column 8 gives the ratio of the volume of a pound of

steam at the given pressure, and the volume of a pound of water at 39.2°. The values in column 8 may be obtained by dividing 62.425, the weight of a cubic foot of water at 39.2°, by the numbers in column 6.

EXAMPLES ON THE USE OF THE STEAM TABLE.

EXAMPLE 1.—Calculate the heat required to change 5 lb. of water at 32° into steam at 92 lb. pressure above vacuum.

SOLUTION.—From column 5, the total heat of 1 lb. at 92 lb. pressure is 1,180.045 B. T. U.

$$1,180.045 \times 5 = 5,900.225 \text{ B. T. U.}$$

EXAMPLE 2.—How many heat units are required to raise 8¼ lb. of water from 32° to 250° F.?

SOLUTION.—Looking in column 8, the heat of the liquid of 1 lb. at 250.293° is 219.261 B. T. U. 219.261 − .293 = 218.968 B. T. U. = heat of liquid for 250°. Then, for 8¼ lb. it is 218.968 × 8¼ = 1,861.228 B. T. U.

EXAMPLE 3.—How many foot-pounds of work will it require to change 60 lb. of boiling water at 80 lb. pressure, absolute, into steam of the same pressure?

SOLUTION.—Looking under column 4, the latent heat of vaporization is 895.108; that is, it takes 895.108 B. T. U. to change 1 lb. of water at 80 lb. pressure into steam of the same pressure. Therefore, it takes 895.108 × 60 = 53,706.48 B. T. U. to perform the same operation on 60 lb. of water.

$$53,706.48 \times 778 = 41,783,641.44 \text{ ft.-lb.}$$

EXAMPLE 4.—Find the volume occupied by 14 lb. of steam at 30 lb., *gauge* pressure.

SOLUTION.— 30 lb., gauge pressure = 30 + 14.7 = 44.7, absolute pressure. The nearest pressure in the table is 44 lb., and the volume of a pound of steam at that pressure is 9.403 cu. ft. The volume of a pound at 46 lb. pressure is 9.018 cu. ft. 9.403 − 9.018 = .385 cu. ft., the difference in volume for a difference in pressure of 2 lb. $\frac{.385}{2}$ = .1925 cu. ft., the difference in volume for a difference in pressure of 1 lb. .1925 × .7 = .135 cu. ft., the difference in volume for a difference in pressure of .7 lb. Therefore, 9.403 − .135 = 9.268 cu. ft. is the volume of 1 lb. of steam at 44.7 lb. pressure. The .135 cu. ft.

is subtracted from 9.403 cu. ft., since the volume is less for a pressure of 44.7 lb. than for a pressure of 44 lb.

$$9.268 \times 14 = 129.752 \text{ cu. ft.}$$

EXAMPLE 5.—Find the weight of 40 cu. ft. of steam at a temperature of 254° F.

SOLUTION.—The weight of 1 cu. ft. of steam at 254.002°, from the table, is .078839 lb. Neglecting the .002°, the weight of 40 cu. ft. is, therefore,

$$.078839 \times 40 = 3.15356 \text{ lb.}$$

EXAMPLE 6.—How many pounds of steam at 64 lb. pressure, absolute, are required to raise the temperature of 300 lb. of water from 40° to 130° F., the water and steam being mixed?

SOLUTION.—The number of heat units required to raise 1 lb. from 40° to 130° is 130 − 40 = 90 B. T. U. (Actually a little more than 90 would be required, but the above is near enough for all practical purposes.) Then, to raise 300 lb. from 40° to 130° requires 90 × 300 = 27,000 B. T. U. This quantity of heat must necessarily come from the steam. Now, 1 lb. of steam at 64 lb. pressure gives up, in condensing, its latent heat of vaporization, or 905.9 B. T. U. But, in addition to its latent heat, each pound of steam on condensing must give up an additional amount of heat in falling to 130°. Since the original temperature of the steam was 296.805° F. (see table), each pound gives up by its fall of temperature 296.805 − 130 = 166.805 B. T. U. Therefore, each pound of the steam gives up a total of

$$905.9 + 166.805 = 1,072.705 \text{ B. T. U.}$$

It will, therefore, take $\dfrac{27,000}{1,072.705} = 25.17$ lb. of steam to accomplish the desired result.

With the steam tables a reliable thermometer may be used for ascertaining the pressure of saturated steam or for testing the accuracy of a steam gauge. The temperature of the steam being measured by the thermometer, the corresponding absolute pressure is found from the steam tables; the gauge pressure is then found by subtracting 14.7 from the absolute pressure. Thus, the temperature of the steam in a condenser being 142°, we find from the steam tables that the corresponding absolute pressure is 3 lb. per sq. in., nearly.

THE PROPERTIES OF SATURATED STEAM.

Pressure Above Vacuum in Pounds per Square Inch.	Temperature, Fahrenheit Degrees.	Quantity of Heat in British Thermal Units.			Weight of a Cubic Foot of Steam in Pounds.	Volume of a Pound of Steam in Cubic Feet.	Ratio of Vol. of Steam to Vol. of Equal Weight of Dist. Water at Temp. of Maximum Density.
		Required to Raise Temperature of the Water From 32° to t°.	Total Latent Heat at Pressure p.	Total Heat Above 32°.			
1	2	3	4	5	6	7	8
p	t	q	L	H	W	V	R
1	102.018	70.040	1,043.015	1,113.055	.003027	330.4	20,623
2	126.302	94.368	1,026.094	1,120.462	.005818	171.9	10,730
3	141.654	109.764	1,015.380	1,125.144	.008522	117.3	7,325
4	153.122	121.271	1,007.370	1,128.641	.011172	89.51	5,588
5	162.370	130.563	1,000.899	1,131.462	.013781	72.56	4,530
6	170.173	138.401	995.441	1,133.842	.016357	61.14	3.816
7	176.945	145.213	990.695	1,135.908	.018908	52.89	3,302
8	182.952	151.255	986.485	1,137.740	.021436	46.65	2,912
9	188.357	156.699	982.690	1,139.389	.023944	41.77	2,607
10	193.284	161.660	979.232	1,140.892	.026437	37.83	2,361
11	197.814	166.225	976.050	1,142.275	.028911	34.59	2,159
12	202.012	170.457	973.098	1,143.555	.031376	31.87	1,990
13	205.929	174.402	970.346	1,144.748	.033828	29.56	1,845
14	209.604	178.112	967.757	1,145.869	.036265	27.58	1,721
14.69	212.000	180.531	966.069	1,146.600	.037928	26.37	1,646
15	213.067	181.608	965.318	1,146.926	.038688	25.85	1,614
16	216.847	184.919	963.007	1,147.926	.041109	24.33	1,519
17	219.452	188.056	960.818	1,148.874	.043519	22.98	1,434
18	222.424	191.058	958.721	1,149.779	.045920	21.78	1,359
19	225.255	193.918	956.725	1,150.643	.048312	20.70	1,292

TABLE—(*Continued*).

1	2	3	4	5	6	7	8
p	t	q	L	H	W	V	R
20	227.964	196.655	954.814	1,151.469	.050696	19.730	1,231.0
22	233.069	201.817	951.209	1,153.026	.055446	18.040	1,126.0
24	237.803	206.610	947.861	1,154.471	.060171	16.620	1,038.0
26	242.225	211.089	944.730	1,155.819	.064870	15.420	962.3
28	246.376	215.293	941.791	1,157.084	.069545	14.380	897.6
30	250.293	219.261	939.019	1,158.280	.074201	13.480	841.3
32	254.002	223.021	936.389	1,159.410	.078839	12.680	791.8
34	257.523	226.594	933.891	1,160.485	.083461	11.980	948.0
36	260.883	230.001	931.508	1,161.509	.088067	11.360	708.8
38	264.093	233.261	929.227	1,162.488	.092657	10.790	673.7
40	267.168	236.386	927.040	1,163.426	.097231	10.280	642.0
42	270.122	239.389	924.940	1,164.329	.101794	9.826	613.3
44	272.965	242.275	922.919	1,165.194	.106345	9.403	587.0
46	275.704	245.061	920.968	1,166.029	.110884	9.018	563.0
48	278.348	247.752	919.084	1,166.836	.115411	8.665	540.9
50	280.904	250.355	917.260	1,167.615	.119927	8.338	520.5
52	283.381	252.875	915.494	1,168.369	.124433	8.037	501.7
54	285.781	255.321	913.781	1,169.102	.128928	7.756	484.2
56	288.111	257.695	912.118	1,169.813	.133414	7.496	467.9
58	290.374	260.002	910.501	1,170.503	.137892	7.252	452.7
60	292.575	262.248	908.928	1,171.176	.142362	7.024	438.5
62	294.717	264.433	907.396	1,171.829	.146824	6.811	425.2
64	296.805	266.566	905.900	1,172.466	.151277	6.610	412.6
66	298.842	268.644	904.443	1,173.087	.155721	6.422	400.8
68	300.831	270.674	903.020	1,173.694	.160157	6.244	389.8
70	302.774	272.657	901.629	1,174.286	.164584	6.076	379.3
72	304.669	274.597	900.269	1,174.866	.169003	5.917	369.4
74	306.526	276.493	898.938	1,175.431	.173417	5.767	360.0
76	308.344	278.350	897.635	1,175.985	.177825	5.624	351.1
78	310.123	280.170	896.359	1,176.529	.182229	5.488	342.6
80	311.866	281.952	895.108	1,177.060	.186627	5.358	334.5
82	313.576	283.701	893.879	1,177.580	.191017	5.235	326.8
84	315.250	285.414	892.677	1,178.091	.195401	5.118	319.5
86	316.893	287.096	891.496	1,178.592	.199781	5.006	312.5
88	318.510	288.750	890.335	1,179.085	.204155	4.898	305.8

TABLE—(*Continued*).

1	2	3	4	5	6	7	8
p	t	q	L	H	W	V	R
90	320.094	290.373	889.196	1,179.569	.208525	4.796	299.4
92	321.653	291.970	888.075	1,180.045	.212892	4.697	293.2
94	323.183	293.539	886.972	1,180.511	.217253	4.603	287.3
96	324.688	295.083	885.887	1,180.970	.221604	4.513	281.7
98	326.169	296.601	884.821	1,181.422	.225950	4.426	276.3
100	327.625	298.098	883.773	1,181.866	.230293	4.342	271.1
105	331.169	301.731	881.214	1,182.945	.241139	4.147	258.9
110	334.582	305.242	878.744	1,183.986	.251947	3.969	247.8
115	337.874	308.621	876.371	1,184.992	.262732	3.806	237.6
120	341.058	311.885	874.076	1,185.961	.273500	3.656	228.3
125	344.136	315.051	871.848	1,186.899	.284243	3.518	219.6
130	347.121	318.121	869.688	1,187.809	.294961	3.390	211.6
135	350.015	321.105	867.590	1,188.695	.305659	3.272	204.2
140	352.827	324.008	865.552	1,189.555	.316338	3.161	197.3
145	355.562	326.823	863.567	1,190.390	.326998	3.058	190.9
150	358.223	329.566	861.634	1,191.200	.337643	2.962	184.9
160	363.346	334.850	857.912	1,192.762	.358886	2.786	173.9
170	368.226	339.892	854.359	1,194.251	.380071	2.631	164.3
180	372.886	344.708	850.963	1,195.671	.401201	2.493	155.6
190	377.352	349.329	847.703	1,197.032	.422280	2.368	147.8
200	381.636	353.766	844.573	1,198.339	.443310	2.256	140.8
210	385.759	358.041	841.556	1,199.597	.464295	2.154	134.5
220	389.736	362.168	838.642	1,200.810	.485237	2.061	128.7
230	393.575	366.152	835.828	1,201.960	.506139	1.976	123.3
240	397.285	370.008	833.103	1,203.111	.527003	1.898	118.5
250	400.883	373.750	830.459	1,204.209	.547831	1.825	114.0
260	404.370	377.377	827.896	1,205.273	.568626	1.759	109.8
270	407.755	380.905	825.401	1,206.306	.589390	1.697	105.9
280	411.048	384.337	822.973	1,207.310	.610124	1.639	102.8
290	414.250	387.677	820.609	1,208.286	.630829	1.585	99.0
300	417.371	390.933	818.305	1,209.238	.651506	1.535	95.8

4

LOGARITHMS.

EXPONENTS.

By the use of logarithms, the processes of multiplication, division, involution, and evolution are greatly shortened, and some operations may be performed that would be impossible without them. Ordinary logarithms cannot be applied to addition and subtraction.

The *logarithm* of a number is that *exponent* by which some fixed number, called the *base*, must be affected in order to equal the number. Any number may be taken as the base. Suppose we choose 4. Then the logarithm of 16 is 2, because 2 is the exponent by which 4 (the base) must be affected in order to equal 16, since $4^2 = 16$. In this case, instead of reading 4^2 as 4 square, read it 4 exponent 2. With the same base, the logarithms of 64 and 8 would be 3 and 1.5, respectively, since $4^3 = 64$, and $4^{1.5} = 4^{\frac{3}{2}} = 8$. In these cases, as in the preceding, read 4^3 and $4^{1.5}$ as 4 exponent 3, and 4 exponent 1.5, respectively.

Although any positive number except 1 *can* be used as a base and a table of logarithms calculated, but two numbers have ever been employed. For all arithmetical operations (except addition and subtraction) the logarithms used are called the *Briggs*, or *common*, logarithms, and the base used is 10. In abstract mathematical analysis, the logarithms used are variously called *hyperbolic, Napierian*, or *natural* logarithms, and the base is 2.718281828+. The common logarithm of any number may be converted into a Napierian logarithm by multiplying the common logarithm by 2.30258509+, which is usually expressed as 2.3026, and sometimes as 2.3. Only the common system of logarithms will be considered here.

Since in the common system the base is 10, it follows that, since $10^1 = 10$, $10^2 = 100$, $10^3 = 1,000$, etc., the logarithm (exponent) of 10 is 1, of 100 is 2, of 1,000 is 3, etc. For the sake of brevity in writing, the words "logarithm of" are abbreviated to "log." Thus, instead of writing logarithm of 100 = 2, write log 100 = 2. When speaking, however, the words for which "log" stands should always be pronounced in full.

From the above it will be seen that, when the base is 10,

since $10^0 = 1$,   the exponent $0 = \log 1$;

since $10^1 = 10$,   the exponent $1 = \log 10$;

since $10^2 = 100$,   the exponent $2 = \log 100$;

since $10^3 = 1,000$, the exponent $3 = \log 1,000$; etc.

Also,

since $10^{-1} = \frac{1}{10} = .1$,   the exponent $-1 = \log .1$;

since $10^{-2} = \frac{1}{100} = .01$,   the exponent $-2 = \log .01$;

since $10^{-3} = \frac{1}{1000} = .001$, the exponent $-3 = \log .001$; etc.

From this it will be seen that the logarithms of exact powers of 10 and of decimals like .1, .01, and .001 are the whole numbers 1, 2, 3, etc. and -1, -2, -3, etc., respectively. *Only numbers consisting of 1 and one or more ciphers have whole numbers for logarithms.*

Now, it is evident that, to produce a number between 1 and 10, the exponent of 10 must be a fraction; to produce a number between 10 and 100, it must be 1 plus a fraction; to produce a number between 100 and 1,000, it must be 2 plus a fraction; etc. Hence, the logarithm of any number between 1 and 10 is a fraction; of any number between 10 and 100, 1 plus a fraction; of any number between 100 and 1,000, 2 plus a fraction, etc. A logarithm, therefore, usually consists of two parts: a whole number, called the *characteristic*, and a fraction, called the *mantissa*. The mantissa is always expressed as a decimal. For example, to produce 20, 10 must have an exponent of *approximately* 1.30103, or $10^{1.30103} = 20$, very nearly, the degree of exactness depending on the number of decimal places used. Hence, $\log 20 = 1.30103$, 1 being the characteristic, and .30103, the mantissa.

Referring to the second part of the preceding table, it is clear that the logarithms of all numbers less than 1 are negative, the logarithms of those between 1 and .1 being -1 plus a fraction. For, since $\log .1 = -1$, the logarithms of .2, .3, etc. (which are all greater than .1, but less than 1) must be greater than -1; i. e., they must equal -1 *plus* a fraction. For the same reason, to produce a number between .1 and .01, the logarithm (exponent of 10) would be equal to -2 plus a fraction, and for a number between .01 and .001, it would be equal to -3 plus a fraction. Hence, the logarithm

of any number between 1 and .1 has a negative characteristic of 1 and a positive mantissa; of a number between
.1 and .01, a negative characteristic of 2 and a positive
mantissa; of a number between .01 and .001, a negative
characteristic of 3 and a positive mantissa; of a number
between .001 and .0001, a negative characteristic of 4 and a
positive mantissa, etc. *The negative characteristics are distinguished from the positive by the — sign written over the characteristic.* Thus, $\overline{3}$ indicates that 3 is negative.

*It must be remembered that in all cases the mantissa is posi-
tive.* Thus, the logarithm 1.30103 means +1 + .30103, and the
logarithm $\overline{1}$.30103 means —1 + .30103. Were the minus sign
written in front of the characteristic, it would indicate that
the entire logarithm was negative. Thus, —1.30103 = —1
—.30103.

Rule for Characteristic.—Starting from the unit figure, count
the number of places to the first (left-hand) digit of the given
number, calling unit's place zero; the number of places thus
counted will be the required characteristic. If the first digit
lies to the left of the unit figure, the characteristic is positive;
if to the right, negative. If the first digit of the number is the
unit figure, the characteristic is 0. Thus, the charactetisic of
the logarithm of 4,826 is 3, since the first digit, 4, lies in the 3d
place to the left of the unit figure, 6. The characteristic of
the logarithm of 0.0000072 is —6 or $\overline{6}$, since the first digit, 7,
lies in the 6th place to the right of the unit figure. The characteristic of the logarithm of 4.391 is 0, since 4 is both the first
digit of the number and also the unit figure.

TO FIND THE LOGARITHM OF A NUMBER.

To aid in obtaining the mantissas of logarithms, *tables of
logarithms* have been calculated, some of which are very
elaborate and convenient. In the Table of Logarithms, the
mantissas of the logarithms of numbers from 1 to 9,999 are
given to five places of decimals. The mantissas of logarithms
of larger numbers can be found by interpolation. The table
contains the *mantissas only;* the characteristics may be easily
found by the preceding rule.

The table depends on the principle, which will be explained later, that all numbers having the same figures in the same order have the same mantissa, without regard to the position of the decimal point, which affects the characteristic only. To illustrate, if log 206 = 2.31387, then,

log 20.6 = 1.31387;   log .206 = $\overline{1}$.31387;
log 2.06 = .31387;   log .0206 = $\overline{2}$.31387; etc.

To find the logarithm of a number not having more than four figures:

Rule.—*Find the first three significant figures of the number whose logarithm is desired, in the left-hand column; find the fourth figure in the column at the top (or bottom) of the page; and in the column under (or above) this figure, and opposite the first three figures previously found, will be the mantissa or decimal part of the logarithm. The characteristic being found as previously described, write it at the left of the mantissa, and the resulting expression will be the logarithm of the required number.*

EXAMPLE.—Find from the table the logarithm (*a*) of 476; (*b*) of 25.47; (*c*) of 1.073; (*d*) of .06313.

SOLUTION.—(*a*) In order to economize space and make the labor of finding the logarithms easier, the first two figures of the mantissa are given only in the column headed 0. The last three figures of the mantissa, opposite 476 in the column headed N (N stands for number), are 761, found in the column headed 0; glancing upwards, we find the first two figures of the mantissa, viz., 67. The characteristic is 2; hence, log 476 = 2.67761.

NOTE.—Since all numbers in the table are decimal fractions, the decimal point is omitted throughout; this is customary in all tables of logarithms.

(*b*) To find the logarithm of 25.47, we find the first three figures, 254, in the column headed N, and on the same horizontal line, under the column headed 7 (the fourth figure of the given number), will be found the last three figures of the mantissa, viz., 603. The first two figures are evidently 40, and the characteristic is 1; hence, log 25.47 = 1.40603.

(*c*) For 1.073; in the column headed 3, opposite 107 in the column headed N, the last three figures of the mantissa are found, in the usual manner, to be 060. It will be noticed

that these figures are printed *060, the star meaning that
instead of glancing *upwards* in the column headed 0, and
taking 02 for the first two figures, we must glance *downwards*
and take the two figures opposite the number 108, in the
left-hand column, i. e., 03.   The characteristic being 0, log
1.073 = 0.03060, or, more simply, .03060.

(d) For .06313; the last three figures of the mantissa are
found opposite 631, in column headed 3, to be 024.   In this
case, the first two figures occur in the same row, and are 80.
Since the characteristic is $\overline{2}$, log .06313 = $\overline{2}$.80024.

If the original number contains but one digit (a cipher is
not a digit), annex mentally two ciphers to the right of the
digit; if the number contains but two digits (with no ciphers
between, as in 4,008), annex mentally one cipher on the
right before seeking the mantissa.   Thus, if the logarithm of
7 is wanted, seek the mantissa for 700, which is .84510; or, if
the logarithm of 48 is wanted, seek the mantissa for 480,
which is .68124.   Or, find the mantissas of logarithms of numbers between 0 and 100, on the first page of the tables.

The process of finding the logarithm of a number from the
table is technically called *taking out the logarithm.*

To take out the logarithm of a number consisting of more
than four figures, it is inexpedient to use more than five
figures of the number when using five-place logarithms (the
logarithms given in the accompanying table are five-place).
Hence, if the number consists of more than five figures and
the sixth figure is less than 5, replace all figures after the fifth
with ciphers; if the sixth figure is 5 or greater, increase the
fifth figure by 1 and replace the remaining figures with
ciphers.   Thus, if the number is 31,415,926, find the logarithm
of 31,416,000; if 31,415,426, find the logarithm of 31,415,000.

EXAMPLE.—Find log 31,416.

SOLUTION.—Find the mantissa of the logarithm of the first
four figures, as explained above.   This is, in the present case,
.49707.   Now, subtract the number in the column headed 1,
opposite 314 (the first three figures of the given number), from
the next greater consecutive number, in this case 721, in the
column headed 2.   721 — 707 = 14: this number is called the
difference.   At the extreme right of the page will be found a

secondary table headed P. P., and at the top of one of these columns, in this table, in bold-face type, will be found the difference. It will be noticed that each column is divided into two parts by a vertical line, and that the figures on the left of this line run in sequence from 1 to 9. Considering the difference column headed 14, we see opposite the number 6 (6 is the last or fifth figure of the number whose logarithm we are taking out) the number 8.4, and we add this number to the mantissa found above, disregarding the decimal point in the mantissa, obtaining 49,707 + 8.4 = 49,715.4. Now, since 4 is less than 5, we reject it, and obtain for our complete mantissa .49715. Since the characteristic of the logarithm of 31,416 is 4, log 31,416 = 4.49715.

EXAMPLE.—Find log 380.93.

SOLUTION.—Proceeding in exactly the same manner as above, the mantissa for 3,809 is 58,081 (the star directs us to take 58 instead of 57 for the first two figures); the next greater mantissa is 58,092, found in the column headed 0, opposite 381 in column headed N. The difference is 092 − 081 = 11. Looking in the section headed P. P. for column headed 11, we find opposite 3, 3.3; neglecting the .3, since it is less than 5, 3 is the amount to be added to the mantissa of the logarithm of 3,809 to form the logarithm of 38,093. Hence, 58,081 + 3 = 58,084, and since the characteristic is 2, log 380.93 = 2.58084.

EXAMPLE.—Find log 1,296,728.

SOLUTION.—Since this number consists of more than five figures and the sixth figure is less than 5, we find the logarithm of 1,296,700 and call it the logarithm of 1,296,728. The mantissa of log 1,296 is found to be 11,261. The difference is 294 − 261 = 33. Looking in the P. P. section for column headed 33, we find opposite 7, on the extreme left, 23.1; neglecting the .1, the amount to be added to the above mantissa is 23. Hence, the mantissa of log 1,296,728 = 11,261 + 23 = 11,284; since the characteristic is 6, log 1,296,728 = 6.11284.

EXAMPLE.—Find log 89.126.

SOLUTION.—Log 89.12 = 1.94998. Difference between this and log 89.13 = 1.95002 − 1.94998 = 4. The P. P. (proportional part) for the fifth figure of the number 6 is 2.4, or 2.

Hence, log 89.126 = 1.94998 + .00002 = 1.95000.

EXAMPLE.—Find log .096725.

SOLUTION.—   Log .09672 = $\overline{2}$.98552. Difference = 4.

P. P. for 5 =   2

Hence,   log .096725 = $\overline{2}$.98554.

To find the logarithm of a number consisting of five or more figures:

Rule.—I. *If the number consists of more than five figures and the sixth figure is 5 or greater, increase the fifth figure by 1 and write ciphers in place of the sixth and remaining figures.*

II. *Find the mantissa corresponding to the logarithm of the first four figures, and substract this mantissa from the next greater mantissa in the table; the remainder is the difference.*

III. *Find in the secondary table headed P. P. a column headed by the same number as that just found for the difference, and in this column, opposite the number corresponding to the fifth figure (or fifth figure increased by 1) of the given number (this figure is always situated at the left of the dividing line of the column), will be found the P. P. (proportional part) for that number. The P. P. thus found is to be added to the mantissa found in II, as in the preceding examples, and the result is the mantissa of the logarithm of the given number, as nearly as may be found with five-place tables.*

.O FIND A NUMBER WHOSE LOGARITHM IS GIVEN.

Rule.—I. *Consider the mantissa first. Glance along the different columns of the table which are headed 0, until the first two figures of the mantissa are found. Then, glance down the same column until the third figure is found (or 1 less than the third figure). Having found the first three figures, glance to the right along the row in which they are situated until the last three figures of the mantissa are found. Then, the number that heads the column in which the last three figures of the mantissa are found is the fourth figure of the required number, and the first three figures lie in the column headed N, and in the same row in which lie the last three figures of the mantissa.*

II. *If the mantissa cannot be found in the table, find the mantissa that is nearest to, but less than, the given mantissa, and which call the next less mantissa. Subtract the next less mantissa from the next greater mantissa in the table to obtain the difference. Also, subtract the next less mantissa from the mantissa of the given logarithm, and call the remainder the P. P. Looking in the secondary table headed P. P. for the column headed by the difference just found, find the number opposite the P. P. just found (or the P. P. corresponding most nearly to that just found); this number is the fifth figure of the required number; the fourth figure will be found at the top of the column containing the next less mantissa, and the first three figures in the column headed N and in the same row that contains the next less mantissa.*

III. *Having found the figures of the number as above directed, locate the decimal point by the rules for the characteristic, annexing ciphers to bring the number up to the required number of figures if the characteristic is greater than 4.*

EXAMPLE.—Find the number whose logarithm is 3.56867.

SOLUTION.—The first two figures of the mantissa are 56; glancing down the column, we find the third figure, 8 (in connection with 820), opposite 370 in the N column. Glancing to the right along the row containing 820, the last three figures of the mantissa, 867, are found in the column headed 4; hence, the fourth figure of the required number is 4, and the first three figures are 370, making the figures of the required number 3,704. Since the characteristic is 3, there are three figures to the left of the unit figure, and the number whose logarithm is 3.56867 is 3,704.

EXAMPLE.—Find the number whose logarithm is 3.56871.

SOLUTION.—The mantissa is not found in the table. The next less mantissa is 56,867; the difference between this and the next greater mantissa is 879 — 867 = 12, and the P. P. is 56,871 — 56,867 = 4. Looking in the P. P. section for the column headed 12, we do not find 4, but we do find 3.6 and 4.8. Since 3.6 is nearer 4 than 4.8, we take the number opposite 3.6 for the fifth figure of the required number; this is 3. Hence, the fourth figure is 4; the first three figures 370, and the figures of the number are 37,043. The characteristic being 3, the number is 3,704.3.

EXAMPLE.—Find the number whose logarithm is 5.95424.

SOLUTION.—The mantissa is found in the column headed 0, opposite 900 in the column headed N. Hence, the fourth figure is 0, and the number is 900,000, the characteristic being 5. Had the logarithm been $\overline{5}$.95424, the number would have been .00009.

EXAMPLE.—Find the number whose logarithm is .93036.

SOLUTION.—The first three figures of the mantissa, 930, are found in the 0 column, opposite 852 in the N column; but since the last two figures of all the mantissas in this row are greater than 36, we must seek the next less mantissa in the preceding row. We find it to be 93,034 (the star directing us to use 93 instead of 92 for the first two figures), in the column headed 8. The difference for this case is 039 — 034 = 5, and the P. P. is 036 — 034 = 2. Looking in the P. P. section for the column headed 5, we find the P. P., 2, opposite 4. Hence, the fifth figure is 4; the fourth figure is 8; the first three figures 851, and the number is 8.5184, the characteristic being 0.

EXAMPLE.—Find the number whose logarithm is $\overline{2}$.05753.

SOLUTION.—The next less mantissa is found in column headed 1, opposite 114 in the N column; hence, the first four figures are 1,141. The difference for this case is 767 — 729 = 38, and the P. P. is 753 — 729 = 24. Looking in the P. P. section for the column headed 38, we find that 24 falls between 22.8 and 26.6. The difference between 24 and 22.8 is 1.2, and between 24 and 26.6 is 2.6; hence, 24 is nearer 22.8 than it is to 26.6, and 6, opposite 22.8, is the fifth figure of the number. Hence, the number whose logarithm is $\overline{2}$.05753 is .011416.

In order to calculate by means of logarithms, a table is absolutely necessary. Hence, for this reason, we do not explain the method of calculating a logarithm. The work involved in calculating even a single logarithm is very great, and no method has yet been demonstrated, of which we are aware, by which the logarithm of a number like 121 can be calculated directly. Moreover, even if the logarithm could be readily obtained, it would be useless without a complete table, such as that which is here given, for the reason that after having used it, say to extract a root, the number corresponding to the logarithm of the result could not be found.

MULTIPLICATION BY LOGARITHMS.

The principle upon which the process is based may be illustrated as follows: Let X and Y represent two numbers whose logarithms are x and y. To find the logarithm of their product, we have, from the definition of a logarithm,

$$10^x = X, \quad (1)$$

and $$10^y = Y. \quad (2)$$

Since both members of (1) may be multiplied by the same quantity without destroying the equality, they evidently may be multiplied by equal quantities like 10^y and Y. Hence, multiplying (1) by (2), member by member,

$$10^x \times 10^y = 10^{x+y} = XY,$$

or, by the definition of a logarithm, $x + y = \log XY$. But XY is the product of X and Y, and $x + y$ is the sum of their logarithms; from which it follows that the sum of the logarithms of two numbers is equal to the logarithm of their product. Hence,

To multiply two or more numbers by using logarithms:

Rule.—*Add the logarithms of the several numbers, and the sum will be the logarithm of the product. Find the number corresponding to this logarithm, and the result will be the number sought.*

EXAMPLE.—Multiply 4.38, 5.217, and 83 together.

SOLUTION.— Log 4.38 = .64147
           Log 5.217 = .71742
           Log 83 = 1.91908
           ――――――――――――――――
Adding,   3.27797 = log (4.38 × 5.217 × 83).

Number corresponding to 3.27797 = 1,896.6. Hence, 4.38 × 5.217 × 83 = 1,896.6, nearly. By actual multiplication, the product is 1,896.5818, showing that the result obtained by using logarithms was correct to five figures.

When adding logarithms, their *algebraic* sum is always to be found. Hence, if some of their numbers multiplied together are wholly decimal, the algebraic sum of the characteristics will be the characteristic of the product. It must be remembered that the mantissas are always positive.

EXAMPLE.—Multiply 49.82, .00243, 17, and .97 together.

SOLUTION.—

Log 49.82 = 1.69740
Log .00243 = $\overline{3}$.38561
Log   17 = 1.23045
Log   .97 = $\overline{1}$.98677
_____

Adding,   0.30023 = log (49.82 × .00243 × 17 × .97).
Number corresponding to 0.30023 = 1.9963. Hence, 49.82
× .00243 × 17 × .97 = 1.9963.

In this case the sum of the mantissas was 2.30023. The
integral 2 added to the positive characteristics makes their
sum = 2 + 1 + 1 = 4; sum of negative characteristics = $\overline{3}$
+ $\overline{1}$ = $\overline{4}$, whence 4 + (− 4) = 0. If, instead of 17, the number
had been .17 in the above example, the logarithm of .17 would
have been $\overline{1}$.23045, and the sum of the logarithms would have
been $\overline{2}$.30023; the product would then have been .019963.

It can now be shown why all numbers with figures in the
same order have the same mantissa, without regard to the
decimal point. Thus, suppose it were known that log 2.06
= .31387. Then, log 20.6 = log (2.06 × 10) = log 2.06 + log 10
= .31387 + 1 = 1.31387. And so it might be proved with the
decimal point in any other position.

_____

DIVISION BY LOGARITHMS.

As before, let X and Y represent two numbers whose logarithms are x and y. To find the logarithm of their quotient,
we have, from the definition of a logarithm,

$$10^x = X, \quad (1)$$
and
$$10^y = Y. \quad (2)$$

Dividing (1) by (2), $10^{x-y} = \dfrac{X}{Y}$, or, by the definition of a

logarithm, $x - y = \log \dfrac{X}{Y}$. But $\dfrac{X}{Y}$ is the quotient of $X ÷ Y$,

and $x - y$ is the difference of their logarithms, from which it
follows that *the difference between the logarithms of two numbers
is equal to the logarithm of their quotient.* Hence, to divide
one number by another by means of logarithms:

Rule.—*Subtract the logarithm of the divisor from the logarithm
of the dividend, and the result will be the logarithm of the quotient.*

EXAMPLE.—Divide 6,784.2 by 27.42.

SOLUTION.—   Log 6,784.2 = 3.83150
                Log   27.42 = 1.43807

$$difference = 2.39343 = \log(6,784.2 \div 27.42).$$

Number corresponding to 2.39343 = 247.42. Hence, 6,784.2 ÷ 27.42 = 247.42.

When subtracting logarithms, their *algebraic* difference is to be found. The operation may sometimes be confusing, because the mantissa is always positive, and the characteristic may be either positive or negative. *When the logarithm to be subtracted is greater than the logarithm from which it is to be taken, or when negative characteristics appear, subtract the mantissa first, and then the characteristic, by changing its sign and adding.*

EXAMPLE.—Divide 274.2 by 6,784.2.

SOLUTION.—   Log   274.2 = 2.43807
                Log 6,784.2 = 3.83150
                            ——————
                            $\overline{2}$.60657

First subtracting the mantissa .83150 gives .60657 for the mantissa of the quotient. In subtracting, 1 had to be taken from the characteristic of the minuend, leaving a characteristic of 1. Subtract the characteristic 3 from this, by changing its sign and adding $1 - 3 = \overline{2}$, the characteristic of the quotient. Number corresponding to $\overline{2}$.60657 = .040418. Hence, 274.2 ÷ 6,784.2 = .040418.

EXAMPLE.—Divide .067842 by .002742.

SOLUTION.—   Log .067842 = $\overline{2}$.83150
                Log .002742 = $\overline{3}$.43807
                            ——————
                $difference = 1.39343$

Since .83150 − .43807 = .39343 and −2 + 3 = 1, number corresponding to 1.39343 = 24.742. Hence, .067842 ÷ .002742 = 24.742.

The only case that is likely to cause trouble in subtracting is that in which the logarithm of the minuend has a negative characteristic, or none at all, and a mantissa less than the mantissa of the subtrahend. For example, let it be required to subtract the logarithm 3.74036 from the logarithm

$\bar{3}.55145$. The logarithm $\bar{3}.55145$ is equivalent to $-3+.55145$. Now, if we add both $+1$ and -1 to this logarithm, it will not change its value. Hence, $\bar{3}.55145 = -3-1+1+.55145 = \bar{4} +1.55145$. Therefore, $\bar{3}.55145 - 3.74036 =$

$$\bar{4}+1.55145$$
$$3+.74036$$
$$\text{\emph{difference}} = \bar{7}+.81109 = \bar{7}.81109.$$

Had the characteristic of the above logarithm been 0 instead of $\bar{3}$, the process would have been exactly the same. Thus, $.55145 = \bar{1}+1.55145$; hence,

$$\bar{1}+1.55145$$
$$3+.74036$$
$$\text{\emph{difference}} = \bar{4}+.81109 = \bar{4}.81109.$$

EXAMPLE.—Divide .02742 by 67.842.

SOLUTION.— Log $.02742 = \bar{2}.43807 = \bar{3}+1.43807$
Log $67.842 = 1.83150 = 1+.83150$

$$\text{\emph{difference}} = \bar{4}+.60657 = \bar{4}.60657.$$

Number corresponding to $\bar{4}.60657 = .00040417$. Hence, $.02742 \div 67.842 = .00040417$.

EXAMPLE.—What is the reciprocal of 3.1416?

SOLUTION.—Reciprocal of $3.1416 = \dfrac{1}{3.1416}$, and log $\dfrac{1}{3.1416}$ $= \log 1 - \log 3.1416 = 0 - .49715$. Since $0 = -1+1$,

$$\bar{1}+1.00000$$
$$\phantom{\bar{1}+1}.49715$$
$$\text{\emph{difference}} = \bar{1}+.50285 = \bar{1}.50285.$$

Number whose logarithm is $\bar{1}.50285 = .31831$.

INVOLUTION BY LOGARITHMS.

If X represents a number whose logarithm is x, we have, from the definition of a logarithm,

$$10^x = X.$$

Raising both numbers to some power, as the nth, the equation becomes

$$10^{xn} = X^n.$$

But X^n is the required power of X, and xn is its logarithm, from which it follows that the logarithm of a number multiplied by the exponent of the power to which it is raised is equal to the logarithm of the power. Hence, to raise a number to any power by the use of logarithms:

Rule.—*Multiply the logarithm of the number by the exponent that denotes the power to which the number is to be raised, and the result will be the logarithm of the required power.*

EXAMPLE.—What is (a) the square of 7.92? (b) the cube of 94.7? (c) the 1.6 power of 512, that is, the value of $512^{1.6}$?

SOLUTION.—(a) Log 7.92 = .89873; exponent of power = 2. Hence, .89873 × 2 = 1.79746 = log 7.92^2. Number corresponding to 1.79746 = 62.727. Hence, 7.92^2 = 62.727, nearly.

(b) Log 94.7 = 1.97635; 1.97635 × 3 = 5.92905 = log 94.7^3. Number corresponding to 5.92905 = 849,280, nearly. Hence, 94.7^3 = 849,280, nearly.

(c) Log $512^{1.6}$ = 1.6 × log 512 = 1.6 × 2.70927 = 4.334832, or 4.33483 (when using five-place logarithms) = log 21,619. Hence, $512^{1.6}$ = 21,619 nearly.

If the number is wholly decimal, so that the characteristic is negative, *multiply the two parts of the logarithm separately by the exponent of the number. If, after multiplying the mantissa, the product has a characteristic, add it, algebraically, to the negative characteristic multiplied by the exponent, and the result will be the negative characteristic of the required power.*

EXAMPLE.—Raise .0751 to the fourth power.

SOLUTION.—Log $.0751^4$ = 4 × log .0751 = 4 × $\overline{2}$.87564. Multiplying the parts separately, 4 × $\overline{2}$ = $\overline{8}$ and 4 × .87564 = 3.50256. Adding the 3 and $\overline{8}$, 3 + (−8) = −5; therefore, log $.0751^4$ = $\overline{5}$.50256. Number corresponding to this = .00003181. Hence, $.0751^4$ = .00003181.

A decimal may be raised to a power whose exponent contains a decimal as follows:

EXAMPLE.—Raise .8 to the 1.21 power.

SOLUTION.—Log $.8^{1.21}$ = 1.21 × $\overline{1}$.90309. There are several ways of performing the multiplication.

First Method.—Adding the characteristic and mantissa algebraically, the result is −.09691. Multiplying this by 1.21 gives −.1172611, or −.11726, when using five-place logarithms. To obtain a positive mantissa, add +1 and −1; whence, log $.8^{1.21}$ = −1 + 1 − .11726 = $\overline{1}$.88274.

Second Method.—Multiplying the characteristic and mantissa separately gives −1.21 + 1.09274. Adding characteristic and mantissa algebraically, gives −.11726; then, adding +1 and −1, log .8$^{1.21}$ = $\overline{1}$.88274.

Third Method.—Multiplying the characteristic and mantissa separately gives −1.21 + 1.09274. Adding the decimal part of the characteristic to the mantissa gives −1 + (−.21 + 1.09274) = $\overline{1}$.88274 = log .8$^{1.21}$. The number corresponding to the logarithm $\overline{1}$.88274 = .76338.

Any one of the above three methods may be used, but we recommend the first or the third. The third is the most elegant and saves figures, but requires the exercise of more caution than the first method does. Below will be found the entire work of multiplication for both .8$^{1.21}$ and .8$^{.21}$.

$\overline{1}$.90309	$\overline{1}$.90309
1.21	.21
90309	90309
180618	180618
90309	+1.1896489
1.0927389	−1 − .21
−1.21	$\overline{1}$.9796489, or $\overline{1}$.97965.

$\overline{1}$.8827389, or $\overline{1}$.88274.

In the second case, the negative decimal obtained by multiplying −1 and .21 was greater than the positive decimal obtained by multiplying .90309 and .21; hence, +1 and −1 were added, as shown.

EVOLUTION BY LOGARITHMS.

If X represents a number whose logarithm is x, we have, from the definition of a logarithm,

$$10^x = X.$$

Extracting some root of both members, as the nth, the equation becomes

$$10^{\frac{x}{n}} = \sqrt[n]{X}.$$

But $\sqrt[n]{X}$ is the required root of X, and $\frac{x}{n}$ is its logarithm, from which it follows that the logarithm of a number divided by the index of the root to be extracted is equal to the logarithm of the root. Hence, to extract any root of a number by means of logarithms:

Rule.—*Divide the logarithm of the number by the index of the root; the result will be the logarithm of the root.*

EXAMPLE.—Extract (a) the square root of 77,851; (b) the cube root of 698,970; (c) the 2.4 root of 8,964,300.

SOLUTION.—(a) Log 77,851 = 4.89127; the index of the root is 2; hence, log $\sqrt{77,851}$ = 4.89127 ÷ 2 = 2.44564; number corresponding to this = 279.02. Hence, $\sqrt{77,851}$ = 279.02, nearly.

(b) Log $\sqrt[3]{698,970}$ = 5.84446 ÷ 3 = 1.94815 = log 88.746; or, $\sqrt[3]{698,970}$ = 88.747, nearly.

(c) Log $\sqrt[2.4]{8,964,300}$ = 6.95251 ÷ 2.4 = 2.89688 = log 788.64; or, $\sqrt[2.4]{8,964,300}$ = 788.64, nearly.

If it is required to extract a root of a number wholly decimal, and the negative characteristic will not exactly contain the index of the root, without a remainder, proceed as follows:

Separate the two parts of the logarithm; add as many units (or parts of a unit) to the negative characteristic as will make it exactly contain the index of the root. Add the same number to the mantissa, and divide both parts by the index. The result will be the characteristic and mantissa of the root.

EXAMPLE.—Extract the cube root of .0003181.

SOLUTION.—Log $\sqrt[3]{.0003181}$ = $\dfrac{\log .0003181}{3}$ = $\dfrac{\overline{4}.50256}{3}$.

$$(\overline{4} + \overline{2} = \overline{6}) + (2 + .50256 = 2.50256).$$
$$(\overline{6} + 3 = \overline{2}) + (2.50256 + 3 = .83419);$$

or, log $\sqrt[3]{.0003181}$ = $\overline{2}.83419$ = log .068263.

Hence, $\sqrt[3]{.0003181}$ = .068263.

EXAMPLE.—Find the value of $\sqrt[1.41]{.0003181}$.

SOLUTION.—Log $\sqrt[1.41]{.0003181}$ = $\dfrac{\log .0003181}{1.41}$ = $\dfrac{\overline{4}.50256}{1.41}$.

If −.23 be added to the characteristic, it will contain 1.41 exactly 3 times. Hence,

$$[-4 + (-.23) = -4.23] + (.23 + .50256 = .73256).$$
$$(-4.23 + 1.41 = \overline{3}) + (.73256 + 1.41 = .51955);$$

or, log $\sqrt[1.41]{.0003181}$ = $\overline{3}.51955$ = log .0033079.

Hence, $\sqrt[1.41]{.0003181}$ = .0033079.

5

EXAMPLE.—Solve this expression by logarithms:

$$\frac{497 \times .0181 \times 762}{3,300 \times .6517} = ?$$

SOLUTION.—

Log   497 = 2.69636
Log   .0181 = $\overline{2}$.25768
Log   762 = 2.88195

Log product = 3.83599
Log   3,300 = 3.51851
Log   .6517 = $\overline{1}$.81405

Log product = 3.33256

3.83599 − 3.33256 = .50343 = log 3.1874.

Hence, $\dfrac{497 \times .0181 \times 762}{3,300 \times .6517} = 3.1874.$

EXAMPLE.—Solve $\sqrt[3]{\dfrac{504,203 \times 507}{1.75 \times 71.4 \times 87}}$ by logarithms.

SOLUTION.—

Log   504,203 = 5.70260
Log   507 = 2.70501

Log product = 8.40761
Log   1.75 = .24304
Log   71.4 = 1.85370
Log   87 = 1.93952

Log product = 4.03626

$\dfrac{8.40761 - 4.03626}{3} = 1.45712 = \log 28.65.$

Hence, $\sqrt[3]{\dfrac{504,203 \times 507}{1.75 \times 71.4 \times 87}} = 28.65.$

Logarithms can often be applied to the solution of equations.

EXAMPLE.—Solve the equation $2.43x^5 = \sqrt[6]{.0648}.$

SOLUTION.— $2.43x^5 = \sqrt[6]{.0648}.$

Dividing by 2.43, $x^5 = \dfrac{\sqrt[6]{.0648}}{2.43}.$

Taking the logarithm of both numbers,

$$5 \times \log x = \frac{\log .0648}{6} - \log 2.43;$$

or
$$5 \log x = \frac{\overline{2}.81158}{6} - .38561$$
$$= \overline{1}.80193 - .38561$$
$$= \overline{1}.41632.$$

Dividing by 5, $\log x = \overline{1}.88326$;

whence, $x = .7643$.

EXAMPLE.—Solve the equation $4.5^x = 8$.

SOLUTION.—Taking the logarithms of both numbers,
$$x \log 4.5 = \log 8,$$

whence, $x = \dfrac{\log 8}{\log 4.5} = \dfrac{.90309}{.65321}.$

Taking logarithms again,
$$\log x = \log .90309 - \log .65321 = \overline{1}.95573 - \overline{1}.81505$$
$$= .14068; \text{ whence, } x = 1.3825.$$

REMARK.—Logarithms are particularly useful in those cases when the unknown quantity is an exponent, as in the last example, or when the exponent contains a decimal, as in several instances in the examples given on pages 45–49. Such examples can be solved without the use of logarithms, but the process is very long and somewhat involved, and the arithmetical work required is enormous. To solve the example last given without using the logarithmic table and obtain the value of x correct to five figures would require, perhaps, 100 times as many figures as were used in the solution given, and the resulting liability to error would be correspondingly increased; indeed, to confine the work to this number of figures would also require a good knowledge of short-cut methods in multiplication and division, and judgment and skill on the part of the calculator that can only be acquired by practice and experience.

Formulas containing quantities affected with decimal exponents are generally of an empirical nature; that is, the constants or exponents or both are given such values as will make the results obtained by the formulas agree with those obtained by experiment. Such formulas occur frequently in works treating on thermodynamics, strength of materials, machine design, etc.

COMMON LOGARITHMS

N.	L. 0	1	2	3	4	5	6	7	8	9
100	00 000	043	087	130	173	217	260	303	346	389
101	432	475	518	561	604	647	689	732	775	817
102	860	903	945	988	*030	*072	*115	*157	*199	*242
103	01 284	326	368	410	452	494	536	578	620	662
104	703	745	787	828	870	912	953	995	*036	*078
105	02 119	160	202	243	284	325	366	407	449	490
106	531	572	612	653	694	735	776	816	857	898
107	938	979	*019	*060	*100	*141	*181	*222	*262	*302
108	03 342	383	423	463	503	543	583	623	663	703
109	743	782	822	862	902	941	981	*021	*060	*100
110	04 139	179	218	258	297	336	376	415	454	493
111	532	571	610	650	689	727	766	805	844	883
112	922	961	999	*038	*077	*115	*154	*192	*231	*269
113	05 308	346	385	423	461	500	538	576	614	652
114	690	729	767	805	843	881	918	956	994	*032
115	06 070	108	145	183	221	258	296	333	371	408
116	446	483	521	558	595	633	670	707	744	781
117	819	856	893	930	967	*004	*041	*078	*115	*151
118	07 188	225	262	298	335	372	408	445	482	518
119	555	591	628	664	700	737	773	809	846	882
120	918	954	990	*027	*063	*099	*135	*171	*207	*243
121	08 279	314	350	386	422	458	493	529	565	600
122	636	672	707	743	778	814	849	884	920	955
123	991	*026	*061	*096	*132	*167	*202	*237	*272	*307
124	09 342	377	412	447	482	517	552	587	621	656
125	691	726	760	795	830	864	899	934	968	*003
126	10 037	072	106	140	175	209	243	278	312	346
127	380	415	449	483	517	551	585	619	653	687
128	721	755	789	823	857	890	924	958	992	*025
129	11 059	093	126	160	193	227	261	294	327	361
130	394	428	461	494	528	561	594	628	661	694
131	727	760	793	826	860	893	926	959	992	*024
132	12 057	090	123	156	189	222	254	287	320	352
133	385	418	450	483	516	548	581	613	646	678
134	710	743	775	808	840	872	905	937	969	*001
135	13 033	066	098	130	162	194	226	258	290	322
136	354	386	418	450	481	513	545	577	609	640
137	672	704	735	767	799	830	862	893	925	956
138	988	*019	*051	*052	*114	*145	*176	*208	*239	*270
139	14 301	333	364	395	426	457	489	520	551	582
140	613	644	675	706	737	768	799	829	860	891
141	922	953	983	*014	*045	*076	*106	*137	*168	*198
142	15 229	259	290	320	351	381	412	442	473	503
143	534	564	594	625	655	685	715	746	776	806
144	836	866	897	927	957	987	*017	*047	*077	*107
145	16 137	167	197	227	256	286	316	346	376	406
146	435	465	495	524	554	584	613	643	673	702
147	732	761	791	820	850	879	909	938	967	997
148	17 026	056	085	114	143	173	202	231	260	289
149	319	348	377	406	435	464	493	522	551	580
150	609	638	667	696	725	754	782	811	840	869
N.	L. 0	1	2	3	4	5	6	7	8	9

P. P.

	44	43	42
1	4.4	4.3	4.2
2	8.8	8.6	8.4
3	13.2	12.9	12.6
4	17.6	17.2	16.8
5	22.0	21.5	21.0
6	26.4	25.8	25.2
7	30.8	30.1	29.4
8	35.2	34.4	33.6
9	39.6	38.7	37.8

	41	40	39
1	4.1	4.0	3.9
2	8.2	8.0	7.8
3	12.3	12.0	11.7
4	16.4	16.0	15.6
5	20.5	20.0	19.5
6	24.6	24.0	23.4
7	28.7	28.0	27.3
8	32.8	32.0	31.2
9	36.9	36.0	35.1

	38	37	36
1	3.8	3.7	3.6
2	7.6	7.4	7.2
3	11.4	11.1	10.8
4	15.2	14.8	14.4
5	19.0	18.5	18.0
6	22.8	22.2	21.6
7	26.6	25.9	25.2
8	30.4	29.6	28.8
9	34.2	33.3	32.4

	35	34	33
1	3.5	3.4	3.3
2	7.0	6.8	6.6
3	10.5	10.2	9.9
4	14.0	13.6	13.2
5	17.5	17.0	16.5
6	21.0	20.4	19.8
7	24.5	23.8	23.1
8	28.0	27.2	26.4
9	31.5	30.6	29.7

	32	31	30
1	3.2	3.1	3.0
2	6.4	6.2	6.0
3	9.6	9.3	9.0
4	12.8	12.4	12.0
5	16.0	15.5	15.0
6	19.2	18.6	18.0
7	22.4	21.7	21.0
8	25.6	24.8	24.0
9	28.8	27.9	27.0

P. P.

N.	L.0	1	2	3	4	5	6	7	8	9
150	17 609	638	667	696	725	754	782	811	840	869
151	898	926	955	984	*013	*041	*070	*099	*127	*156
152	18 184	213	241	270	298	327	355	384	412	441
153	469	498	526	554	583	611	639	667	696	724
154	752	780	808	837	865	893	921	949	977	*005
155	19 033	061	089	117	145	173	201	229	257	285
156	312	340	368	396	424	451	479	507	535	562
157	590	618	645	673	700	728	756	783	811	838
158	866	893	921	948	976	*003	*030	*058	*085	*112
159	20 140	167	194	222	249	276	303	330	358	385
160	412	439	466	493	520	548	575	602	629	656
161	683	710	737	763	790	817	844	871	898	925
162	952	978	*005	*032	*059	*085	*112	*139	*165	*192
163	21 219	245	272	299	325	352	378	405	431	458
164	484	511	537	564	590	617	643	669	696	722
165	748	775	801	827	854	880	906	932	958	985
166	22 011	037	063	089	115	141	167	194	220	246
167	272	298	324	350	376	401	427	453	479	505
168	531	557	583	608	634	660	686	712	737	763
169	789	814	840	866	891	917	943	968	994	*019
170	23 045	070	096	121	147	172	198	223	249	274
171	300	325	350	376	401	426	452	477	502	528
172	553	578	603	629	654	679	704	729	754	779
173	805	830	855	880	905	930	955	980	*005	*030
174	24 055	080	105	130	155	180	204	229	254	279
175	304	329	353	378	403	428	452	477	502	527
176	551	576	601	625	650	674	699	724	748	773
177	797	822	846	871	895	920	944	969	993	*018
178	25 042	066	091	115	139	164	188	212	237	261
179	285	310	334	358	382	406	431	455	479	503
180	527	551	575	600	624	648	672	696	720	744
181	768	792	816	840	864	888	912	935	959	983
182	26 007	031	055	079	102	126	150	174	198	221
183	245	269	293	316	340	364	387	411	435	458
184	482	505	529	553	576	600	623	647	670	694
185	717	741	764	788	811	834	858	881	905	928
186	951	975	998	*021	*045	*068	*091	*114	*138	*161
187	27 184	207	231	254	277	300	323	346	370	393
188	416	439	462	485	508	531	554	577	600	623
189	646	669	692	715	738	761	784	807	830	852
190	875	898	921	944	967	989	*012	*035	*058	*081
191	28 103	126	149	171	194	217	240	262	285	307
192	330	353	375	398	421	443	466	488	511	533
193	556	578	601	623	646	668	691	713	735	758
194	780	803	825	847	870	892	914	937	959	981
195	29 003	026	048	070	092	115	137	159	181	203
196	226	248	270	292	314	336	358	380	403	425
197	447	469	491	513	535	557	579	601	623	645
198	667	688	710	732	754	776	798	820	842	863
199	885	907	929	951	973	994	*016	*038	*060	*081
200	30 103	125	146	168	190	211	233	255	276	298
N.	L.0	1	2	3	4	5	6	7	8	9

P.P.

	29	28
1	2.9	2.8
2	5.8	5.6
3	8.7	8.4
4	11.6	11.2
5	14.5	14.0
6	17.4	16.8
7	20.3	19.6
8	23.2	22.4
9	26.1	25.2

	27	26
1	2.7	2.6
2	5.4	5.2
3	8.1	7.8
4	10.8	10.4
5	13.5	13.0
6	16.2	15.6
7	18.9	18.2
8	21.6	20.8
9	24.3	23.4

	25
1	2.5
2	5.0
4	7.5
4	10.0
5	12.5
6	15.0
7	17.5
8	20.0
9	22.5

	24	23
1	2.4	2.3
2	4.8	4.6
3	7.2	6.9
4	9.6	9.2
5	12.0	11.5
6	14.4	13.8
7	16.8	16.1
8	19.2	18.4
9	21.6	20.7

	22	21
1	2.2	2.1
2	4.4	4.2
3	6.6	6.3
4	8.8	8.4
5	11.0	10.5
6	13.2	12.6
7	15.4	14.7
8	17.6	16.8
9	19.8	18.9

P.P.

TABLE—(Continued).

N.	L.0	1	2	3	4	5	6	7	8	9
200	30 103	125	146	168	190	211	233	255	276	298
201	320	341	363	384	406	428	449	471	492	514
202	535	557	578	600	621	643	664	685	707	728
203	750	771	792	814	835	856	878	899	920	942
204	963	984	*006	*027	*048	*069	*091	*112	*133	*154
205	31 175	197	218	239	260	281	302	323	345	366
206	387	408	429	450	471	492	513	534	555	576
207	597	618	639	660	681	702	723	744	765	785
208	806	827	848	869	890	911	931	952	973	994
209	32 015	035	056	077	098	118	139	160	181	201
210	222	243	263	284	305	325	346	366	387	408
211	428	449	469	490	510	531	552	572	593	613
212	634	654	675	695	715	736	756	777	797	818
213	838	858	879	899	919	940	960	980	*001	*021
214	33 041	062	082	102	122	143	163	183	203	224
215	244	264	284	304	325	345	365	385	405	425
216	445	465	486	506	526	546	566	586	606	626
217	646	666	686	706	726	746	766	786	806	826
218	846	866	885	905	925	945	965	985	*005	*025
219	34 044	064	084	104	124	143	163	183	203	223
220	242	262	282	301	321	341	361	380	400	420
221	439	459	479	498	518	537	557	577	596	616
222	635	655	674	694	713	733	753	772	792	811
223	830	850	869	889	908	928	947	967	986	*005
224	35 025	044	064	083	102	122	141	160	180	199
225	218	238	257	276	295	315	334	353	372	392
226	411	430	449	468	488	507	526	545	564	583
227	603	622	641	660	679	698	717	736	755	774
228	793	813	832	851	870	889	908	927	946	965
229	984	*003	*021	*040	*059	*078	*097	*116	*135	*154
230	36 173	192	211	229	248	267	286	305	324	342
231	361	380	399	418	436	455	474	493	511	530
232	549	568	586	605	624	642	661	680	698	717
233	736	754	773	791	810	829	847	866	884	903
234	922	940	959	977	996	*014	*033	*051	*070	*088
235	37 107	125	144	162	181	199	218	236	254	273
236	291	310	328	346	365	383	401	420	438	457
237	475	493	511	530	548	566	585	603	621	639
238	658	676	694	712	731	749	767	785	803	822
239	840	858	876	894	912	931	949	967	985	*003
240	38 021	039	057	075	093	112	130	148	166	184
241	202	220	238	256	274	292	310	328	346	364
242	382	399	417	435	453	471	489	507	525	543
243	561	578	596	614	632	650	668	686	703	721
244	739	757	775	792	810	828	846	863	881	899
245	917	934	952	970	987	*005	*023	*041	*058	*076
246	39 094	111	129	146	164	182	199	217	235	252
247	270	287	305	322	340	358	375	393	410	428
248	445	463	480	498	515	533	550	568	585	602
249	620	637	655	672	690	707	724	742	759	777
250	794	811	829	846	863	881	898	915	933	950
N.	L.0	1	2	3	4	5	6	7	8	9

P. P.

	22	21
1	2.2	2.1
2	4.4	4.2
3	6.6	6.3
4	8.8	8.4
5	11.0	10.5
6	13.2	12.6
7	15.4	14.7
8	17.6	16.8
9	19.8	18.9

	20
1	2.0
2	4.0
3	6.0
4	8.0
5	10.0
6	12.0
7	14.0
8	16.0
9	18.0

	19
1	1.9
2	3.8
3	5.7
4	7.6
5	9.5
6	11.4
7	13.3
8	15.2
9	17.1

	18
1	1.8
2	3.6
3	5.4
4	7.2
5	9.0
6	10.8
7	12.6
8	14.4
9	16.2

	17
1	1.7
2	3.4
3	5.1
4	6.8
5	8.5
6	10.2
7	11.9
8	13.6
9	15.3

P. P.

TABLE—(Continued).

N.	L.0	1	2	3	4	5	6	7	8	9
250	39 794	811	829	846	863	881	898	915	933	950
251	967	985	*002	*019	*037	*054	*071	*088	*106	*123
252	40 140	157	175	192	209	226	243	261	278	295
253	312	329	346	364	381	398	415	432	449	466
254	483	500	518	535	552	569	586	603	620	637
255	654	671	688	705	722	739	756	773	790	807
256	824	841	858	875	892	909	926	943	960	976
257	993	*010	*027	*044	*061	*078	*095	*111	*128	*145
258	41 162	179	196	212	229	246	263	280	296	313
259	330	347	363	380	397	414	430	447	464	481
260	497	514	531	547	564	581	597	614	631	647
261	664	681	697	714	731	747	764	780	797	814
262	830	847	863	880	896	913	929	946	963	979
263	996	*012	*029	*045	*062	*078	*095	*111	*127	*144
264	42 160	177	193	210	226	243	259	275	292	308
265	325	341	357	374	390	406	423	439	455	472
266	488	504	521	537	553	570	586	602	619	635
267	651	667	684	700	716	732	749	765	781	797
268	813	830	846	862	878	894	911	927	943	959
269	975	991	*008	*024	*040	*056	*072	*088	*104	*120
270	43 136	152	169	185	201	217	233	249	265	281
271	297	313	329	345	361	377	393	409	425	441
272	457	473	489	505	521	537	553	569	584	600
273	616	632	648	664	680	696	712	727	743	759
274	775	791	807	823	838	854	870	886	902	917
275	933	949	965	981	996	*012	*028	*044	*059	*075
276	44 091	107	122	138	154	170	185	201	217	232
277	248	264	279	295	311	326	342	358	373	389
278	404	420	436	451	467	483	498	514	529	545
279	560	576	592	607	623	638	654	669	685	700
280	716	731	747	762	778	793	809	824	840	855
281	871	886	902	917	932	948	963	979	994	*010
282	45 025	040	056	071	086	102	117	133	148	163
283	179	194	209	225	240	255	271	286	301	317
284	332	347	362	378	393	408	423	439	454	469
285	484	500	515	530	545	561	576	591	606	621
286	637	652	667	682	697	712	728	743	758	773
287	788	803	818	834	849	864	879	894	909	924
288	939	954	969	984	*000	*015	*030	*045	*060	*075
289	46 090	105	120	135	150	165	180	195	210	225
290	240	255	270	285	300	315	330	345	359	374
291	389	404	419	434	449	464	479	494	509	523
292	538	553	568	583	598	613	627	642	657	672
293	687	702	716	731	746	761	776	790	805	820
294	835	850	864	879	894	909	923	938	953	967
295	982	997	*012	*026	*041	*056	*070	*085	*100	*114
296	47 129	144	159	173	188	202	217	232	246	261
297	276	290	305	319	334	349	363	378	392	407
298	422	436	451	465	480	494	509	524	538	553
299	567	582	596	611	625	640	654	669	683	698
300	712	727	741	756	770	784	799	813	828	842

N.	L.0	1	2	3	4	5	6	7	8	9

P. P.

18

1	1.8
2	3.6
3	5.4
4	7.2
5	9.0
6	10.8
7	12.6
8	14.4
9	16.2

17

1	1.7
2	3.4
3	5.1
4	6.8
5	8.5
6	10.2
7	11.9
8	13.6
9	15.3

16

1	1.6
2	3.2
3	4.8
4	6.4
5	8.0
6	9.6
7	11.2
8	12.8
9	14.4

15

1	1.5
2	3.0
3	4.5
4	6.0
5	7.5
6	9.0
7	10.5
8	12.0
9	13.5

14

1	1.4
2	2.8
3	4.2
4	5.6
5	7.0
6	8.4
7	9.8
8	11.2
9	12.6

TABLE—(Continued).

N.	L. 0	1	2	3	4	5	6	7	8	9
300	47 712	727	741	756	770	784	799	813	828	842
301	857	871	885	900	914	929	943	958	972	986
302	48 001	015	029	044	058	073	087	101	116	130
303	144	159	173	187	202	216	230	244	259	273
304	287	302	316	330	344	359	373	387	401	416
305	430	444	458	473	487	501	515	530	544	558
306	572	586	601	615	629	643	657	671	686	700
307	714	728	742	756	770	785	799	813	827	841
308	855	869	883	897	911	926	940	954	968	982
309	996	*010	*024	*038	*052	*066	*080	*094	*108	*122
310	49 136	150	164	178	192	206	220	234	248	262
311	276	290	304	318	332	346	360	374	388	402
312	415	429	443	457	471	485	499	513	527	541
313	554	568	582	596	610	624	638	651	665	679
314	693	707	721	734	748	762	776	790	803	817
315	831	845	859	872	886	900	914	927	941	955
316	969	982	996	*010	*024	*037	*051	*065	*079	*092
317	50 106	120	133	147	161	174	188	202	215	229
318	243	256	270	284	297	311	325	338	352	365
319	379	393	406	420	433	447	461	474	488	501
320	515	529	542	556	569	583	596	610	623	637
321	651	664	678	691	705	718	732	745	759	772
322	786	799	813	826	840	853	866	880	893	907
323	920	934	947	961	974	987	*001	*014	*028	*041
324	51 055	068	081	095	108	121	135	148	162	175
325	188	202	215	228	242	255	268	282	295	308
326	322	335	348	362	375	388	402	415	428	441
327	455	468	481	495	508	521	534	548	561	574
328	587	601	614	627	640	654	667	680	693	706
329	720	733	746	759	772	786	799	812	825	838
330	851	865	878	891	904	917	930	943	957	970
331	983	996	*009	*022	*035	*048	*061	*075	*088	*101
332	52 114	127	140	153	166	179	192	205	218	231
333	244	257	270	284	297	310	323	336	349	362
334	375	388	401	414	427	440	453	466	479	492
335	504	517	530	543	556	569	582	595	608	621
336	634	647	660	673	686	699	711	724	737	750
337	763	776	789	802	815	827	840	853	866	879
338	892	905	917	930	943	956	969	982	994	*007
339	53 020	033	046	058	071	084	097	110	122	135
340	148	161	173	186	199	212	224	237	250	263
341	275	288	301	314	326	339	352	364	377	390
342	403	415	428	441	453	466	479	491	504	517
343	529	542	555	567	580	593	605	618	631	643
344	656	668	681	694	706	719	732	744	757	769
345	782	794	807	820	832	845	857	870	882	895
346	908	920	933	945	958	970	983	995	*008	*020
347	54 033	045	058	070	083	095	108	120	133	145
348	158	170	183	195	208	220	233	245	258	270
349	283	295	307	320	332	345	357	370	382	394
350	407	419	432	444	456	469	481	494	506	518
N.	L. 0	1	2	3	4	5	6	7	8	9

P. P.

15
1 | 1.5
2 | 3.0
3 | 4.5
4 | 6.0
5 | 7.5
6 | 9.0
7 | 10.5
8 | 12.0
9 | 13.5

14
1 | 1.4
2 | 2.8
3 | 4.2
4 | 5.6
5 | 7.0
6 | 8.4
7 | 9.8
8 | 11.2
9 | 12.6

13
1 | 1.3
2 | 2.6
3 | 3.9
4 | 5.2
5 | 6.5
6 | 7.8
7 | 9.1
8 | 10.4
9 | 11.7

12
1 | 1.2
2 | 2.4
3 | 3.6
4 | 4.8
5 | 6.0
6 | 7.2
7 | 8.4
8 | 9.6
9 | 10.8

TABLE—(*Continued*).

N.	L.0	1	2	3	4	5	6	7	8	9	P. P.
350	54 407	419	432	444	456	469	481	494	506	518	
351	531	543	555	568	580	593	605	617	630	642	
352	654	667	679	691	704	716	728	741	753	765	
353	777	790	802	814	827	839	851	864	876	888	**13**
354	900	913	925	937	949	962	974	986	998	*011	1 1.3
355	55 023	035	047	060	072	084	096	108	121	133	2 2.6
356	145	157	169	182	194	206	218	230	242	255	3 3.9
357	267	279	291	303	315	328	340	352	364	376	4 5.2
358	388	400	413	425	437	449	461	473	485	497	5 6.5
359	509	522	534	546	558	570	582	594	606	618	6 7.8
360	630	642	654	666	678	691	703	715	727	739	7 9.1 / 8 10.4
361	751	763	775	787	799	811	823	835	847	859	9 11.7
362	871	883	895	907	919	931	943	955	967	979	
363	991	*003	*015	*027	*038	*050	*062	*074	*086	*098	
364	56 110	122	134	146	158	170	182	194	205	217	
365	229	241	253	265	277	289	301	312	324	336	**12**
366	348	360	372	384	396	407	419	431	443	455	1 1.2
367	467	478	490	502	514	526	538	549	561	573	2 2.4
368	585	597	608	620	632	644	656	667	679	691	3 3.6
369	703	714	726	738	750	761	773	785	797	808	4 4.8
370	820	832	844	855	867	879	891	902	914	926	5 6.0 / 6 7.2
371	937	949	961	972	984	996	*008	*019	*031	*043	7 8.4
372	57 054	066	078	089	101	113	124	136	148	159	8 9.6
373	171	183	194	206	217	229	241	252	264	276	9 10.8
374	287	299	310	322	334	345	357	368	380	392	
375	403	415	426	438	449	461	473	484	496	507	
376	519	530	542	553	565	576	588	600	611	623	
377	634	646	657	669	680	692	703	715	726	738	**11**
378	749	761	772	784	795	807	818	830	841	852	1 1.1
379	864	875	887	898	910	921	933	944	955	967	2 2.2
380	978	990	*001	*013	*024	*035	*047	*058	*070	*081	3 3.3 / 4 4.4
381	58 092	104	115	127	138	149	161	172	184	195	5 5.5
382	206	218	229	240	252	263	274	286	297	309	6 6.6
383	320	331	343	354	365	377	388	399	410	422	7 7.7
384	433	444	456	467	478	490	501	512	524	535	8 8.8
385	546	557	569	580	591	602	614	625	636	647	9 9.9
386	659	670	681	692	704	715	726	737	749	760	
387	771	782	794	805	816	827	838	850	861	872	
388	883	894	906	917	928	939	950	961	973	984	
389	995	*006	*017	*028	*040	*051	*062	*073	*084	*095	**10**
390	59 106	118	129	140	151	162	173	184	195	207	1 1.0
391	218	229	240	251	262	273	284	295	306	318	2 2.0
392	329	340	351	362	373	384	395	406	417	428	3 3.0
393	439	450	461	472	483	494	506	517	528	539	4 4.0
394	550	561	572	583	594	605	616	627	638	649	5 5.0
395	660	671	682	693	704	715	726	737	748	759	6 6.0
396	770	780	791	802	813	824	835	846	857	868	7 7.0
397	879	890	901	912	923	934	945	956	966	977	8 8.0
398	988	999	*010	*021	*032	*043	*054	*065	*076	*086	9 9.0
399	60 097	108	119	130	141	152	163	173	184	195	
400	206	217	228	239	249	260	271	282	293	304	
N.	L.0	1	2	3	4	5	6	7	8	9	P. P.

TABLE—(*Continued*).

N.	L.0	1	2	3	4	5	6	7	8	9
400	60 206	217	228	239	249	260	271	282	293	304
401	314	325	336	347	358	369	379	390	401	412
402	423	433	444	455	466	477	487	498	509	520
403	531	541	552	563	574	584	595	606	617	627
404	638	648	660	670	681	692	703	713	724	735
405	746	756	767	778	788	799	810	821	831	842
406	853	863	874	885	895	906	917	927	938	949
407	959	970	981	991	*002	*013	*023	*034	*045	*055
408	61 066	077	087	098	109	119	130	140	151	162
409	172	183	194	204	215	225	236	247	257	268
410	278	289	300	310	321	331	342	352	363	374
411	384	395	405	416	426	437	448	458	469	479
412	490	500	511	521	532	542	553	563	574	584
413	595	606	616	627	637	648	658	669	679	690
414	700	711	721	731	742	752	763	773	784	794
415	805	815	826	836	847	857	868	878	888	899
416	909	920	930	941	951	962	972	982	993	*003
417	62 014	024	034	045	055	066	076	086	097	107
418	118	128	138	149	159	170	180	190	201	211
419	221	232	242	252	263	273	284	294	304	315
420	325	335	346	356	366	377	387	397	408	418
421	428	439	449	459	469	480	490	500	511	521
422	531	542	552	562	572	583	593	603	613	624
423	634	644	655	665	675	685	696	706	716	726
424	737	747	757	767	778	788	798	808	818	829
425	839	849	859	870	880	890	900	910	921	931
426	941	951	961	972	982	992	*002	*012	*022	*033
427	63 043	053	063	073	083	094	104	114	124	134
428	144	155	165	175	185	195	205	215	225	236
429	246	256	266	276	286	296	306	317	327	337
430	347	357	367	377	387	397	407	417	428	438
431	448	458	468	478	488	498	508	518	528	538
432	548	558	568	579	589	599	609	619	629	639
433	649	659	669	679	689	699	709	719	729	739
434	749	759	769	779	789	799	809	819	829	839
435	849	859	869	879	889	899	909	919	929	939
436	949	959	969	979	988	998	*008	*018	*028	*038
437	64 048	058	068	078	088	098	108	118	128	137
438	147	157	167	177	187	197	207	217	227	237
439	246	256	266	276	286	296	306	316	326	335
440	345	355	365	375	385	395	404	414	424	434
441	444	454	464	473	483	493	503	513	523	532
442	542	552	562	572	582	591	601	611	621	631
443	640	650	660	670	680	689	699	709	719	729
444	738	748	758	768	777	787	797	807	816	826
445	836	846	856	865	875	885	895	904	914	924
446	933	943	953	963	972	982	992	*002	*011	*021
447	65 031	040	050	060	070	079	089	099	108	118
448	128	137	147	157	167	176	186	196	205	215
449	225	234	244	254	263	273	283	292	302	312
450	321	331	341	350	360	369	379	389	398	408
N.	L.0	1	2	3	4	5	6	7	8	9

P. P.

11
1	1.1
2	2.2
3	3.3
4	4.4
5	5.5
6	6.6
7	7.7
8	8.8
9	9.9

10
1	1.0
2	2.0
3	3.0
4	4.0
5	5.0
6	6.0
7	7.0
8	8.0
9	9.0

9
1	0.9
2	1.8
3	2.7
4	3.6
5	4.5
6	5.4
7	6.3
8	7.2
9	8.1

TABLE—(*Continued*).

N.	L.0	1	2	3	4	5	6	7	8	9
450	65 321	331	341	350	360	369	379	389	398	408
451	418	427	437	447	456	466	475	485	495	504
452	514	523	533	543	552	562	571	581	591	600
453	610	619	629	639	648	658	667	677	686	696
454	706	715	725	734	744	753	763	772	782	792
455	801	811	820	830	839	849	858	868	877	887
456	896	906	916	925	935	944	954	963	973	982
457	992	*001	*011	*020	*030	*039	*049	*058	*068	*077
458	66 087	096	106	115	124	134	143	153	162	172
459	181	191	200	210	219	229	238	247	257	266
460	276	285	295	304	314	323	332	342	351	361
461	370	380	389	398	408	417	427	436	445	455
462	464	474	483	492	502	511	521	530	539	549
463	558	567	577	586	596	605	614	624	633	642
464	652	661	671	680	689	699	708	717	727	736
465	745	755	764	773	783	792	801	811	820	829
466	839	848	857	867	876	885	894	904	913	922
467	932	941	950	960	969	978	987	997	*006	*015
468	67 025	034	043	052	062	071	080	089	099	108
469	117	127	136	145	154	164	173	182	191	201
470	210	219	228	237	247	256	265	274	284	293
471	302	311	321	330	339	348	357	367	376	385
472	394	403	413	422	431	440	449	459	468	477
473	486	495	504	514	523	532	541	550	560	569
474	578	587	596	605	614	624	633	642	651	660
475	669	679	688	697	706	715	724	733	742	752
476	761	770	779	788	797	806	815	825	834	843
477	852	861	870	879	888	897	906	916	925	934
478	943	952	961	970	979	988	997	*006	*015	*024
479	68 034	043	052	061	070	079	088	097	106	115
480	124	133	142	151	160	169	178	187	196	205
481	215	224	233	242	251	260	269	278	287	296
482	305	314	323	332	341	350	359	368	377	386
483	395	404	413	422	431	440	449	458	467	476
484	485	494	502	511	520	529	538	547	556	565
485	574	583	592	601	610	619	628	637	646	655
486	664	673	681	690	699	708	717	726	735	744
487	753	762	771	780	789	797	806	815	824	833
488	842	851	860	869	878	886	895	904	913	922
489	931	940	949	958	966	975	984	993	*002	*011
490	69 020	028	037	046	055	064	073	082	090	099
491	108	117	126	135	144	152	161	170	179	188
492	197	205	214	223	232	241	249	258	267	276
493	285	294	302	311	320	329	338	346	355	364
494	373	381	390	399	408	417	425	434	443	452
495	461	469	478	487	496	504	513	522	531	539
496	548	557	566	574	583	592	601	609	618	627
497	636	644	653	662	671	679	688	697	705	714
498	723	732	740	749	758	767	775	784	793	801
499	810	819	827	836	845	854	862	871	880	888
500	897	906	914	923	932	940	949	958	966	975
N.	L.0	1	2	3	4	5	6	7	8	9

P. P.

10		9		8	
1	1.0	1	0.9	1	0.8
2	2.0	2	1.8	2	1.6
3	3.0	3	2.7	3	2.4
4	4.0	4	3.6	4	3.2
5	5.0	5	4.5	5	4.0
6	6.0	6	5.4	6	4.8
7	7.0	7	6.3	7	5.6
8	8.0	8	7.2	8	6.4
9	9.0	9	8.1	9	7.2

TABLE—(Continued).

N.	L.0	1	2	3	4	5	6	7	8	9
500	69 897	906	914	923	932	940	949	958	966	975
501	984	992	*001	*010	*018	*027	*036	*044	*053	*062
502	70 070	079	088	096	105	114	122	131	140	148
503	157	165	174	183	191	200	209	217	226	234
504	243	252	260	269	278	286	295	303	312	321
505	329	338	346	355	364	372	381	389	398	406
506	415	424	432	441	449	458	467	475	484	492
507	501	509	518	526	535	544	552	561	569	578
508	586	595	603	612	621	629	638	646	655	663
509	672	680	689	697	706	714	723	731	740	749
510	757	766	774	783	791	800	808	817	825	834
511	842	851	859	868	876	885	893	902	910	919
512	927	935	944	952	961	969	978	986	995	*003
513	71 012	020	029	037	046	054	063	071	079	088
514	096	105	113	122	130	139	147	155	164	172
515	181	189	198	206	214	223	231	240	248	257
516	265	273	282	290	299	307	315	324	332	341
517	349	357	366	374	383	391	399	408	416	425
518	433	441	450	458	466	475	483	492	500	508
519	517	525	533	542	550	559	567	575	584	592
520	600	609	617	625	634	642	650	659	667	675
521	684	692	700	709	717	725	734	742	750	759
522	767	775	784	792	800	809	817	825	834	842
523	850	858	867	875	883	892	900	908	917	925
524	933	941	950	958	966	975	983	991	999	*008
525	72 016	024	032	041	049	057	066	074	082	090
526	099	107	115	123	132	140	148	156	165	173
527	181	189	198	206	214	222	230	239	247	255
528	263	272	280	288	296	304	313	321	329	337
529	346	354	362	370	378	387	395	403	411	419
530	428	436	444	452	460	469	477	485	493	501
531	509	518	526	534	542	550	558	567	575	583
532	591	599	607	616	624	632	640	648	656	665
533	673	681	689	697	705	713	722	730	738	746
534	754	762	770	779	787	795	803	811	819	827
535	835	843	852	860	868	876	884	892	900	908
536	916	925	933	941	949	957	965	973	981	989
537	997	*006	*014	*022	*030	*038	*046	*054	*062	*070
538	73 078	086	094	102	111	119	127	135	143	151
539	159	167	175	183	191	199	207	215	223	231
540	239	247	255	263	272	280	288	296	304	312
541	320	328	336	344	352	360	368	376	384	392
542	400	408	416	424	432	440	448	456	464	472
543	480	488	496	504	512	520	528	536	544	552
544	560	568	576	584	592	600	608	616	624	632
545	640	648	656	664	672	679	687	695	703	711
546	719	727	735	743	751	759	767	775	783	791
547	799	807	815	823	830	838	846	854	862	870
548	878	886	894	902	910	918	926	933	941	949
549	957	965	973	981	989	997	*005	*013	*020	*028
550	74 036	044	052	060	068	076	084	092	099	107
N.	L.0	1	2	3	4	5	6	7	8	9

P. P.

	9
1	0.9
2	1.8
3	2.7
4	3.6
5	4.5
6	5.4
7	6.3
8	7.2
9	8.1

	8
1	0.8
2	1.6
3	2.4
4	3.2
5	4.0
6	4.8
7	5.6
8	6.4
9	7.2

	7
1	0.7
2	1.4
3	2.1
4	2.8
5	3.5
6	4.2
7	4.9
8	5.6
9	6.3

TABLE—(*Continued*).

N.	L.0	1	2	3	4	5	6	7	8	9
550	74 036	044	052	060	068	076	084	092	099	107
551	115	123	131	139	147	155	162	170	178	186
552	194	202	210	218	225	233	241	249	257	265
553	273	280	288	296	304	312	320	327	335	343
554	351	359	367	374	382	390	398	406	414	421
555	429	437	445	453	461	468	476	484	492	500
556	507	515	523	531	539	547	554	562	570	578
557	586	593	601	609	617	624	632	640	648	656
558	663	671	679	687	695	702	710	718	726	733
559	741	749	757	764	772	780	788	796	803	811
560	819	827	834	842	850	858	865	873	881	889
561	896	904	912	920	927	935	943	950	958	966
562	974	981	989	997	*005	*012	*020	*028	*035	*043
563	75 051	059	066	074	082	089	097	105	113	120
564	128	136	143	151	159	166	174	182	189	197
565	205	213	220	228	236	243	251	259	266	274
566	282	289	297	305	312	320	328	335	343	351
567	358	366	374	381	389	397	404	412	420	427
568	435	442	450	458	465	473	481	488	496	504
569	511	519	526	534	542	549	557	565	572	580
570	587	595	603	610	618	626	633	641	648	656
571	664	671	679	686	694	702	709	717	724	732
572	740	747	755	762	770	778	785	793	800	808
573	815	823	831	838	846	853	861	868	876	884
574	891	899	906	914	921	929	937	944	952	959
575	967	974	982	989	997	*005	*012	*020	*027	*035
576	76 042	050	057	065	072	080	087	095	103	110
577	118	125	133	140	148	155	163	170	178	185
578	193	200	208	215	223	230	238	245	253	260
579	268	275	283	290	298	305	313	320	328	335
580	343	350	358	365	373	380	388	395	403	410
581	418	425	433	440	448	455	462	470	477	485
582	492	500	507	515	522	530	537	545	552	559
583	567	574	582	589	597	604	612	619	626	634
584	641	649	656	664	671	678	686	693	701	708
585	716	723	730	738	745	753	760	768	775	782
586	790	797	805	812	819	827	834	842	849	856
587	864	871	879	886	893	901	908	916	923	930
588	938	945	953	960	967	975	982	989	997	*004
589	77 012	019	026	034	041	048	056	063	070	078
590	085	093	100	107	115	122	129	137	144	151
591	159	166	173	181	188	195	203	210	217	225
592	232	240	247	254	262	269	276	283	291	298
593	305	313	320	327	335	342	349	357	364	371
594	379	386	393	401	408	415	422	430	437	444
595	452	459	466	474	481	488	495	503	510	517
596	525	532	539	546	554	561	568	576	583	590
597	597	605	612	619	627	634	641	648	656	663
598	670	677	685	692	699	706	714	721	728	735
599	743	750	757	764	772	779	786	793	801	808
600	815	822	830	837	844	851	*859	866	873	880
N.	L.0	1	2	3	4	5	6	7	8	9

P. P.

8

1	0.8
2	1.6
3	2.4
4	3.2
5	4.0
6	4.8
7	5.6
8	6.4
9	7.2

7

1	0.7
2	1.4
3	2.1
4	2.8
5	3.5
6	4.2
7	4.9
8	5.6
9	6.3

TABLE—(Continued).

N.	L.0	1	2	3	4	5	6	7	8	9
600	77 815	822	830	837	844	851	859	866	873	880
601	887	895	902	909	916	924	931	938	945	952
602	960	967	974	981	988	996	*003	*010	*017	*025
603	78 032	039	046	053	061	068	075	082	089	097
604	104	111	118	125	132	140	147	154	161	168
605	176	183	190	197	204	211	219	226	233	240
606	247	254	262	269	276	283	290	297	305	312
607	319	326	333	340	347	355	362	369	376	383
608	390	398	405	412	419	426	433	440	447	455
609	462	469	476	483	490	497	504	512	519	526
610	533	540	547	554	561	569	576	583	590	597
611	604	611	618	625	633	640	647	654	661	668
612	675	682	689	696	704	711	718	725	732	739
613	746	753	760	767	774	781	789	796	803	810
614	817	824	831	838	845	852	859	866	873	880
615	888	895	902	909	916	923	930	937	944	951
616	958	965	972	979	986	993	*000	*007	*014	*021
617	79 029	036	043	050	057	064	071	078	085	092
618	099	106	113	120	127	134	141	148	155	162
619	169	176	183	190	197	204	211	218	225	232
620	239	246	253	260	267	274	281	288	295	302
621	309	316	323	330	337	344	351	358	365	372
622	379	386	393	400	407	414	421	428	435	442
623	449	456	463	470	477	484	491	498	505	511
624	518	525	532	539	546	553	560	567	574	581
625	588	595	602	609	616	623	630	637	644	650
626	657	664	671	678	685	692	699	706	713	720
627	727	734	741	748	754	761	768	775	782	789
628	796	803	810	817	824	831	837	844	851	858
629	865	872	879	886	893	900	906	913	920	927
630	934	941	948	955	962	969	975	982	989	996
631	80 003	010	017	024	030	037	044	051	058	065
632	072	079	085	092	099	106	113	120	127	134
633	140	147	154	161	168	175	182	188	195	202
634	209	216	223	229	236	243	250	257	264	271
635	277	284	291	298	305	312	318	325	332	339
636	346	353	359	366	373	380	387	393	400	407
637	414	421	428	434	441	448	455	462	468	475
638	482	489	496	502	509	516	523	530	536	543
639	550	557	564	570	577	584	591	598	604	611
640	618	625	632	638	645	652	659	665	672	679
641	686	693	699	706	713	720	726	733	740	747
642	754	760	767	774	781	787	794	801	808	814
643	821	828	835	841	848	855	862	868	875	882
644	889	895	902	909	916	922	929	936	943	949
645	956	963	969	976	983	990	996	*003	*010	*017
646	81 023	030	037	043	050	057	064	070	077	084
647	090	097	104	111	117	124	131	137	144	151
648	158	164	171	178	184	191	198	204	211	218
649	224	231	238	245	251	258	265	271	278	285
650	291	298	305	311	318	325	331	338	345	351
N.	L.0	1	2	3	4	5	6	7	8	9

P. P.

	8		7		6
1	0.8	1	0.7	1	0.6
2	1.6	2	1.4	2	1.2
3	2.4	3	2.1	3	1.8
4	3.2	4	2.8	4	2.4
5	4.0	5	3.5	5	3.0
6	4.8	6	4.2	6	3.6
7	5.6	7	4.9	7	4.2
8	6.4	8	5.6	8	4.8
9	7.2	9	6.8	9	5.4

TABLE—(Continued).

N.	L.0	1	2	3	4	5	6	7	8	9
650	81 291	298	305	311	318	325	331	338	345	351
651	358	365	371	378	385	391	398	405	411	418
652	425	431	438	445	451	458	465	471	478	485
653	491	498	505	511	518	525	531	538	544	551
654	558	564	571	578	584	591	598	604	611	617
655	624	631	637	644	651	657	664	671	677	684
656	690	697	704	710	717	723	730	737	743	750
657	757	763	770	776	783	790	796	803	809	816
658	823	829	836	842	849	856	862	869	875	882
659	889	895	902	908	915	921	928	935	941	948
660	954	961	968	974	981	987	994	*000	*007	*014
661	82 020	027	033	040	046	053	060	066	073	079
662	086	092	099	105	112	119	125	132	138	145
663	151	158	164	171	178	184	191	197	204	210
664	217	223	230	236	243	249	256	263	269	276
665	282	289	295	302	308	315	321	328	334	341
666	347	354	360	367	373	380	387	393	400	406
667	413	419	426	432	439	445	452	458	465	471
668	478	484	491	497	504	510	517	523	530	536
669	543	549	556	562	569	575	582	588	595	601
670	607	614	620	627	633	640	646	653	659	666
671	672	679	685	692	698	705	711	718	724	730
672	737	743	750	756	763	769	776	782	789	795
673	802	808	814	821	827	834	840	847	853	860
674	866	872	879	885	892	898	905	911	918	924
675	930	937	943	950	956	963	969	975	982	988
676	995	*001	*008	*014	*020	*027	*033	*040	*046	*052
677	83 059	065	072	078	085	091	097	104	110	117
678	123	129	136	142	149	155	161	168	174	181
679	187	193	200	206	213	219	225	232	238	245
680	251	257	264	270	276	283	289	296	302	308
681	315	321	327	334	340	347	353	359	366	372
682	378	385	391	398	404	410	417	423	429	436
683	442	448	455	461	467	474	480	487	493	499
684	506	512	518	525	531	537	544	550	556	563
685	569	575	582	588	594	601	607	613	620	626
686	632	639	645	651	658	664	670	677	683	689
687	696	702	708	715	721	727	734	740	746	753
688	759	765	771	778	784	790	797	803	809	816
689	822	828	835	841	847	853	860	866	872	879
690	885	891	897	904	910	916	923	929	935	942
691	948	954	960	967	973	979	985	992	998	*004
692	84 011	017	023	029	036	042	048	055	061	067
693	073	080	086	092	098	105	111	117	123	130
694	136	142	148	155	161	167	173	180	186	192
695	198	205	211	217	223	230	236	242	248	255
696	261	267	273	280	286	292	298	305	311	317
697	323	330	336	342	348	354	361	367	373	379
698	386	392	398	404	410	417	423	429	435	442
699	448	454	460	466	473	479	485	491	497	504
700	510	516	522	528	535	541	547	553	559	566
N.	L.0	1	2	3	4	5	6	7	8	9

P. P.

7	
1	0.7
2	1.4
3	2.1
4	2.8
5	3.5
6	4.2
7	4.9
8	5.6
9	6.3

6	
1	0.6
2	1.2
3	1.8
4	2.4
5	3.0
6	3.6
7	4.2
8	4.8
9	5.4

TABLE—(*Continued*).

N.	L.0	1	2	3	4	5	6	7	8	9
700	84 510	516	522	528	535	541	547	553	559	566
701	572	578	584	590	597	603	609	615	621	628
702	634	640	646	652	658	665	671	677	683	689
703	696	702	708	714	720	726	733	739	745	751
704	757	763	770	776	782	788	794	800	807	813
705	819	825	831	837	844	850	856	862	868	874
706	880	887	893	899	905	911	917	924	930	936
707	942	948	954	960	967	973	979	985	991	997
708	85 003	009	016	022	028	034	040	046	052	058
709	065	071	077	083	089	095	101	107	114	120
710	126	132	138	144	150	156	163	169	175	181
711	187	193	199	205	211	217	224	230	236	242
712	248	254	260	266	272	278	285	291	297	303
713	309	315	321	327	333	339	345	352	358	364
714	370	376	382	388	394	400	406	412	418	425
715	431	437	443	449	455	461	467	473	479	485
716	491	497	503	509	516	522	528	534	540	546
717	552	558	564	570	576	582	588	594	600	606
718	612	618	625	631	637	643	649	655	661	667
719	673	679	685	691	697	703	709	715	721	727
720	733	739	745	751	757	763	769	775	781	788
721	794	800	806	812	818	824	830	836	842	848
722	854	860	866	872	878	884	890	896	902	908
723	914	920	926	932	938	944	950	956	962	968
724	974	980	986	992	998	*004	*010	*016	*022	*028
725	86 034	040	046	052	058	064	070	076	082	088
726	094	100	106	112	118	124	130	136	141	147
727	153	159	165	171	177	183	189	195	201	207
728	213	219	225	231	237	243	249	255	261	267
729	273	279	285	291	297	303	308	314	320	326
730	332	338	344	350	356	362	368	374	380	386
731	392	398	404	410	415	421	427	433	439	445
732	451	457	463	469	475	481	487	493	499	504
733	510	516	522	528	534	540	546	552	558	564
734	570	576	581	587	593	599	605	611	617	623
735	629	635	641	646	652	658	664	670	676	682
736	688	694	700	705	711	717	723	729	735	741
737	747	753	759	764	770	776	782	788	794	800
738	806	812	817	823	829	835	841	847	853	859
739	864	870	876	882	888	894	900	906	911	917
740	925	929	935	941	947	953	958	964	970	976
741	982	988	994	999	*005	*011	*017	*023	*029	*035
742	87 040	046	052	058	064	070	075	081	087	093
743	099	105	111	116	122	128	134	140	146	151
744	157	163	169	175	181	186	192	198	204	210
745	216	221	227	233	239	245	251	256	262	268
746	274	280	286	291	297	303	309	315	320	326
747	332	338	344	349	355	361	367	373	379	384
748	390	396	402	408	413	419	425	431	437	442
749	448	454	460	466	471	477	483	489	495	500
750	506	512	518	523	529	535	541	547	552	558
N.	L.0	1	2	3	4	5	6	7	8	9

P. P.

	7
1	0.7
2	1.4
3	2.1
4	2.8
5	3.5
6	4.2
7	4.9
8	5.6
9	6.3

	6
1	0.6
2	1.2
3	1.8
4	2.4
5	3.0
6	3.6
7	4.2
8	4.8
9	5.4

	5
1	0.5
2	1.0
3	1.5
4	2.0
5	2.5
6	3.0
7	3.5
8	4.0
9	4.5

TABLE—(*Continued*).

N.	L.0	1	2	3	4	5	6	7	8	9
750	87 506	512	518	523	529	535	541	547	552	558
751	564	570	576	581	587	593	599	604	610	616
752	622	628	633	639	645	651	656	662	668	674
753	679	685	691	697	703	708	714	720	726	731
754	737	743	749	754	760	766	772	777	783	789
755	795	800	806	812	818	823	829	835	841	846
756	852	858	864	869	875	881	887	892	898	904
757	910	915	921	927	933	938	944	950	955	961
758	967	973	978	984	990	996	*001	*007	*013	*018
759	88 024	030	036	041	047	053	058	064	070	076
760	081	087	093	098	104	110	116	121	127	133
761	138	144	150	156	161	167	173	178	184	190
762	195	201	207	213	218	224	230	235	241	247
763	252	258	264	270	275	281	287	292	298	304
764	309	315	321	326	332	338	343	349	355	360
765	366	372	377	383	389	395	400	406	412	417
766	423	429	434	440	446	451	457	463	468	474
767	480	485	491	497	502	508	513	519	525	530
768	536	542	547	553	559	564	570	576	581	587
769	593	598	604	610	615	621	627	632	638	643
770	649	655	660	666	672	677	683	689	694	700
771	705	711	717	722	728	734	739	745	750	756
772	762	767	773	779	784	790	795	801	807	812
773	818	824	829	835	840	846	852	857	863	868
774	874	880	885	891	897	902	908	913	919	925
775	930	936	941	947	953	958	964	969	975	981
776	986	992	997	*003	*009	*014	*020	*025	*031	*037
777	89 042	048	053	059	064	070	076	081	087	092
778	098	104	109	115	120	126	131	137	143	148
779	154	159	165	170	176	182	187	193	198	204
780	209	215	221	226	232	237	243	248	254	260
781	265	271	276	282	287	293	298	304	310	315
782	321	326	332	337	343	348	354	360	365	371
783	376	382	387	393	398	404	409	415	421	426
784	432	437	443	448	454	459	465	470	476	481
785	487	492	498	504	509	515	520	526	531	537
786	542	548	553	559	564	570	575	581	586	592
787	597	603	609	614	620	625	631	636	642	647
788	653	658	664	669	675	680	686	691	697	702
789	708	713	719	724	730	735	741	746	752	757
790	763	768	774	779	785	790	796	801	807	812
791	818	823	829	834	840	845	851	856	862	867
792	873	878	883	889	894	900	905	911	916	922
793	927	933	938	944	949	955	960	966	971	977
794	982	988	993	998	*004	*009	*015	*020	*026	*031
795	90 037	042	048	053	059	064	069	075	080	086
796	091	097	102	108	113	119	124	129	135	140
797	146	151	157	162	168	173	179	184	189	195
798	200	206	211	217	222	227	233	238	244	249
799	255	260	266	271	276	282	287	293	298	304
800	309	314	320	325	331	336	342	347	352	358
N.	L.0	1	2	3	4	5	6	7	8	9

P. P.

	6
1	0.6
2	1.2
3	1.8
4	2.4
5	3.0
6	3.6
7	4.2
8	4.8
9	5.4

	5
1	0.5
2	1.0
3	1.5
4	2.0
5	2.5
6	3.0
7	3.5
8	4.0
9	4.5

Table—(*Continued*).

N.	L.0	1	2	3	4	5	6	7	8	9
800	90 309	314	320	325	331	336	342	347	352	358
801	363	369	374	380	385	390	396	401	407	412
802	417	423	428	434	439	445	450	455	461	466
803	472	477	482	488	493	499	504	509	515	520
804	526	531	536	542	547	553	558	563	569	574
805	580	585	590	596	601	607	612	617	623	628
806	634	639	644	650	655	660	666	671	677	682
807	687	693	698	703	709	714	720	725	730	736
808	741	747	752	757	763	768	773	779	784	789
809	795	800	806	811	816	822	827	832	838	843
810	849	854	859	865	870	875	881	886	891	897
811	902	907	913	918	924	929	934	940	945	950
812	956	961	966	972	977	982	988	993	998	*004
813	91 009	014	020	025	030	036	041	046	052	057
814	062	068	073	078	084	089	094	100	105	110
815	116	121	126	132	137	142	148	153	158	164
816	169	174	180	185	190	196	201	206	212	217
817	222	228	233	238	243	249	254	259	265	270
818	275	281	286	291	297	302	307	312	318	323
819	328	334	339	344	350	355	360	365	371	376
820	381	387	392	397	403	408	413	418	424	429
821	434	440	445	450	455	461	466	471	477	482
822	487	492	498	503	508	514	519	524	529	535
823	540	545	551	556	561	566	572	577	582	587
824	593	598	603	609	614	619	624	630	635	640
825	645	651	656	661	666	672	677	682	687	693
826	698	703	709	714	719	724	730	735	740	745
827	751	756	761	766	772	777	782	787	793	798
828	803	808	814	819	824	829	834	840	845	850
829	855	861	866	871	876	882	887	892	897	903
830	908	913	918	924	929	934	939	944	950	955
831	960	965	971	976	981	986	991	997	*002	*007
832	92 012	018	023	028	033	038	044	049	054	059
833	065	070	075	080	085	091	096	101	106	111
834	117	122	127	132	137	143	148	153	158	163
835	169	174	179	184	189	195	200	205	210	215
836	221	226	231	236	241	247	252	257	262	267
837	273	278	283	288	293	298	304	309	314	319
838	324	330	335	340	345	350	355	361	366	371
839	376	381	387	392	397	402	407	412	418	423
840	428	433	438	443	449	454	459	464	469	474
841	480	485	490	495	500	505	511	516	521	526
842	531	536	542	547	552	557	562	567	572	578
843	583	588	593	598	603	609	614	619	624	629
844	634	639	645	650	655	660	665	670	675	681
845	686	691	696	701	706	711	716	722	727	732
846	737	742	747	752	758	763	768	773	778	783
847	788	793	799	804	809	814	819	824	829	334
848	840	845	850	855	860	865	870	875	881	886
849	891	896	901	906	911	916	921	927	932	937
850	942	947	952	957	962	967	973	978	983	988
N.	L.0	1	2	3	4	5	6	7	8	9

P. P.

8

1	0.6
2	1.2
3	1.8
4	2.4
5	3.0
6	3.6
7	4.2
8	4.8
9	5.4

5

1	0.5
2	1.0
3	1.5
4	2.0
5	2.5
6	3.0
7	3.5
8	4.0
9	4.5

Table—(Continued).

N.	L.0	1	2	3	4	5	6	7	8	9
850	92 942	947	952	957	962	967	973	978	983	988
851	993	998	*003	*008	*013	*018	*024	*029	*034	*039
852	93 044	049	054	059	064	069	075	080	085	090
853	095	100	105	110	115	120	125	131	136	141
854	146	151	156	161	166	171	176	181	186	192
855	197	202	207	212	217	222	227	232	237	242
856	247	252	258	263	268	273	278	283	288	293
857	298	303	308	313	318	323	328	334	339	344
858	349	354	359	364	369	374	379	384	389	394
859	399	404	409	414	420	425	430	435	440	445
860	450	455	460	465	470	475	480	485	490	495
861	500	505	510	515	520	526	531	536	541	546
862	551	556	561	566	571	576	581	586	591	596
863	601	606	611	616	621	626	631	636	641	646
864	651	656	661	666	671	676	682	687	692	697
865	702	707	712	717	722	727	732	737	742	747
866	752	757	762	767	772	777	782	787	792	797
867	802	807	812	817	822	827	832	837	842	847
868	852	857	862	867	872	877	882	887	892	897
869	902	907	912	917	922	927	932	937	942	947
870	952	957	962	967	972	977	982	987	992	997
871	94 002	007	012	017	022	027	032	037	042	047
872	052	057	062	067	072	077	082	086	091	096
873	101	106	111	116	121	126	131	136	141	146
874	151	156	161	166	171	176	181	186	191	196
875	201	206	211	216	221	226	231	236	240	245
876	250	255	260	265	270	275	280	285	290	295
877	300	305	310	315	320	325	330	335	340	345
878	349	354	359	364	369	374	379	384	389	394
879	399	404	409	414	419	424	429	433	438	443
880	448	453	458	463	468	473	478	483	488	493
881	498	503	507	512	517	522	527	532	537	542
882	547	552	557	562	567	571	576	581	586	591
883	596	601	606	611	616	621	626	630	635	640
884	645	650	655	660	665	670	675	680	685	689
885	694	699	704	709	714	719	724	729	734	738
886	743	748	753	758	763	768	773	778	783	787
887	792	797	802	807	812	817	822	827	832	836
888	841	846	851	856	861	866	871	876	880	885
889	890	895	900	905	910	915	919	924	929	934
890	939	944	949	954	959	963	968	973	978	983
891	988	993	998	*002	*007	*012	*017	*022	*027	*032
892	95 036	041	046	051	056	061	066	071	075	080
893	085	090	095	100	105	109	114	119	124	129
894	134	139	143	148	153	158	163	168	173	177
895	182	187	192	197	202	207	211	216	221	226
896	231	236	240	245	250	255	260	265	270	274
897	279	284	289	294	299	303	308	313	318	323
898	328	332	337	342	347	352	357	361	366	371
899	376	381	386	390	395	400	405	410	415	419
900	424	429	434	439	444	448	453	458	463	468
N.	L.0	1	2	3	4	5	6	7	8	9

P. P.

8
1 0.6
2 1.2
3 1.8
4 2.4
5 3.0
6 3.6
7 4.2
8 4.8
9 5.4

5
1 0.5
2 1.0
3 1.5
4 2.0
5 2.5
6 3.0
7 3.5
8 4.0
9 4.5

4
1 0.4
2 0.8
3 1.2
4 1.6
5 2.0
6 2.4
7 2.8
8 3.2
9 3.6

TABLE—(Continued).

N.	L.0	1	2	3	4	5	6	7	8	9	P. P.
900	95 424	429	434	439	444	448	453	458	463	468	
901	472	477	482	487	492	497	501	506	511	516	
902	521	525	530	535	540	545	550	554	559	564	
903	569	574	578	583	588	593	598	602	607	612	
904	617	622	626	631	636	641	646	650	655	660	
905	665	670	674	679	684	689	694	698	703	708	
906	713	718	722	727	732	737	742	746	751	756	
907	761	766	770	775	780	785	789	794	799	804	
908	809	813	818	823	828	832	837	842	847	852	
909	856	861	866	871	875	880	885	890	895	899	
910	904	909	914	918	923	928	933	938	942	947	
911	952	957	961	966	971	976	980	985	990	995	5
912	999	*004	*009	*014	*019	*023	*028	*033	*038	*042	1 | 0.5
913	96 047	052	057	061	066	071	076	080	085	090	2 | 1.0
914	095	099	104	109	114	118	123	128	133	137	3 | 1.5
915	142	147	152	156	161	166	171	175	180	185	4 | 2.0
916	190	194	199	204	209	213	218	223	227	232	5 | 2.5
917	237	242	246	251	256	261	265	270	275	280	6 | 3.0
918	284	289	294	298	303	308	313	317	322	327	7 | 3.5
919	332	336	341	346	350	355	360	365	369	374	8 | 4.0
920	379	384	388	393	398	402	407	412	417	421	9 | 4.5
921	426	431	435	440	445	450	454	459	464	468	
922	473	478	483	487	492	497	501	506	511	515	
923	520	525	530	534	539	544	548	553	558	562	
924	567	572	577	581	586	591	595	600	605	609	
925	614	619	624	628	633	638	642	647	652	656	
926	661	666	670	675	680	685	689	694	699	703	
927	708	713	717	722	727	731	736	741	745	750	
928	755	759	764	769	774	778	783	788	792	797	
929	802	806	811	816	820	825	830	834	839	844	
930	848	853	858	862	867	872	876	881	886	890	
931	895	900	904	909	914	918	923	928	932	937	4
932	942	946	951	956	960	965	970	974	979	984	1 | 0.4
933	988	993	997	*002	*007	*011	*016	*021	*025	*030	2 | 0.8
934	97 035	039	044	049	053	058	063	067	072	077	3 | 1.2
935	081	086	090	095	100	104	109	114	118	123	4 | 1.6
936	128	132	137	142	146	151	155	160	165	169	5 | 2.0
937	174	179	183	188	192	197	202	206	211	216	6 | 2.4
938	220	225	230	234	239	243	248	253	257	262	7 | 2.8
939	267	271	276	280	285	290	294	299	304	308	8 | 3.2
940	313	317	322	327	331	336	340	345	350	354	9 | 3.6
941	359	364	368	373	377	382	387	391	396	400	
942	405	410	414	419	424	428	433	437	442	447	
943	451	456	460	465	470	474	479	483	488	493	
944	497	502	506	511	516	520	525	529	534	539	
945	543	548	552	557	562	566	571	575	580	585	
946	589	594	598	603	607	612	617	621	626	630	
947	635	640	644	649	653	658	663	667	672	676	
948	681	685	690	695	699	704	708	713	717	722	
949	727	731	736	740	745	749	754	759	763	768	
950	772	777	782	786	791	795	800	804	809	813	
N.	L.0	1	2	3	4	5	6	7	8	9	P. P.

TABLE—(*Continued*).

N.	L.0	1	2	3	4	5	6	7	8	9	P. P.
950	97 772	777	782	786	791	795	800	804	809	813	
951	818	823	827	832	836	841	845	850	855	859	
952	864	868	873	877	882	886	891	896	900	905	
953	909	914	918	923	928	932	937	941	946	950	
954	955	959	964	968	973	978	982	987	991	996	
955	98 000	005	009	014	019	023	028	032	037	041	
956	046	050	055	059	064	068	073	078	082	087	
957	091	096	100	105	109	114	118	123	127	132	
958	137	141	146	150	155	159	164	168	173	177	
959	182	186	191	195	200	204	209	214	218	223	
960	227	232	236	241	245	250	254	259	263	268	
961	272	277	281	286	290	295	299	304	308	313	**5**
962	318	322	327	331	336	340	345	349	354	358	1 0.5
963	363	367	372	376	381	385	390	394	399	403	2 1.0
964	408	412	417	421	426	430	435	439	444	448	3 1.5
965	453	457	462	466	471	475	480	484	489	493	4 2.0
966	498	502	507	511	516	520	525	529	534	538	5 2.5
967	543	547	552	556	561	565	570	574	579	583	6 3.0
968	588	592	597	601	605	610	614	619	623	628	7 3.5
969	632	637	641	646	650	655	659	664	668	673	8 4.0
970	677	682	686	691	695	700	704	709	713	717	9 4.5
971	722	726	731	735	740	744	749	753	758	762	
972	767	771	776	780	784	789	793	798	802	807	
973	811	816	820	825	829	834	838	843	847	851	
974	856	860	865	869	874	878	883	887	892	896	
975	900	905	909	914	918	923	927	932	936	941	
976	945	949	954	958	963	967	972	976	981	985	
977	989	994	998	*003	*007	*012	*016	*021	*025	*029	
978	99 034	038	043	047	052	056	061	065	069	074	
979	078	083	087	092	096	100	105	109	114	118	
980	123	127	131	136	140	145	149	154	158	162	
981	167	171	176	180	185	189	193	198	202	207	**4**
982	211	216	220	224	229	233	238	242	247	251	1 0.4
983	255	260	264	269	273	277	282	286	291	295	2 0.8
984	300	304	308	313	317	322	326	330	335	339	3 1.2
985	344	348	352	357	361	366	370	374	379	383	4 1.6
986	388	392	396	401	405	410	414	419	423	427	5 2.0
987	432	436	441	445	449	454	458	463	467	471	6 2.4
988	476	480	484	489	493	498	502	506	511	515	7 2.8
989	520	524	528	533	537	542	546	550	555	559	8 3.2
990	564	568	572	577	581	585	590	594	599	603	9 3.6
991	607	612	616	621	625	629	634	638	642	647	
992	651	656	660	664	669	673	677	682	686	691	
993	695	699	704	708	712	717	721	726	730	734	
994	739	743	747	752	756	760	765	769	774	778	
995	782	787	791	795	800	804	808	813	817	822	
996	826	830	835	839	843	848	852	856	861	865	
997	870	874	878	883	887	891	896	900	904	909	
998	913	917	922	926	930	935	939	944	948	952	
999	957	961	965	970	974	978	983	987	991	996	
1000	00 000	004	009	013	017	022	026	030	035	039	
N.	L.0	1	2	3	4	5	6	7	8	9	P. P.

TRIGONOMETRIC FUNCTIONS.

DIRECTIONS FOR USING THE TABLE.

The table given on pages 74–78 contains the natural sines, cosines, tangents, and cotangents of angles from 0° to 90°. Angles less than 45° are given in the first column at the left-hand side of the page, and the names of the functions are given at the top of the page; angles greater than 45° appear at the right-hand side of the page, and the names of the functions are given at the bottom. Thus, the second column contains the sines of angles less than 45° and the cosines of angles greater than 45°; the sixth column contains the cotangents of angles less than 45° and the tangents of angles greater than 45°. To find the function of an angle less than 45°, look in the column of angles at the left of the page for the angle, and at the top of the page for the name of the function; to find a function of an angle greater than 45°, look in the column at the right of the page for the angle and at the bottom of the page for the name of the function. The successive angles differ by an interval of 10′; they increase downwards in the left-hand column and upwards in the right-hand column. Thus, for angles less than 45° read down from top of page, and for angles greater than 45° read up from bottom of page.

The third, fifth, seventh, and ninth columns, headed *d*, contain the differences between the successive functions; for example, in the second column we find that the sine of 32° 10′ is .5324 and that the sine of 32° 20′ is .5348; the difference is .5348 − .5324 = .0024, and the 24 is written in the third column, just opposite the space between .5324 and .5348. In like manner the differences between the successive tabular values of the tangents are given in the fifth column, those between the cotangents in the seventh column, and those for the cosines in the ninth column. These differences in the functions correspond to a difference of 10′ in the angle; thus, when the angle 32° 10′ is increased by 10′, that is, to 32° 20′, the increase of the sine is .0024, or, as given in the table, 24. It will be observed that in the tabular difference no attention is paid to the decimal point, it being understood that the difference is

merely the number obtained by subtracting the last two or
three figures of the smaller function from those of the larger.
These differences are used to obtain the sines, cosines, etc.
of angles not given in the table; the method employed may
be illustrated by an example. Required, the tangent of
27° 34'. Looking in the table, we see that the tangent of 27° 30'
is .5206, and (in column 5) the difference for 10' is 37. Difference for 1' is 37 ÷ 10 = 3.7, and difference for 4' is 3.7 × 4 = 14.8.
Adding this difference to the value of the tan 27° 30', we have

$$\text{tan } 27° 30' = .5206$$
$$\text{difference for } 4' = \quad 14.8$$
$$\overline{\qquad\qquad\qquad\qquad}$$
$$\text{tan } 27° 34' = .5220.8 \text{ or } .5221, \text{ to four places.}$$

Since only four decimal places are retained, the 8 in the
fifth place is dropped and the figure in the fourth place is
increased by 1, because 8 is greater than 5.

To avoid multiplication, the column of proportional parts,
headed P. P., at the extreme right of the page, is used. At
the head of each table in this column is the difference for 10',
and below are the differences for any intermediate number
of minutes from 1' to 9'. In the above example, the difference for 10' was 37; looking in the table with 37 at the head,
the difference opposite 4 is 14.8; that opposite 7 is 25.9; and so
on. For want of space, the differences for the cotangents for
angles less than 45° (or the tangents of angles greater than
45°) have been omitted from the tables of proportional parts.
The use of these functions should be avoided, if possible,
since the differences change very rapidly, and the computation is therefore likely to be inexact. The method to be
employed when dealing with these functions may be shown
by an example: Required, the tangent of 76° 34'. Since this
angle is greater than 45°, we look for it in the column at the
right, and read up; opposite the 76° 30', we find, in sixth column, the number 4.1653, and corresponding to it in seventh
column is the difference 540. Since 540 is the difference for 10',
the difference for 4' is 540 × $\frac{4}{10}$ = 216. Adding this difference:

$$\text{tan } 76° 30' = 4.1653$$
$$\text{difference for } 4' = \quad 216$$
$$\overline{\qquad\qquad\qquad\qquad}$$
$$\text{tan } 76° 34' = 4.1869$$

When the angle contains a certain number of seconds, divide the number by 6, and take the whole number nearest to the quotient; look out this number in the table of proportional parts (under the proper *difference*), and take out the number that is opposite to it. Shift the decimal point one place to the left, and then add it to the partial function already found.

Find the sine of 34° 26′ 44″.

sine 34° 20′ = .5640   Difference for 10′ = 24.

difference for 6′ =   14.4

difference for 44″ =   1.7 $\frac{44}{6}$ = 7¼. Look out in the P. P.

sine 34° 26′ 44″ = .5656   table the number under 24 and opposite 7. It is 16.8. Shifting the decimal point one place to the left, we get 1.68, or, say, 1.7.

The tangent is found in the same way as the sine.

To find the cosine of an angle:

As the angle increases, the value of the cosine decreases, so that, instead of adding the values corresponding to 6′ and 44″ to the function already found, we subtract them from it.

Thus, find cos 34° 26′ 44″.

cos 34° 20′ = .8258   Difference for 10′ = 17.

difference for 6′ =   10.2

difference for 44″ =   1.2   The number under the 17 and opposite the 7, in the P. P.

total difference =   11.4   table, is 11.9. Therefore, take

.8247   1.19, or, say, 1.2.

Therefore, cos 34° 26′ 44″ = .8258 — .0011 = .8240.

Only four decimal places are kept; therefore, the figure of the difference following the decimal point is dropped before subtracting.

The cotangent is found in the same manner.

We will now consider angles greater than 45°.

Find the sine of 68° 47′ 22″.

In obtaining the *difference*, it must be remembered to choose the one between the sine of 68° 40′ and the next angle above it, namely, 68° 50′.

sine 68° 40' = .9815 Difference for 10' = 10.

difference for 7' = 7

difference for 19" = .4 $\frac{19}{50}$ = 3⅘, say 4. Under the 10

sine 68° 47' 22" = .9822   and opposite the 4 is the number 4.0; shifting the decimal point, we get .4.

As usual, only four decimal places are kept.

The tangent is found in the same manner.

Find cos 68° 47' 22".

As before, the cosine decreases as the angle increases; therefore, we subtract the successive sine values corresponding to the increments in the angle.

cos 68° 40' = .3638 Difference for 10' = 27.

difference for 7' = 18.9

difference for 22" = 1.1   Under the 27 and opposite the

total difference = 20   4 is the number 10.8; there-

.3618   fore, take 1.08 in this case, or, say, 1.1.

Therefore, cos 68° 47' 22" = .3638 − .002 = .3618.

The cotangent is found in the same way.

In finding the functions of an angle, the only difficulty likely to be encountered is to determine whether the difference obtained from the table of proportional parts is to be added or subtracted. This can be told in every case by observing whether the function is increasing or decreasing as the angle increases. For example, take the angle 21°; its sine is .3584, and the following sines, reading downwards, are .3611, .3638, etc. It is plain, therefore, that the sine of say 21° 6' is greater than that of 21°, and that the difference for 6' must be added. On the other hand, the cosine of 21° is .9336, and the following cosines, reading downwards, are .9325, .9315, etc.; that is, as the angle grows larger the cosine decreases. The cosine of an angle between 21° and 21° 10', say 21° 6', must therefore lie between .9325 and .9315; that is, it must be smaller than .9325, which shows that in this case the difference for 6' must be subtracted from the cosine of 21°.

We will now consider the case in which the function, i. e., the sine, cosine, tangent, or cotangent, is given and the corresponding angle is to be found.

Find the angle whose sine is .4943. The operation is arranged as follows:

$$
\begin{array}{rl}
.4943 & \text{Difference for } 10' = 26. \\
.4924 & = \sin 29^\circ 30'. \\
\hline
\text{1st remainder} \quad 19 & \\
18.2 & = \text{difference for } 7'. \\
\hline
\text{2d remainder} \quad .8 & \\
.78 & = \text{difference for .3' or 18''.}
\end{array}
$$

.4943 = sin 29° 37′ 18″.

Looking down the second column, we find the sine next *smaller* than .4943 to be .4924, and the difference for 10′ to be 26. The angle corresponding to .4924 is 29° 30′. Subtracting the .4924 from .4943, the first remainder is 19; looking in the table of proportional parts, the part next lower than this difference is 18.2, opposite which is 7′. Subtracting this difference from the remainder, we get .8, and, looking in the table, we see that 7.8 with its decimal point moved one place to the left is nearest to the second difference. This is the difference for .3′ or 18″. Hence, the angle is 29° 30′ + 7′ + 18 = 29° 37′ 18″.

Find the angle whose tangent is .8824.

$$
\begin{array}{rl}
.8824 & \text{Difference for } 10' = 51. \\
.8796 & = \tan 41^\circ 20'. \\
\hline
\text{1st remainder} \quad .28 & \\
25.5 & = \text{difference for } 5'. \\
\hline
\text{2d remainder} \quad 2.5 & \\
2.55 & = \text{difference for .5' or 30''.}
\end{array}
$$

.8824 = tan 41° 25′ 30″.

In the two examples just given, the minutes and seconds corresponding to the 1st and 2d remainders are added to the angle taken from the table. Thus, in the first example, an inspection of the table shows that the angle increases as the sine increases; hence, the angle whose sine is .4943 must be greater than 29° 30′, whose sine is .4924. For this reason the correction must be *added* to 29° 30′. The same reasoning applies to the second example.

Find the angle whose cosine is .7742.

```
              .7742            Difference for 10′ = 18.
              .7735    = cos 39° 20′.
1st remainder    7
              5.4   = difference for 3′.
2d remainder   1.6
              1.62  = difference for .9′ or 54″.
```

39° 20′ − 3′ 54″ = 39° 16′ 6″, which is the angle whose cosine is .7742.

Looking down the eighth column, headed cos, the next smaller cosine is .7735, to which corresponds the angle 39° 20′. The difference for 10′ is 18. Subtracting, the remainder is 7, and the next lower number in the table of proportional parts is 5.4, which is the difference for 3′. Subtracting this from 1st remainder, 2d remainder is 1.6, which is nearest 16.2 of table of proportional parts, if the decimal point of the latter is moved to the left one place. Since 16.2 corresponds to a difference of 9′, 1.62 corresponds to a difference of .9′, or 54″. Hence, the correction for the angle 39° 20′ is 3′ 54″. From the table, it appears that, as the cosine increases, the angle grows smaller; therefore, the angle whose cosine is .7742 must be smaller than the angle whose cosine is .7735, and the correction for the angle must be subtracted.

Find the angle whose cotangent is .9847.

```
              .9847            Difference for 10′ = 57.
              .9827    = cos 45° 30′.
- 1st remainder   20
              17.1  = difference for 3′.
2d remainder   2.9
              2.85  = difference for .5′ or 30″.
```

45° 30′ − 3′ 30″ = 45° 26′ 30″, the angle whose cotangent is .9847.

In finding the angle corresponding to a function, as in the above examples, the angles obtained may vary from the true angle by 2 or 3 seconds; in order to obtain the number of seconds accurately, the functions should contain six or seven decimal places.

°	′	Sin.	d.	Tan.	d.	Cot.	d.	Cos.	d.		P. P.
0	0	0.0000		0.0000		infinit.		1.0000	0	0 **90**	
			29		29						
	10	0.0029	29	0.0029	29	343.7737		1.0000	0	50	
	20	0.0058	29	0.0058	29	171.8854		1.0000	0	40	**30**
	30	0.0087	29	0.0087	29	114.5887		1.0000	0	30	1 | 3.0
	40	0.0116	29	0.0116	29	85.9398		0.9999	1	20	2 | 6.0
	50	0.0145		0.0145		68.7501		0.9999	0	10	3 | 9.0
			30		30				1		4 | 12.0
1	0	0.0175	29	0.0175	29	57.2900		0.9998	0	0 **89**	5 | 15.0
	10	0.0204	29	0.0204	29	49.1039	81861	0.9998	0	50	6 | 18.0
	20	0.0233	29	0.0233	29	42.9641	61398	0.9997	1	40	7 | 21.0
	30	0.0262	29	0.0262	29	38.1885	47756	0.9997	0	30	8 | 24.0
	40	0.0291	29	0.0291	29	34.3678	38207	0.9996	1	20	9 | 27.0
	50	0.0320	29	0.0320	29	31.2416	31262	0.9995	1	10	
2	0	0.0349	29	0.0349	29	28.6363	26053	0.9994	1	0 **88**	
	10	0.0378	29	0.0378	29	26.4316	22047	0.9993	1	50	**29**
	20	0.0407	29	0.0407	30	24.5418	18598	0.9992	1	40	1 | 2.9
	30	0.0436	29	0.0437	29	22.9038	16380	0.9990	1	30	2 | 5.8
	40	0.0465	29	0.0466	29	21.4704	14334	0.9989	1	20	3 | 8.7
	50	0.0494	29	0.0495	29	20.2056	12648	0.9988	2	10	4 | 11.6
3	0	0.0523	29	0.0524	29	19.0811	11245	0.9986	1	0 **87**	5 | 14.5
	10	0.0552	29	0.0553	29	18.0750	10061	0.9985	2	50	6 | 17.4
	20	0.0581	29	0.0582	30	17.1693	9057	0.9983	2	40	7 | 20.3
	30	0.0610	30	0.0612	29	16.3199	8194	0.9981	1	30	8 | 23.2
	40	0.0640	29	0.0641	29	15.6048	7451	0.9980	2	20	9 | 26.1
	50	0.0669	29	0.0670	29	14.9244	6804	0.9978	2	10	
4	0	0.0698	29	0.0699	30	14.3007	6237	0.9976	2	0 **86**	**28**
	10	0.0727	29	0.0729	29	13.7267	5710	0.9974	3	50	1 | 2.8
	20	0.0756	29	0.0758	29	13.1969	5298	0.9971	2	40	2 | 5.6
	30	0.0785	29	0.0787	29	12.7062	4907	0.9969	2	30	3 | 8.4
	40	0.0814	29	0.0816	30	12.2505	4557	0.9967	3	20	4 | 11.2
	50	0.0843	29	0.0846		11.8262	4243	0.9964	2	10	5 | 14.0
5	0	0.0872	29	0.0875		11.4301	3961	0.9962	3	0 **85**	6 | 16.8
	10	0.0901	28	0.0904	30	11.0594	3707	0.9959	2	50	7 | 19.6
	20	0.0929	29	0.0934	29	10.7119	3475	0.9957	3	40	8 | 22.4
	30	0.0958	29	0.0963	29	10.3854	3265	0.9954	3	30	9 | 25.2
	40	0.0987	29	0.0992	30	10.0780	3074	0.9951	3	20	
	50	0.1016		0.1022		9.7882	2898	0.9948	3	10	**5**
6	0	0.1045	29	0.1051	29	9.5144	2738	0.9945	3	0 **84**	1 | 0.5
	10	0.1074	29	0.1080	30	9.2553	2591	0.9942	3	50	2 | 1.0
	20	0.1103	29	0.1110	29	9.0098	2455	0.9939	3	40	3 | 1.5
	30	0.1132	29	0.1139	30	8.7769	2329	0.9936	4	30	4 | 2.0
	40	0.1161	29	0.1169	29	8.5555	2214	0.9932	3	20	5 | 2.5
	50	0.1190	29	0.1198	30	8.3450	2105	0.9929	4	10	6 | 3.0
7	0	0.1219	29	0.1228	29	8.1443	2007	0.9925	3	0 **83**	7 | 3.5
	10	0.1248	28	0.1257	30	7.9530	1913	0.9922	4	50	8 | 4.0
	20	0.1276	29	0.1287	30	7.7704	1826	0.9918	4	40	9 | 4.5
	30	0.1305	29	0.1317	29	7.5958	1746	0.9914	3	30	
	40	0.1334	29	0.1346	30	7.4287	1671	0.9911	4	20	**4**
	50	0.1363	29	0.1376	29	7.2687	1600	0.9907	4	10	1 | 0.4
8	0	0.1392	29	0.1405	30	7.1154	1533	0.9903	4	0 **82**	2 | 0.8
	10	0.1421	28	0.1435	30	6.9682	1472	0.9899	5	50	3 | 1.2
	20	0.1449	29	0.1465	30	6.8269	1413	0.9894	4	40	4 | 1.6
	30	0.1478	29	0.1495	29	6.6912	1357	0.9890	4	30	5 | 2.0
	40	0.1507	29	0.1524	30	6.5606	1306	0.9886	5	20	6 | 2.4
	50	0.1536	28	0.1554	30	6.4348	1258	0.9881	4	10	7 | 2.8
9	0	0.1564		0.1584		6.3138	1210	0.9877		0 **81**	8 | 3.2
		Cos.	d.	Cot.	d.	Tan.	d.	Sin.	d.	′ °	9 | 3.6

°	′	Sin.	d.	Tan.	d.	Cot.	d.	Cos.	d.	′	°
9	0	0.1564	29	0.1584	30	6.3138	1168	0.9877	5	0	81
	10	0.1593	29	0.1614	30	6.1970	1126	0.9872	4	50	
	20	0.1622	28	0.1644	29	6.0844	1086	0.9868	5	40	
	30	0.1650	29	0.1673	30	5.9758	1050	0.9863	5	30	
	40	0.1679	29	0.1703	30	5.8708	1014	0.9858	5	20	
	50	0.1708	28	0.1733	30	5.7694	981	0.9853	5	10	
10	0	0.1736	29	0.1763	30	5.6713	949	0.9848	5	0	80
	10	0.1765	29	0.1793	30	5.5764	919	0.9843	5	50	
	20	0.1794	28	0.1823	30	5.4845	890	0.9838	5	40	
	30	0.1822	29	0.1853	30	5.3955	862	0.9833	6	30	
	40	0.1851	29	0.1883	31	5.3093	836	0.9827	5	20	
	50	0.1880	28	0.1914	30	5.2257	811	0.9822	6	10	
11	0	0.1908	29	0.1944	30	5.1446	788	0.9816	5	0	79
	10	0.1937	28	0.1974	30	5.0658	764	0.9811	6	50	
	20	0.1965	29	0.2004	31	4.9894	742	0.9805	6	40	
	30	0.1994	28	0.2035	30	4.9152	722	0.9799	6	30	
	40	0.2022	29	0.2065	30	4.8430	701	0.9793	6	20	
	50	0.2051	28	0.2095	31	4.7729	683	0.9787	6	10	
12	0	0.2079	29	0.2126	30	4.7046	664	0.9781	6	0	78
	10	0.2108	28	0.2156	30	4.6382	646	0.9775	6	50	
	20	0.2136	28	0.2186	31	4.5736	629	0.9769	6	40	
	30	0.2164	29	0.2217	30	4.5107	613	0.9763	6	30	
	40	0.2193	28	0.2247	31	4.4494	597	0.9757	7	20	
	50	0.2221	29	0.2278	31	4.3897	582	0.9750	6	10	
13	0	0.2250	28	0.2309	30	4.3315	568	0.9744	7	0	77
	10	0.2278	28	0.2339	31	4.2747	554	0.9737	7	50	
	20	0.2306	28	0.2370	31	4.2193	540	0.9730	6	40	
	30	0.2334	29	0.2401	31	4.1653	527	0.9724	7	30	
	40	0.2363	28	0.2432	30	4.1126	515	0.9717	7	20	
	50	0.2391	28	0.2462	31	4.0611	503	0.9710	7	10	
14	0	0.2419	28	0.2493	31	4.0108	491	0.9703	7	0	76
	10	0.2447	29	0.2524	31	3.9617	481	0.9696	7	50	
	20	0.2476	28	0.2555	31	3.9136	469	0.9689	8	40	
	30	0.2504	28	0.2586	31	3.8667	459	0.9681	7	30	
	40	0.2532	28	0.2617	31	3.8208	448	0.9674	7	20	
	50	0.2560	28	0.2648	31	3.7760	439	0.9667	8	10	
15	0	0.2588	28	0.2679	32	3.7321	430	0.9659	7	0	75
	10	0.2616	28	0.2711	31	3.6891	421	0.9652	8	50	
	20	0.2644	28	0.2742	31	3.6470	411	0.9644	8	40	
	30	0.2672	28	0.2773	32	3.6059	403	0.9636	8	30	
	40	0.2700	28	0.2805	31	3.5656	395	0.9628	7	20	
	50	0.2728	28	0.2836	31	3.5261	387	0.9621	8	10	
16	0	0.2756	28	0.2867	32	3.4874	379	0.9613	8	0	74
	10	0.2784	28	0.2899	32	3.4495	371	0.9605	9	50	
	20	0.2812	28	0.2931	31	3.4124	365	0.9596	8	40	
	30	0.2840	28	0.2962	32	3.3759	357	0.9588	8	30	
	40	0.2868	28	0.2994	32	3.3402	350	0.9580	8	20	
	50	0.2896	28	0.3026	31	3.3052	343	0.9572	9	10	
17	0	0.2924	28	0.3057	32	3.2709	338	0.9563	8	0	73
	10	0.2952	27	0.3089	32	3.2371	330	0.9555	9	50	
	20	0.2979	28	0.3121	32	3.2041	325	0.9546	9	40	
	30	0.3007	28	0.3153	32	3.1716	319	0.9537	9	30	
	40	0.3035	27	0.3185	32	3.1397	313	0.9528	8	20	
	50	0.3062	28	0.3217	32	3.1084	307	0.9520	9	10	
18	0	0.3090		0.3249		3.0777		0.9511		0	72
		Cos.	d.	Cot.	d.	Tan.	d.	Sin.	d.	′	°

P. P.

	32	31	30
1	3.2	3.1	3.0
2	6.4	6.2	6.0
3	9.6	9.3	9.0
4	12.8	12.4	12.0
5	16.0	15.5	15.0
6	19.2	18.6	18.0
7	22.4	21.7	21.0
8	25.6	24.8	24.0
9	28.8	27.9	27.0

	29	28	27
1	2.9	2.8	2.7
2	5.8	5.6	5.4
3	8.7	8.4	8.1
4	11.6	11.2	10.8
5	14.5	14.0	13.5
6	17.4	16.8	16.2
7	20.3	19.6	18.9
8	23.2	22.4	21.6
9	26.1	25.2	24.3

	9	8
1	0.9	0.8
2	1.8	1.6
3	2.7	2.4
4	3.6	3.2
5	4.5	4.0
6	5.4	4.8
7	6.3	5.6
8	7.2	6.4
9	8.1	7.2

	7	6
1	0.7	0.6
2	1.4	1.2
3	2.1	1.8
4	2.8	2.4
5	3.5	3.0
6	4.2	3.6
7	4.9	4.2
8	5.6	4.8
9	6.3	5.4

	5	4
1	0.5	0.4
2	1.0	0.8
3	1.5	1.2
4	2.0	1.6
5	2.5	2.0
6	3.0	2.4
7	3.5	2.8
8	4.0	3.2
9	4.5	3.6

P. P.

° '	Sin.	d.	Tan.	d.	Cot.	d.	Cos.	d.	'	°
18 0	0.3090	28	0.3249	32	3.0777	302	0.9511	9	0	**72**
10	0.3118	27	0.3281	33	3.0475	297	0.9502	10	50	
20	0.3145	28	0.3314	32	3.0178	291	0.9492	9	40	
30	0.3173	28	0.3346	32	2.9887	287	0.9483	9	30	
40	0.3201	27	0.3378	33	2.9600	281	0.9474	9	20	
50	0.3228	28	0.3411	32	2.9319	277	0.9465	10	10	
19 0	0.3256	27	0.3443	33	2.9042	272	0.9455	9	0	**71**
10	0.3283	28	0.3476	32	2.8770	268	0.9446	10	50	
20	0.3311	27	0.3508	33	2.8502	263	0.9436	10	40	
30	0.3338	27	0.3541	33	2.8239	259	0.9426	9	30	
40	0.3365	28	0.3574	33	2.7980	255	0.9417	10	20	
50	0.3393	27	0.3607	33	2.7725	250	0.9407	10	10	
20 0	0.3420	28	0.3640	33	2.7475	247	0.9397	10	0	**70**
10	0.3448	27	0.3673	33	2.7228	243	0.9387	10	50	
20	0.3475	27	0.3706	33	2.6985	239	0.9377	10	40	
30	0.3502	27	0.3739	33	2.6746	235	0.9367	11	30	
40	0.3529	28	0.3772	33	2.6511	232	0.9356	10	20	
50	0.3557	27	0.3805	34	2.6279	228	0.9346	10	10	
21 0	0.3584	27	0.3839	34	2.6051	225	0.9336	11	0	**69**
10	0.3611	27	0.3872	34	2.5826	221	0.9325	10	50	
20	0.3638	27	0.3906	33	2.5605	219	0.9315	11	40	
30	0.3665	27	0.3939	34	2.5386	214	0.9304	11	30	
40	0.3692	27	0.3973	33	2.5172	212	0.9293	10	20	
50	0.3719	27	0.4006	34	2.4960	209	0.9283	11	10	
22 0	0.3746	27	0.4040	34	2.4751	206	0.9272	11	0	**68**
10	0.3773	27	0.4074	34	2.4545	203	0.9261	11	50	
20	0.3800	27	0.4108	34	2.4342	200	0.9250	11	40	
30	0.3827	27	0.4142	34	2.4142	197	0.9239	11	30	
40	0.3854	27	0.4176	34	2.3945	195	0.9228	12	20	
50	0.3881	26	0.4210	35	2.3750	191	0.9216	11	10	
23 0	0.3907	27	0.4245	34	2.3559	190	0.9205	11	0	**67**
10	0.3934	27	0.4279	35	2.3369	186	0.9194	12	50	
20	0.3961	26	0.4314	34	2.3183	185	0.9182	11	40	
30	0.3987	27	0.4348	35	2.2998	181	0.9171	12	30	
40	0.4014	27	0.4383	34	2.2817	180	0.9159	12	20	
50	0.4041	26	0.4417	35	2.2637	177	0.9147	12	10	
24 0	0.4067	27	0.4452	35	2.2460	174	0.9135	11	0	**66**
10	0.4094	26	0.4487	35	2.2286	173	0.9124	12	50	
20	0.4120	27	0.4522	35	2.2113	170	0.9112	12	40	
30	0.4147	26	0.4557	35	2.1943	168	0.9100	12	30	
40	0.4173	27	0.4592	36	2.1775	166	0.9088	13	20	
50	0.4200	26	0.4628	35	2.1609	164	0.9075	12	10	
25 0	0.4226	27	0.4663	36	2.1445	162	0.9063	12	0	**65**
10	0.4253	26	0.4699	35	2.1283	160	0.9051	13	50	
20	0.4279	26	0.4734	36	2.1123	158	0.9038	12	40	
30	0.4305	26	0.4770	36	2.0965	156	0.9026	13	30	
40	0.4331	27	0.4806	35	2.0809	154	0.9013	12	20	
50	0.4358	26	0.4841	36	2.0655	152	0.9001	13	10	
26 0	0.4384	26	0.4877	36	2.0503	150	0.8988	13	0	**64**
10	0.4410	26	0.4913	37	2.0353	149	0.8975	13	50	
20	0.4436	26	0.4950	36	2.0204	147	0.8962	13	40	
30	0.4462	26	0.4986	36	2.0057	145	0.8949	13	30	
40	0.4488	26	0.5022	37	1.9912	144	0.8936	13	20	
50	0.4514	26	0.5059	36	1.9768	142	0.8923	13	10	
27 0	0.4540		0.5095		1.9626		0.8910		0	**63**
	Cos.	d.	Cot.	d.	Tan.	d.	Sin.	d.	'	°

P. P.

	37	36	35
1	3.7	3.6	3.5
2	7.4	7.2	7.0
3	11.1	10.8	10.5
4	14.8	14.4	14.0
5	18.5	18.0	17.5
6	22.2	21.6	21.0
7	25.9	25.2	24.5
8	29.6	28.8	28.0
9	33.3	32.4	31.5

	34	33	32
1	3.4	3.3	3.2
2	6.8	6.6	6.4
3	10.2	9.9	9.6
4	13.6	13.2	12.8
5	17.0	16.5	16.0
6	20.4	19.8	19.2
7	23.8	23.1	22.4
8	27.2	26.4	25.6
9	30.6	29.7	28.8

	28	27	26
1	2.8	2.7	2.6
2	5.6	5.4	5.2
3	8.4	8.1	7.8
4	11.2	10.8	10.4
5	14.0	13.5	13.0
6	16.8	16.2	15.6
7	19.6	18.9	18.2
8	22.4	21.6	20.8
9	25.2	24.3	23.4

	13	12
1	1.3	1.2
2	2.6	2.4
3	3.9	3.6
4	5.2	4.8
5	6.5	6.0
6	7.8	7.2
7	9.1	8.4
8	10.4	9.6
9	11.7	10.8

	11	10	9
1	1.1	1.0	0.9
2	2.2	2.0	1.8
3	3.3	3.0	2.7
4	4.4	4.0	3.6
5	5.5	5.0	4.5
6	6.6	6.0	5.4
7	7.7	7.0	6.3
8	8.8	8.0	7.2
9	9.9	9.0	8.1

P. P.

°	'	Sin.	d.	Tan.	d.	Cot.	d.	Cos.	d.		P. P.
27	0	0.4540	26	0.5095	37	1.9626	140	0.8910	13	0 63	
	10	0.4566	26	0.5132	37	1.9486	139	0.8897	13	50	
	20	0.4592	25	0.5169	37	1.9347	137	0.8884	14	40	
	30	0.4617	26	0.5206	37	1.9210	136	0.8870	13	30	
	40	0.4643	26	0.5243	37	1.9074	134	0.8857	14	20	
	50	0.4669	26	0.5280	37	1.8940	133	0.8843	14	10	
28	0	0.4695	25	0.5317	37	1.8807	131	0.8829	13	0 62	
	10	0.4720	26	0.5354	38	1.8676	130	0.8816	14	50	
	20	0.4746	26	0.5392	38	1.8546	128	0.8802	14	40	
	30	0.4772	25	0.5430	37	1.8418	127	0.8788	14	30	
	40	0.4797	26	0.5467	38	1.8291	126	0.8774	14	20	
	50	0.4823	25	0.5505	38	1.8165	125	0.8760	14	10	
29	0	0.4848	26	0.5543	38	1.8040	123	0.8746	14	0 61	
	10	0.4874	25	0.5581	38	1.7917	121	0.8732	14	50	
	20	0.4899	25	0.5619	39	1.7796	121	0.8718	14	40	
	30	0.4924	26	0.5658	38	1.7675	119	0.8704	15	30	
	40	0.4950	25	0.5696	39	1.7556	119	0.8689	14	20	
	50	0.4975	25	0.5735	39	1.7437	116	0.8675	15	10	
30	0	0.5000	25	0.5774	38	1.7321	116	0.8660	14	0 60	
	10	0.5025	25	0.5812	39	1.7205	115	0.8646	15	50	
	20	0.5050	25	0.5851	39	1.7090	113	0.8631	15	40	
	30	0.5075	25	0.5890	40	1.6977	113	0.8616	15	30	
	40	0.5100	25	0.5930	39	1.6864	111	0.8601	14	20	
	50	0.5125	25	0.5969	40	1.6753	110	0.8587	15	10	
31	0	0.5150	25	0.6009	39	1.6643	109	0.8572	15	0 59	
	10	0.5175	25	0.6048	40	1.6534	108	0.8557	15	50	
	20	0.5200	25	0.6088	40	1.6426	107	0.8542	16	40	
	30	0.5225	25	0.6128	40	1.6319	107	0.8526	15	30	
	40	0.5250	25	0.6168	40	1.6212	105	0.8511	15	20	
	50	0.5275	24	0.6208	41	1.6107	104	0.8496	16	10	
32	0	0.5299	25	0.6249	40	1.6003	103	0.8480	15	0 58	
	10	0.5324	24	0.6289	41	1.5900	102	0.8465	15	50	
	20	0.5348	25	0.6330	41	1.5798	101	0.8450	16	40	
	30	0.5373	25	0.6371	41	1.5697	100	0.8434	16	30	
	40	0.5398	24	0.6412	41	1.5597	100	0.8418	15	20	
	50	0.5422	24	0.6453	41	1.5497	98	0.8403	16	10	
33	0	0.5446	25	0.6494	42	1.5399	98	0.8387	16	0 57	
	10	0.5471	24	0.6536	41	1.5301	97	0.8371	16	50	
	20	0.5495	24	0.6577	42	1.5204	96	0.8355	16	40	
	30	0.5519	25	0.6619	42	1.5108	95	0.8339	16	30	
	40	0.5544	24	0.6661	42	1.5013	94	0.8323	16	20	
	50	0.5568	24	0.6703	42	1.4919	93	0.8307	17	10	
34	0	0.5592	24	0.6745	42	1.4826	93	0.8290	16	0 56	
	10	0.5616	24	0.6787	43	1.4733	92	0.8274	16	50	
	20	0.5640	24	0.6830	43	1.4641	91	0.8258	17	40	
	30	0.5664	24	0.6873	43	1.4550	90	0.8241	16	30	
	40	0.5688	24	0.6916	43	1.4460	90	0.8225	17	20	
	50	0.5712	24	0.6959	43	1.4370	89	0.8208	16	10	
35	0	0.5736	24	0.7002	44	1.4281	88	0.8192	17	0 55	
	10	0.5760	23	0.7046	43	1.4193	87	0.8175	17	50	
	20	0.5783	24	0.7089	44	1.4106	87	0.8158	17	40	
	30	0.5807	24	0.7133	44	1.4019	85	0.8141	17	30	
	40	0.5831	23	0.7177	44	1.3934	86	0.8124	17	20	
	50	0.5854	24	0.7221	44	1.3848	84	0.8107	17	10	
36	0	0.5878		0.7265		1.3764		0.8090		0 54	
		Cos.	d.	Cot.	d.	Tan.	d.	Sin.	d.	' °	P. P.

P. P.

	44	43	42
1	4.4	4.3	4.2
2	8.8	8.6	8.4
3	13.2	12.9	12.6
4	17.6	17.2	16.8
5	22.0	21.5	21.0
6	26.4	25.8	25.2
7	30.8	30.1	29.4
8	35.2	34.4	33.6
9	39.6	38.7	37.8

	41	40	39
1	4.1	4.0	3.9
2	8.2	8.0	7.8
3	12.3	12.0	11.7
4	16.4	16.0	15.6
5	20.5	20.0	19.5
6	24.6	24.0	23.4
7	28.7	28.0	27.3
8	32.8	32.0	31.2
9	36.9	36.0	35.1

	38	37
1	3.8	3.7
2	7.6	7.4
3	11.4	11.1
4	15.2	14.8
5	19.0	18.5
6	22.8	22.2
7	26.6	25.9
8	30.4	29.6
9	34.2	33.3

	26	25	24
1	2.6	2.5	2.4
2	5.2	5.0	4.8
3	7.8	7.5	7.2
4	10.4	10.0	9.6
4	13.0	12.5	12.0
6	15.6	15.0	14.4
7	18.2	17.5	16.8
8	20.8	20.0	19.2
9	23.4	22.5	21.6

	23	17	16
1	2.3	1.7	1.6
2	4.6	3.4	3.2
3	6.9	5.1	4.8
4	9.2	6.8	6.4
5	11.5	8.5	8.0
6	13.8	10.2	9.6
7	16.1	11.9	11.2
8	18.4	13.6	12.8
9	20.7	15.3	14.4

	15	14	13
1	1.5	1.4	1.3
2	3.0	2.8	2.6
3	4.5	4.2	3.9
4	6.0	5.6	5.2
5	7.5	7.0	6.5
6	9.0	8.4	7.8
7	10.5	9.8	9.1
8	12.0	11.2	10.4
9	13.5	12.6	11.7

° '	Sin.	d.	Tan.	d.	Cot.	d.	Cos.	d.	
36 0	0.5878	23	0.7265	45	1.3764	84	0.8090	17	0 **54**
10	0.5901	24	0.7310	45	1.3680	83	0.8073	17	50
20	0.5925	23	0.7355	45	1.3597	83	0.8056	17	40
30	0.5948	24	0.7400	45	1.3514	82	0.8039	18	30
40	0.5972	23	0.7445	45	1.3432	81	0.8021	17	20
50	0.5995	23	0.7490	46	1.3351	81	0.8004	18	10
37 0	0.6018	23	0.7536	45	1.3270	80	0.7986	17	0 **53**
10	0.6041	24	0.7581	46	1.3190	79	0.7969	18	50
20	0.6065	23	0.7627	46	1.3111	79	0.7951	18	40
30	0.6088	23	0.7673	47	1.3032	78	0.7934	18	30
40	0.6111	23	0.7720	46	1.2954	78	0.7916	18	20
50	0.6134	23	0.7766	47	1.2876	77	0.7898	18	10
38 0	0.6157	23	0.7813	47	1.2799	76	0.7880	18	0 **52**
10	0.6180	22	0.7860	47	1.2723	76	0.7862	18	50
20	0.6202	23	0.7907	47	1.2647	75	0.7844	18	40
30	0.6225	23	0.7954	48	1.2572	75	0.7826	18	30
40	0.6248	23	0.8002	48	1.2497	74	0.7808	18	20
50	0.6271	22	0.8050	48	1.2423	74	0.7790	19	10
39 0	0.6293	23	0.8098	48	1.2349	73	0.7771	18	0 **51**
10	0.6316	22	0.8146	49	1.2276	73	0.7753	18	50
20	0.6338	23	0.8195	48	1.2203	72	0.7735	19	40
30	0.6361	22	0.8243	49	1.2131	72	0.7716	18	30
40	0.6383	23	0.8292	50	1.2059	71	0.7698	19	20
50	0.6406	22	0.8342	49	1.1988	70	0.7679	19	10
40 0	0.6428	22	0.8391	50	1.1918	71	0.7660	18	0 **50**
10	0.6450	22	0.8441	50	1.1847	69	0.7642	19	50
20	0.6472	22	0.8491	50	1.1778	70	0.7623	19	40
30	0.6494	23	0.8541	50	1.1708	68	0.7604	19	30
40	0.6517	22	0.8591	51	1.1640	69	0.7585	19	20
50	0.6539	22	0.8642	51	1.1571	67	0.7566	19	10
41 0	0.6561	22	0.8693	51	1.1504	68	0.7547	19	0 **49**
10	0.6583	21	0.8744	52	1.1436	67	0.7528	19	50
20	0.6604	22	0.8796	51	1.1369	66	0.7509	19	40
30	0.6626	22	0.8847	52	1.1303	66	0.7490	20	30
40	0.6648	22	0.8899	53	1.1237	66	0.7470	19	20
50	0.6670	21	0.8952	52	1.1171	65	0.7451	20	10
42 0	0.6691	22	0.9004	53	1.1106	65	0.7431	19	0 **48**
10	0.6713	21	0.9057	53	1.1041	64	0.7412	20	50
20	0.6734	22	0.9110	53	1.0977	64	0.7392	19	40
30	0.6756	21	0.9163	54	1.0913	63	0.7373	20	30
40	0.6777	22	0.9217	54	1.08 0	64	0.7353	20	20
50	0.6799	21	0.9271	54	1.0786	62	0.7333	19	10
43 0	0.6820	21	0.9325	54	1.0724	62	0.7314	19	0 **47**
10	0.6841	21	0.9380	55	1.0661	63	0.7294	20	50
20	0.6862	22	0.9435	55	1.0599	62	0.7274	20	40
30	0.6884	21	0.9490	55	1.0538	61	0.7254	20	30
40	0.6905	21	0.9545	56	1.0477	61	0.7234	20	20
50	0.6926	21	0.9601	56	1.0416	61	0.7214	20	10
44 0	0.6947	20	0.9657	56	1.0355	61	0.7193	21	0 **46**
10	0.6967	21	0.9713	57	1.0295	60	0.7178	20	50
20	0.6988	21	0.9770	57	1.0235	59	0.7153	20	40
30	0.7009	21	0.9827	57	1.0176	59	0.7133	21	30
40	0.7030	20	0.9884	58	1.0117	59	0.7112	20	20
50	0.7050	21	0.9942	58	1.0058	58	0.7092	20	10
45 0	0.7071	21	1.0000	58	1.0000	58	0.7071	21	0 **45**
	Cos.	d.	Cot.	d.	Tan.	d.	Sin.	d.	' °

P. P.

	58	57	56	55
1	5.8	5.7	5.6	5.5
2	11.6	11.4	11.2	11.0
3	17.4	17.1	16.8	16.5
4	23.2	22.8	22.4	22.0
5	29.0	28.5	28.0	27.5
6	34.8	34.2	33.6	33.0
7	40.6	39.9	39.2	38.5
8	46.4	45.6	44.8	44.0
9	52.2	51.3	50.4	49.5

	54	53	52	51
1	5.4	5.3	5.2	5.1
2	10.8	10.6	10.4	10.2
3	16.2	15.9	15.6	15.3
4	21.6	21.2	20.8	20.4
5	27.0	26.5	26.0	25.5
6	32.4	31.8	31.2	30.6
7	37.8	37.1	36.4	35.7
8	43.2	42.4	41.6	40.8
9	48.6	47.7	46.8	45.9

	50	49	48
1	5.0	4.9	4.8
2	10.0	9.8	9.6
3	15.0	14.7	14.4
4	20.0	19.6	19.2
5	25.0	24.5	24.0
6	30.0	29.4	28.8
7	35.0	34.3	33.6
8	40.0	39.2	38.4
9	45.0	44.1	43.2

	47	46	45
1	4.7	4.6	4.5
2	9.4	9.2	9.0
3	14.1	13.8	13.5
4	18.8	18.4	18.0
5	23.5	23.0	22.5
6	28.2	27.6	27.0
7	32.9	32.2	31.5
8	37.6	36.8	36.0
9	42.3	41.4	40.5

	24	23	22	21
1	2.4	2.3	2.2	2.1
2	4.8	4.6	4.4	4.2
3	7.2	6.9	6.6	6.3
4	9.6	9.2	8.8	8.4
5	12.0	11.5	11.0	10.5
6	14.4	13.8	13.2	12.6
7	16.8	16.1	15.4	14.7
8	19.2	18.4	17.6	16.8
9	21.6	20.7	19.8	18.9

	20	19	18	17
1	2.0	1.9	1.8	1.7
2	4.0	3.8	3.6	3.4
3	6.0	5.7	5.4	5.1
4	8.0	7.6	7.2	6.8
5	10.0	9.5	9.0	8.5
6	12.0	11.4	10.8	10.2
7	14.0	13.3	12.6	11.9
8	16.0	15.2	14.4	13.6
9	18.0	17.1	16.2	15.3

P. P.

PRIME NUMBERS.

Every prime number is an odd number and has for its unit figure 1, 3, 7, or 9; any odd number that has 5 for its unit figure is divisible by 5, and is not a prime number. The prime factors of any number less than 1,000 may be found from the following table. If the number is odd and does not end with 5, the factors are given directly; thus, the prime factors of 357 are 3, 7, and 17; those of 931 are 7, 7, and 19, the exponent 2 of the 7 indicating that 7 is used twice as a factor. If a number is a prime number, the space beside it is blank; thus, 317 and 859 are prime numbers. To find the prime factors of an odd number that has 5 for the unit figure, divide by 5 until a quotient is obtained which does not have 5 for a unit figure; the factors of this quotient are then found from the table, and with the 5's already used as divisors constitute the prime factors. For example, to find the prime factors of 5,775 proceed as follows: $5{,}775 \div 5 = 1{,}155$; $1{,}155 \div 5 = 231$; from the table, $231 = 3 \times 7 \times 11$; hence, $5{,}775 = 3 \times 5 \times 5 \times 7 \times 11$. If the number is even, divide it by 2, the quotient by 2, and so on until an odd quotient is reached; then find the prime factors of the quotient from the table. The process of finding the prime factors of 936 is as follows:

$936 \div 2 = 468$; $468 \div 2 = 234$; $234 \div 2 = 117$; $117 = 3^2 \times 13$, from table. Hence, $936 = 2^3 \times 3^2 \times 13 = 2 \times 2 \times 2 \times 3 \times 3 \times 13$.

FACTORS OF 3.1416.

NOT REGARDING DECIMAL POINT, 3.1416 =

2×15708	22×1428	68×462
3×10472	24×1309	77×408
4×7854	28×1122	84×374
6×5236	33×952	88×357
7×4488	34×924	102×308
8×3927	42×748	119×264
11×2856	44×714	132×238
12×2618	51×616	136×231
14×2244	56×561	154×204
17×1848	66×476	168×187
21×1496		

PRIME FACTORS.

PRIME FACTORS OF ALL ODD NUMBERS FROM 1 TO 1,000
THAT ARE NOT DIVISIBLE BY 5.

1		101		201	$3 \cdot 67$	301	$7 \cdot 43$	401	
3		103		203	$7 \cdot 29$	303	$3 \cdot 101$	403	$13 \cdot 31$
7		107		207	$3^2 \cdot 23$	307		407	$11 \cdot 37$
9	3^2	109		209	$11 \cdot 19$	309	$3 \cdot 103$	409	
11		111	$3 \cdot 37$	211		311		411	$3 \cdot 137$
13		113		213	$3 \cdot 71$	313		413	$7 \cdot 59$
17		117	$3^2 \cdot 13$	217	$7 \cdot 31$	317		417	$3 \cdot 139$
19		119	7.17	219	$3 \cdot 73$	319	$11 \cdot 29$	419	
21	$3 \cdot 7$	121	11^2	221	$13 \cdot 17$	321	$3 \cdot 107$	421	
23		123	$3 \cdot 41$	223		323	$17 \cdot 19$	423	$3^2 \cdot 47$
27	3^3	127		227		327	$3 \cdot 109$	427	$7 \cdot 61$
29		129	$3 \cdot 43$	229		329	$7 \cdot 47$	429	$3 \cdot 11 \cdot 13$
31		131		231	$3 \cdot 7 \cdot 11$	331		431	
33	$3 \cdot 11$	133	$7 \cdot 19$	233		333	$3^2 \cdot 37$	433	
37		137		237	$3 \cdot 79$	337		437	$19 \cdot 23$
39	$3 \cdot 13$	139		239		339	$3 \cdot 113$	439	
41		141	$3 \cdot 47$	241		341	$11 \cdot 31$	441	$3^2 \cdot 7^2$
43		143	$11 \cdot 13$	243	3^5	343	7^3	443	
47		147	$3 \cdot 7^2$	247	$13 \cdot 19$	347		447	$3 \cdot 149$
49	7^2	149		249	$3 \cdot 83$	349		449	
51	$3 \cdot 17$	151		251		351	$3^3 \cdot 13$	451	$11 \cdot 41$
53		153	$3^2 \cdot 17$	253	$11 \cdot 23$	353		453	$3 \cdot 151$
57	$3 \cdot 19$	157		257		357	$3 \cdot 7 \cdot 17$	457	
59		159	$3 \cdot 53$	259	$7 \cdot 37$	359		459	$3^3 \cdot 17$
61		161	$7 \cdot 23$	261	$3^2 \cdot 29$	361	19^2	461	
63	$3^2 \cdot 7$	163		263		363	$3 \cdot 11^2$	463	
67		167		267	$3 \cdot 89$	367		467	
69	$3 \cdot 23$	169	13^2	269		369	$3^2 \cdot 41$	469	$7 \cdot 67$
71		171	$3^2 \cdot 19$	271		371	$7 \cdot 53$	471	$3 \cdot 157$
73		173		273	$3 \cdot 7 \cdot 13$	373		473	$11 \cdot 43$
77	$7 \cdot 11$	177	$3 \cdot 59$	277		377	$13 \cdot 29$	477	$3^2 \cdot 53$
79		179		279	$3^2 \cdot 31$	379		479	
81	3^4	181		281		381	$3 \cdot 127$	481	$13 \cdot 37$
83		183	$3 \cdot 61$	283		383		483	$3 \cdot 7 \cdot 23$
87	$3 \cdot 29$	187	$11 \cdot 17$	287	$7 \cdot 41$	387	$3^2 \cdot 43$	487	
89		189	$3^3 \cdot 7$	289	17^2	389		489	$3 \cdot 163$
91	$7 \cdot 13$	191		291	$3 \cdot 97$	391	$17 \cdot 23$	491	
93	$3 \cdot 31$	193		293		393	$3 \cdot 131$	493	$17 \cdot 29$
97		197		297	$3^3 \cdot 11$	397		497	$7 \cdot 71$
99	$3^2 \cdot 11$	199		299	$13 \cdot 23$	399	$3 \cdot 7 \cdot 19$	499	

PRIME FACTORS OF ALL ODD NUMBERS FROM 1 TO 1,000
THAT ARE NOT DIVISIBLE BY 5.

(*Continued*).

501	$3 \cdot 167$	601		701		801	$3^2 \cdot 89$	901	$17 \cdot 53$
503		603	$3^2 \cdot 67$	703	$19 \cdot 37$	803	$11 \cdot 73$	903	$3 \cdot 7 \cdot 43$
507	$3 \cdot 13^2$	607		707	$7 \cdot 101$	807	$3 \cdot 269$	907	
509		609	$3 \cdot 7 \cdot 29$	709		809		909	$3^2 \cdot 101$
511	$7 \cdot 73$	611	$13 \cdot 47$	711	$3^2 \cdot 79$	811		911	
513	$3^3 \cdot 19$	613		713	$23 \cdot 31$	813	$3 \cdot 271$	913	$11 \cdot 83$
517	$11 \cdot 47$	617		717	$3 \cdot 239$	817	$19 \cdot 43$	917	$7 \cdot 131$
519	$3 \cdot 173$	619		719		819	$3^2 \cdot 7 \cdot 13$	919	
521		621	$3^3 \cdot 23$	721	$7 \cdot 103$	821		921	$3 \cdot 307$
523		623	$7 \cdot 89$	723	$3 \cdot 241$	823		923	$13 \cdot 71$
527	$17 \cdot 31$	627	$3 \cdot 11 \cdot 19$	727		827		927	$3^2 \cdot 103$
529	23^2	629	$17 \cdot 37$	729	3^6	829		929	
531	$3^2 \cdot 59$	631		731	$17 \cdot 43$	831	$3 \cdot 277$	931	$7^2 \cdot 19$
533	$13 \cdot 41$	633	$3 \cdot 211$	733		833	$7^2 \cdot 17$	933	$3 \cdot 311$
537	$3 \cdot 179$	637	$7^2 \cdot 13$	737	$11 \cdot 67$	837	$3^3 \cdot 31$	937	
539	$7^2 \cdot 11$	639	$3^2 \cdot 71$	739		839		939	$3 \cdot 313$
541		641		741	$3 \cdot 13 \cdot 19$	841	29^2	941	
543	$3 \cdot 181$	643		743		843	$3 \cdot 281$	943	$23 \cdot 41$
547		647		747	$3^2 \cdot 83$	847	$7 \cdot 11^2$	947	
549	$3^2 \cdot 61$	649	$11 \cdot 59$	749	$7 \cdot 107$	849	$3 \cdot 283$	949	$13 \cdot 73$
551	$19 \cdot 29$	651	$3 \cdot 7 \cdot 31$	751		851	$23 \cdot 37$	951	$3 \cdot 317$
553	$7 \cdot 79$	653		753	$3 \cdot 251$	853		953	
557		657	$3^2 \cdot 73$	757		857		957	$3 \cdot 11 \cdot 29$
559	$13 \cdot 43$	659		759	$3 \cdot 11 \cdot 23$	859		959	$7 \cdot 137$
561	$3 \cdot 11 \cdot 17$	661		761		861	$3 \cdot 7 \cdot 41$	961	31^2
563		663	$3 \cdot 13 \cdot 17$	763	$7 \cdot 109$	863		963	$3^2 \cdot 107$
567	$3^4 \cdot 7$	667	$23 \cdot 29$	767	$13 \cdot 59$	867	$3 \cdot 17^2$	967	
569		669	$3 \cdot 223$	769		869	$11 \cdot 79$	969	$3 \cdot 17 \cdot 19$
571		671	$11 \cdot 61$	771	$3 \cdot 257$	871	$13 \cdot 67$	971	
573	$3 \cdot 191$	673		773		873	$3^2 \cdot 97$	973	$7 \cdot 139$
577		677		777	$3 \cdot 7 \cdot 37$	877		977	
579	$3 \cdot 193$	679	$7 \cdot 97$	779	$19 \cdot 41$	879	$3 \cdot 293$	979	$11 \cdot 89$
581	$7 \cdot 83$	681	$3 \cdot 227$	781	$11 \cdot 71$	881		981	$3^2 \cdot 109$
583	$11 \cdot 53$	683		783	$3^3 \cdot 29$	883		983	
587		687	$3 \cdot 229$	787		887		987	$3 \cdot 7 \cdot 47$
589	$19 \cdot 31$	689	$13 \cdot 53$	789	$3 \cdot 263$	889	$7 \cdot 127$	989	$23 \cdot 43$
591	$3 \cdot 197$	691		791	$7 \cdot 113$	891	$3^4 \cdot 11$	991	
593		693	$3^2 \cdot 7 \cdot 11$	793	$13 \cdot 61$	893	$19 \cdot 47$	993	$3 \cdot 331$
597	$3 \cdot 199$	697	$17 \cdot 41$	797		897	$3 \cdot 13 \cdot 23$	997	
599		699	$3 \cdot 233$	799	$17 \cdot 47$	899	$29 \cdot 31$	999	$3^3 \cdot 37$

CIRCUMFERENCES AND AREAS OF CIRCLES FROM 1-64 TO 100.

Diam.	Circum.	Area.	Diam.	Circum.	Area.
1/64	.0491	.0002	4 3/8	13.7445	15.0330
1/32	.0982	.0008	4 1/2	14.1372	15.9043
1/16	.1963	.0031	4 5/8	14.5299	16.8002
1/8	.3927	.0123	4 3/4	14.9226	17.7206
3/16	.5890	.0276	4 7/8	15.3153	18.6555
1/4	.7854	.0491	5	15.7080	19.6350
5/16	.9817	.0767	5 1/8	16.1007	20.6290
3/8	1.1781	.1104	5 1/4	16.4934	21.6476
7/16	1.3744	.1503	5 3/8	16.8861	22.6907
1/2	1.5708	.1963	5 1/2	17.2788	23.7583
9/16	1.7671	.2485	5 5/8	17.6715	24.8505
5/8	1.9635	.3068	5 3/4	18.0642	25.9673
11/16	2.1598	.3712	5 7/8	18.4569	27.1086
3/4	2.3562	.4418	6	18.8496	28.2744
13/16	2.5525	.5185	6 1/8	19.2423	29.4648
7/8	2.7489	.6013	6 1/4	19.6350	30.6797
15/16	2.9452	.6903	6 3/8	20.0277	31.9191
1	3.1416	.7854	6 1/2	20.4204	33.1831
1 1/8	3.5343	.9940	6 5/8	20.8131	34.4717
1 1/4	3.9270	1.2272	6 3/4	21.2058	35.7848
1 3/8	4.3197	1.4849	6 7/8	21.5985	37.1224
1 1/2	4.7124	1.7671	7	21.9912	38.4846
1 5/8	5.1051	2.0739	7 1/8	22.3839	39.8713
1 3/4	5.4978	2.4053	7 1/4	22.7766	41.2826
1 7/8	5.8905	2.7612	7 3/8	23.1693	42.7184
2	6.2832	3.1416	7 1/2	23.5620	44.1787
2 1/8	6.6759	3.5466	7 5/8	23.9547	45.6636
2 1/4	7.0686	3.9761	7 3/4	24.3474	47.1731
2 3/8	7.4613	4.4301	7 7/8	24.7401	48.7071
2 1/2	7.8540	4.9087	8	25.1328	50.2656
2 5/8	8.2467	5.4119	8 1/8	25.5255	51.8487
2 3/4	8.6394	5.9396	8 1/4	25.9182	53.4563
2 7/8	9.0321	6.4918	8 3/8	26.3109	55.0884
3	9.4248	7.0686	8 1/2	26.7036	56.7451
3 1/8	9.8175	7.6699	8 5/8	27.0963	58.4264
3 1/4	10.2102	8.2958	8 3/4	27.4890	60.1322
3 3/8	10.6029	8.9462	8 7/8	27.8817	61.8625
3 1/2	10.9956	9.6211	9	28.2744	63.6174
3 5/8	11.3883	10.3206	9 1/8	28.6671	65.3968
3 3/4	11.7810	11.0447	9 1/4	29.0598	67.2008
3 7/8	12.1737	11.7933	9 3/8	29.4525	69.0293
4	12.5664	12.5664	9 1/2	29.8452	70.8823
4 1/8	12.9591	13.3641	9 5/8	30.2379	72.7599
4 1/4	13.3518	14.1863	9 3/4	30.6306	74.6621

TABLE—(*Continued*).

Diam.	Circum.	Area.	Diam.	Circum.	Area.
9⅞	31.0233	76.589	15⅝	49.0875	191.748
10	31.4160	78.540	15¾	49.4802	194.828
10⅛	31.8087	80.516	15⅞	49.8729	197.933
10¼	32.2014	82.516	16	50.2656	201.062
10⅜	32.5941	84.541	16⅛	50.6583	204.216
10½	32.9868	86.590	16¼	51.0510	207.395
10⅝	33.3795	88.664	16⅜	51.4437	210.598
10¾	33.7722	90.763	16½	51.8364	213.825
10⅞	34.1649	92.886	16⅝	52.2291	217.077
11	34.5576	95.033	16¾	52.6218	220.354
11⅛	34.9503	97.205	16⅞	53.0145	223.655
11¼	35.3430	99.402	17	53.4072	226.981
11⅜	35.7357	101.623	17⅛	53.7999	230.331
11½	36.1284	103.869	17¼	54.1926	233.706
11⅝	36.5211	106.139	17⅜	54.5853	237.105
11¾	36.9138	108.434	17½	54.9780	240.529
11⅞	37.3065	110.754	17⅝	55.3707	243.977
12	37.6992	113.098	17¾	55.7634	247.450
12⅛	38.0919	115.466	17⅞	56.1561	250.948
12¼	38.4846	117.859	18	56.5488	254.470
12⅜	38.8773	120.277	18⅛	56.9415	258.016
12½	39.2700	122.719	18¼	57.3342	261.587
12⅝	39.6627	125.185	18⅜	57.7269	265.183
12¾	40.0554	127.677	18½	58.1196	268.803
12⅞	40.4481	130.192	18⅝	58.5123	272.448
13	40.8408	132.733	18¾	58.9050	276.117
13⅛	41.2335	135.297	18⅞	59.2977	279.811
13¼	41.6262	137.887	19	59.6904	283.529
13⅜	42.0189	140.501	19⅛	60.0831	287.272
13½	42.4116	143.189	19¼	60.4758	291.040
13⅝	42.8043	145.802	19⅜	60.8685	294.832
13¾	43.1970	148.490	19½	61.2612	298.648
13⅞	43.5897	151.202	19⅝	61.6539	302.489
14	43.9824	153.938	19¾	62.0466	306.355
14⅛	44.3751	156.700	19⅞	62.4393	310.245
14¼	44.7678	159.485	20	62.8320	314.160
14⅜	45.1605	162.296	20⅛	63.2247	318.099
14½	45.5582	165.130	20¼	63.6174	322.063
14⅝	45.9459	167.990	20⅜	64.0101	326.051
14¾	46.3386	170.874	20½	64.4028	330.064
14⅞	46.7313	173.782	20⅝	64.7955	334.102
15	47.1240	176.715	20¾	65.1882	338.164
15⅛	47.5167	179.673	20⅞	65.5809	342.250
15¼	47.9094	182.655	21	65.9736	346.361
15⅜	48.3021	185.661	21⅛	66.3663	350.497
15½	48.6948	188.692	21¼	66.7590	354.657

TABLE—(*Continued*).

Diam.	Circum.	Area.	Diam.	Circum.	Area.
21⅜	67.1517	358.842	27⅛	85.2159	577.870
21½	67.5444	363.051	27¼	85.6086	583.209
21⅝	67.9371	367.285	27⅜	86.0013	588.571
21¾	68.3298	371.543	27½	86.3940	593.959
21⅞	68.7225	375.826	27⅝	86.7867	599.371
22	69.1152	380.134	27¾	87.1794	604.807
22⅛	69.5079	384.466	27⅞	87.5721	610.268
22¼	69.9006	388.822	28	87.9648	615.754
22⅜	70.2933	393.203	28⅛	88.3575	621.264
22½	70.6860	397.609	28¼	88.7502	626.798
22⅝	71.0787	402.038	28⅜	89.1429	632.357
22¾	71.4714	406.494	28½	89.5356	637.941
22⅞	71.8641	410.973	28⅝	89.9283	643.549
23	72.2568	415.477	28¾	90.3210	649.182
23⅛	72.6495	420.004	28⅞	90.7137	654.840
23¼	73.0422	424.558	29	91.1064	660.521
23⅜	73.4349	429.135	29⅛	91.4991	666.228
23½	73.8276	433.737	29¼	91.8918	671.959
23⅝	74.2203	438.364	29⅜	92.2845	677.714
23¾	74.6130	443.015	29½	92.6772	683.494
23⅞	75.0057	447.690	29⅝	93.0699	689.299
24	75.3984	452.390	29¾	93.4626	695.128
24⅛	75.7911	457.115	29⅞	93.8553	700.982
24¼	76.1838	461.864	30	94.2480	706.860
24⅜	76.5765	466.638	30⅛	94.6407	712.763
24½	76.9692	471.436	30¼	95.0834	718.690
24⅝	77.3619	476.259	30⅜	95.4261	724.642
24¾	77.7546	481.107	30½	95.8188	730.618
24⅞	78.1473	485.979	30⅝	96.2115	736.619
25	78.5400	490.875	30¾	96.6042	742.645
25⅛	78.9327	495.796	30⅞	96.9969	748.695
25¼	79.3254	500.742	31	97.3896	754.769
25⅜	79.7181	505.712	31⅛	97.7823	760.869
25½	80.1108	510.706	31¼	98.1750	766.992
25⅝	80.5085	515.726	31⅜	98.5677	773.140
25¾	80.8962	520.769	31½	98.9604	779.313
25⅞	81.2889	525.838	31⅝	99.3531	785.510
26	81.6816	530.930	31¾	99.7458	791.732
26⅛	82.0743	536.048	31⅞	100.1385	797.979
26¼	82.4670	541.190	32	100.5312	804.250
26⅜	82.8597	546.356	32⅛	100.9239	810.545
26½	83.2524	551.547	32¼	101.3166	816.865
26⅝	83.6451	556.763	32⅜	101.7093	823.210
26¾	84.0378	562.003	32½	102.1020	829.579
26⅞	84.4305	567.267	32⅝	102.4947	835.972
27	84.8232	572.557	32¾	102.8874	842.391

TABLE—(*Continued*).

Diam.	Circum.	Area.	Diam.	Circum.	Area.
32 7/8	103.280	848.833	38 5/8	121.344	1,171.731
33	103.673	855.301	38 3/4	121.737	1,179.327
33 1/8	104.065	861.792	38 7/8	122.130	1,186.948
33 1/4	104.458	868.309	39	122.522	1,194.593
33 3/8	104.851	874.850	39 1/8	122.915	1,202.263
33 1/2	105.244	881.415	39 1/4	123.308	1,209.958
33 5/8	105.636	888.005	39 3/8	123.700	1,217.677
33 3/4	106.029	894.620	39 1/2	124.093	1,225.420
33 7/8	106.422	901.259	39 5/8	124.486	1,233.188
34	106.814	907.922	39 3/4	124.879	1,240.981
34 1/8	107.207	914.611	39 7/8	125.271	1,248.798
34 1/4	107.600	921.323	40	125.664	1,256.640
34 3/8	107.992	928.061	40 1/8	126.057	1,264.510
34 1/2	108.385	934.822	40 1/4	126.449	1,272.400
34 5/8	108.778	941.609	40 3/8	126.842	1,280.310
34 3/4	109.171	948.420	40 1/2	127.235	1,288.250
34 7/8	109.563	955.255	40 5/8	127.627	1,296.220
35	109.956	962.115	40 3/4	128.020	1,304.210
35 1/8	110.349	969.000	40 7/8	128.413	1,312.220
35 1/4	110.741	975.909	41	128.806	1,320.260
35 3/8	111.134	982.842	41 1/8	129.198	1,328.320
35 1/2	111.527	989.800	41 1/4	129.591	1,336.410
35 5/8	111.919	996.783	41 3/8	129.984	1,344.520
35 3/4	112.312	1,003.790	41 1/2	130.376	1,352.660
35 7/8	112.705	1,010.822	41 5/8	130.769	1,360.820
36	113.098	1,017.878	41 3/4	131.162	1,369.000
36 1/8	113.490	1,024.960	41 7/8	131.554	1,377.210
36 1/4	113.883	1,032.065	42	131.947	1,385.450
36 3/8	114.276	1,039.195	42 1/8	132.340	1,393.700
36 1/2	114.668	1,046.349	42 1/4	132.733	1,401.990
36 5/8	115.061	1,053.528	42 3/8	133.125	1,410.300
36 3/4	115.454	1,060.732	42 1/2	133.518	1,418.630
36 7/8	115.846	1,067.960	42 5/8	133.911	1,426.990
37	116.239	1,075.213	42 3/4	134.303	1,435.370
37 1/8	116.632	1,082.490	42 7/8	134.696	1,443.770
37 1/4	117.025	1,089.792	43	135.089	1,452.200
37 3/8	117.417	1,097.118	43 1/8	135.481	1,460.660
37 1/2	117.810	1,104.469	43 1/4	135.874	1,469.140
37 5/8	118.203	1,111.844	43 3/8	136.267	1,477.640
37 3/4	118.595	1,119.244	43 1/2	136.660	1,486.170
37 7/8	118.988	1,126.669	43 5/8	137.052	1,494.730
38	119.381	1,134.118	43 3/4	137.445	1,503.300
38 1/8	119.773	1,141.591	43 7/8	137.838	1,511.910
38 1/4	120.166	1,149.089	44	138.230	1,520.530
38 3/8	120.559	1,156.612	44 1/8	138.623	1,529.190
38 1/2	120.952	1,164.159	44 1/4	139.016	1,537.860

TABLE—(*Continued*).

Diam.	Circum.	Area.	Diam.	Circum.	Area.
44⅜	139.408	1,546.56	50⅛	157.473	1,973.33
44½	139.801	1,555.29	50¼	157.865	1,983.18
44⅝	140.194	1,564.04	50⅜	158.258	1,993.06
44¾	140.587	1,572.81	50½	158.651	2,002.97
44⅞	140.979	1,581.61	50⅝	159.043	2,012.89
45	141.372	1,590.43	50¾	159.436	2,022.85
45⅛	141.765	1,599.28	50⅞	159.829	2,032.82
45¼	142.157	1,608.16	51	160.222	2,042.83
45⅜	142.550	1,617.05	51⅛	160.614	2,052.85
45½	142.943	1,625.97	51¼	161.007	2,062.90
45⅝	143.335	1,634.92	51⅜	161.400	2,072.98
45¾	143.728	1,643.89	51½	161.792	2,083.08
45⅞	144.121	1,652.89	51⅝	162.185	2,093.20
46	144.514	1,661.91	51¾	162.578	2,103.35
46⅛	144.906	1,670.95	51⅞	162.970	2,113.52
46¼	145.299	1,680.02	52	163.363	2,123.72
46⅜	145.692	1,689.11	52⅛	163.756	2,133.94
46½	146.084	1,698.23	52¼	164.149	2,144.19
46⅝	146.477	1,707.37	52⅜	164.541	2,154.46
46¾	146.870	1,716.54	52½	164.934	2,164.76
46⅞	147.262	1,725.73	52⅝	165.327	2,175.08
47	147.655	1,734.95	52¾	165.719	2,185.42
47⅛	148.048	1,744.19	52⅞	166.112	2,195.79
47¼	148.441	1,753.45	53	166.505	2,206.19
47⅜	148.833	1,762.74	53⅛	166.897	2,216.61
47½	149.226	1,772.06	53¼	167.290	2,227.05
47⅝	149.619	1,781.40	53⅜	167.683	2,237.52
47¾	150.011	1,790.76	53½	168.076	2,248.01
47⅞	150.404	1,800.15	53⅝	168.468	2,258.53
48	150.797	1,809.56	53¾	168.861	2,269.07
48⅛	151.189	1,819.00	53⅞	169.254	2,279.64
48¼	151.582	1,828.46	54	169.646	2,290.23
48⅜	151.975	1,837.95	54⅛	170.039	2,300.84
48½	152.368	1,847.46	54¼	170.432	2,311.48
48⅝	152.760	1,856.99	54⅜	170.824	2,322.15
48¾	153.153	1,866.55	54½	171.217	2,332.83
48⅞	153.546	1,876.14	54⅝	171.610	2,343.55
49	153.938	1,885.75	54¾	172.003	2,354.29
49⅛	154.331	1,895.38	54⅞	172.395	2,365.05
49¼	154.724	1,905.04	55	172.788	2,375.83
49⅜	155.116	1,914.72	55⅛	173.181	2,386.65
49½	155.509	1,924.43	55¼	173.573	2,397.48
49⅝	155.902	1,934.16	55⅜	173.966	2,408.34
49¾	156.295	1,943.91	55½	174.359	2,419.23
49⅞	156.687	1,953.69	55⅝	174.751	2,430.14
50	157.080	1,963.50	55¾	175.144	2,441.07

TABLE—(*Continued*).

Diam.	Circum.	Area.	Diam.	Circum.	Area.
55⅞	175.537	2,452.03	61⅝	193.601	2,982.67
56	175.930	2,463.01	61¾	193.994	2,994.78
56⅛	176.322	2,474.02	61⅞	194.386	3,006.92
56¼	176.715	2,485.05	62	194.779	3,019.08
56⅜	177.108	2,496.11	62⅛	195.172	3,031.26
56½	177.500	2,507.19	62¼	195.565	3,043.47
56⅝	177.893	2,518.30	62⅜	195.957	3,055.71
56¾	178.286	2,529.43	62½	196.350	3,067.97
56⅞	178.678	2,540.58	62⅝	196.743	3,080.25
57	179.071	2,551.76	62¾	197.135	3,092.56
57⅛	179.464	2,562.97	62⅞	197.528	3,104.89
57¼	179.857	2,574.20	63	197.921	3,117.25
57⅜	180.249	2,585.45	63⅛	198.313	3,129.64
57½	180.642	2,596.73	63¼	198.706	3,142.04
57⅝	181.035	2,608.03	63⅜	199.099	3,154.47
57¾	181.427	2,619.36	63½	199.492	3,166.93
57⅞	181.820	2,630.71	63⅝	199.884	3,179.41
58	182.213	2,642.09	63¾	200.277	3,191.91
58⅛	182.605	2,653.49	63⅞	200.670	3,204.44
58¼	182.998	2,664.91	64	201.062	3,217.00
58⅜	183.391	2,676.36	64⅛	201.455	3,229.58
58½	183.784	2,687.84	64¼	201.848	3,242.18
58⅝	184.176	2,699.33	64⅜	202.240	3,254.81
58¾	184.569	2,710.86	64½	202.633	3,267.46
58⅞	184.962	2,722.41	64⅝	203.026	3,280.14
59	185.354	2,733.98	64¾	203.419	3,292.84
59⅛	185.747	2,745.57	64⅞	203.811	3,305.56
59¼	186.140	2,757.20	65	204.204	3,318.31
59⅜	186.532	2,768.84	65⅛	204.597	3,331.09
59½	186.925	2,780.51	65¼	204.989	3,343.89
59⅝	187.318	2,792.21	65⅜	205.382	3,356.71
59¾	187.711	2,803.93	65½	205.775	3,369.56
59⅞	188.103	2,815.67	65⅝	206.167	3,382.44
60	188.496	2,827.44	65¾	206.560	3,395.33
60⅛	188.889	2,839.23	65⅞	206.953	3,408.26
60¼	189.281	2,851.05	66	207.346	3,421.20
60⅜	189.674	2,862.89	66⅛	207.738	3,434.17
60½	190.067	2,874.76	66¼	208.131	3,447.17
60⅝	190.459	2,886.65	66⅜	208.524	3,460.19
60¾	190.852	2,898.57	66½	208.916	3,473.24
60⅞	191.245	2,910.51	66⅝	209.309	3,486.30
61	191.638	2,922.47	66¾	209.702	3,499.40
61⅛	192.030	2,934.46	66⅞	210.094	3,512.52
61¼	192.423	2,946.48	67	210.487	3,525.66
61⅜	192.816	2,958.52	67⅛	210.880	3,538.83
61½	193.208	2,970.58	67¼	211.273	3,552.02

TABLE—(*Continued*).

Diam.	Circum.	Area.	Diam.	Circum.	Area.
67⅜	211.665	3,565.24	73⅛	229.729	4,199.74
67½	212.058	3,578.48	73¼	230.122	4,214.11
67⅝	212.451	3,591.74	73⅜	230.515	4,228.51
67¾	212.843	3,605.04	73½	230.906	4,242.93
67⅞	213.236	3,618.35	73⅝	231.300	4,257.37
68	213.629	3,631.69	73¾	231.693	4,271.84
68⅛	214.021	3,645.05	73⅞	232.086	4,286.33
68¼	214.414	3,658.44	74	232.478	4,300.85
68⅜	214.807	3,671.86	74⅛	232.871	4,315.39
68½	215.200	3,685.29	74¼	233.264	4,329.96
68⅝	215.592	3,698.76	74⅜	233.656	4,344.55
68¾	215.985	3,712.24	74½	234.049	4,359.17
68⅞	216.378	3,725.75	74⅝	234.442	4,373.81
69	216.770	3,739.29	74¾	234.835	4,388.47
69⅛	217.163	3,752.85	74⅞	235.227	4,403.16
69¼	217.556	3,766.43	75	235.620	4,417.87
69⅜	217.948	3,780.04	75⅛	236.013	4,432.61
69½	218.341	3,793.68	75¼	236.405	4,447.38
69⅝	218.734	3,807.34	75⅜	236.798	4,462.16
69¾	219.127	3,821.02	75½	237.191	4,476.98
69⅞	219.519	3,834.73	75⅝	237.583	4,491.81
70	219.912	3,848.46	75¾	237.976	4,506.67
70⅛	220.305	3,862.22	75⅞	238.369	4,521.56
70¼	220.697	3,876.00	76	238.762	4,536.47
70⅜	221.090	3,889.80	76⅛	239.154	4,551.41
70½	221.483	3,903.63	76¼	239.547	4,566.36
70⅝	221.875	3,917.49	76⅜	239.940	4,581.35
70¾	222.268	3,931.37	76½	240.332	4,596.36
70⅞	222.661	3,945.27	76⅝	240.725	4,611.39
71	223.054	3,959.20	76¾	241.118	4,626.45
71⅛	223.446	3,973.15	76⅞	241.510	4,641.53
71¼	223.839	3,987.13	77	241.903	4,656.64
71⅜	224.232	4,001.13	77⅛	242.296	4,671.77
71½	224.624	4,015.16	77¼	242.689	4,686.92
71⅝	225.017	4,029.21	77⅜	243.081	4,702.10
71¾	225.410	4,043.29	77½	243.474	4,717.31
71⅞	225.802	4,057.39	77⅝	243.867	4,732.54
72	226.195	4,071.51	77¾	244.259	4,747.79
72⅛	226.588	4,085.66	77⅞	244.652	4,763.07
72¼	226.981	4,099.84	78	245.045	4,778.87
72⅜	227.373	4,114.04	78⅛	245.437	4,793.70
72½	227.766	4,128.26	78¼	245.830	4,809.05
72⅝	228.159	4,142.51	78⅜	246.223	4,824.43
72¾	228.551	4,156.78	78½	246.616	4,839.83
72⅞	228.944	4,171.08	78⅝	247.008	4,855.26
73	229.337	4,185.40	78¾	247.401	4,870.71

TABLE—(*Continued*).

Diam.	Circum.	Area.	Diam.	Circum.	Area.
78⅞	247.794	4,886.18	84⅝	265.858	5,624.56
79	248.186	4,901.68	84¾	266.251	5,641.18
79⅛	248.579	4,917.21	84⅞	266.643	5,657.84
79¼	248.972	4,932.75	85	267.036	5,674.51
79⅜	249.364	4,948.33	85⅛	267.429	5,691.22
79½	249.757	4,963.92	85¼	267.821	5,707.94
79⅝	250.150	4,979.55	85⅜	268.214	5,724.69
79¾	250.543	4,995.19	85½	268.607	5,741.47
79⅞	250.935	5,010.86	85⅝	268.999	5,758.27
80	251.328	5,026.56	85¾	269.392	5,775.10
80⅛	251.721	5,042.28	85⅞	269.785	5,791.94
80¼	252.113	5,058.03	86	270.178	5,808.82
80⅜	252.506	5,073.79	86⅛	270.570	5,825.72
80½	252.899	5,089.59	86¼	270.963	5,842.64
80⅝	253.291	5,105.41	86⅜	271.356	5,859.59
80¾	253.684	5,121.25	86½	271.748	5,876.56
80⅞	254.077	5,137.12	86⅝	272.141	5,893.55
81	254.470	5,153.01	86¾	272.534	5,910.58
81⅛	254.862	5,168.93	86⅞	272.926	5,927.62
81¼	255.255	5,184.87	87	273.319	5,944.69
81⅜	255.648	5,200.83	87⅛	273.712	5,961.79
81½	256.040	5,216.82	87¼	274.105	5,978.91
81⅝	256.433	5,232.84	87⅜	274.497	5,996.05
81¾	256.826	5,248.88	87½	274.890	6,013.22
81⅞	257.218	5,264.94	87⅝	275.283	6,030.41
82	257.611	5,281.03	87¾	275.675	6,047.63
82⅛	258.004	5,297.14	87⅞	276.068	6,064.87
82¼	258.397	5,313.28	88	276.461	6,082.14
82⅜	258.789	5,329.44	88⅛	276.853	6,099.43
82½	259.182	5,345.63	88¼	277.246	6,116.74
82⅝	259.575	5,361.84	88⅜	277.629	6,134.08
82¾	259.967	5,378.08	88½	278.032	6,151.45
82⅞	260.360	5,394.34	88⅝	278.424	6,168.84
83	260.753	5,410.62	88¾	278.817	6,186.25
83⅛	261.145	5,426.93	88⅞	279.210	6,203.69
83¼	261.538	5,443.26	89	279.602	6,221.15
83⅜	261.931	5,459.62	89⅛	279.995	6,238.64
83½	262.324	5,476.01	89¼	280.388	6,256.15
83⅝	262.716	5,492.41	89⅜	280.780	6,273.69
83¾	263.109	5,508.84	89½	281.173	6,291.25
83⅞	263.502	5,525.30	89⅝	281.566	6,308.84
84	263.894	5,541.78	89¾	281.959	6,326.45
84⅛	264.287	5,558.29	89⅞	282.351	6,344.08
84¼	264.680	5,574.82	90	282.744	6,361.74
84⅜	265.072	5,591.37	90⅛	283.137	6,379.42
84½	265.465	5,607.95	90¼	283.529	6,397.13

TABLE—(*Continued*).

Diam.	Circum.	Area.	Diam.	Circum.	Area.
90⅜	283.922	6,414.86	95¼	299.237	7,125.59
90½	284.315	6,432.62	95⅜	299.630	7,144.31
90⅝	284.707	6,450.40	95½	300.023	7,163.04
90¾	285.100	6,468.21	95⅝	300.415	7,181.81
90⅞	285.493	6,486.04	95¾	300.808	7,200.60
91	285.886	6,503.90	95⅞	301.201	7,219.41
91⅛	286.278	6,521.78	96	301.594	7,238.25
91¼	286.671	6,539.68	96⅛	301.986	7,257.11
91⅜	287.064	6,557.61	96¼	302.379	7,275.99
91½	287.456	6,575.56	96⅜	302.772	7,294.91
91⅝	287.849	6,593.54	96½	303.164	7,313.84
91¾	288.242	6,611.55	96⅝	303.557	7,332.80
91⅞	288.634	6,629.57	96¾	303.950	7,351.79
92	289.027	6,647.63	96⅞	304.342	7,370.79
92⅛	289.420	6,665.70	97	304.735	7,389.83
92¼	289.813	6,683.80	97⅛	305.128	7,408.89
92⅜	290.205	6,701.93	97¼	305.521	7,427.97
92½	290.598	6,720.08	97⅜	305.913	7,447.08
92⅝	290.991	6,738.25	97½	306.306	7,466.21
92¾	291.383	6,756.45	97⅝	306.699	7,485.37
92⅞	291.776	6,774.68	97¾	307.091	7,504.55
93	292.169	6,792.92	97⅞	307.484	7,523.75
93⅛	292.562	6,811.20	98	307.877	7,542.98
93¼	292.954	6,829.49	98⅛	308.270	7,562.24
93⅜	293.347	6,847.82	98¼	308.662	7,581.52
93½	293.740	6,866.16	98⅜	309.055	7,600.82
93⅝	294.132	6,884.53	98½	309.448	7,620.15
93¾	294.525	6,902.93	98⅝	309.840	7,639.50
93⅞	294.918	6,921.35	98¾	310.233	7,658.88
94	295.310	6,939.79	98⅞	310.626	7,678.28
94⅛	295.703	6,958.26	99	311.018	7,697.71
94¼	296.096	6,976.76	99⅛	311.411	7,717.16
94⅜	296.488	6,995.28	99¼	311.804	7,736.63
94½	296.881	7,013.82	99⅜	312.196	7,756.13
94⅝	297.274	7,032.39	99½	312.589	7,775.66
94¾	297.667	7,050.98	99⅝	312.982	7,795.21
94⅞	298.059	7,069.59	99¾	313.375	7,814.78
95	298.452	7,088.24	99⅞	313.767	7,834.38
95⅛	298.845	7,106.90	100	314.160	7,854.00

The preceding table may be used to determine the diameter when the circumference or area is known. Thus, the diameter of a circle having an area of 7,200 sq. in. is, approximately, 95¾ in.

DECIMAL EQUIVALENTS OF PARTS OF ONE INCH.

1-64	.015625	17-64	.265625	33-64	.515625	49-64	.765625
1-32	.031250	9-32	.281250	17-32	.531250	25-32	.781250
3-64	.046875	19-64	.296875	35-64	.546875	51-64	.796875
1-16	.062500	5-16	.312500	9-16	.562500	13-16	.812500
5-64	.078125	21-64	.328125	37-64	.578125	53-64	.828125
3-32	.093750	11-32	.343750	19-32	.593750	27-32	.843750
7-64	.109375	23-64	.359375	39-64	.609375	55-64	.859375
1-8	.125000	3-8	.375000	5-8	.625000	7-8	.875000
9-64	.140625	25-64	.390625	41-64	.640625	57-64	.890625
5-32	.156250	13-32	.406250	21-32	.656250	29-32	.906250
11-64	.171875	27-64	.421875	43-64	.671875	59-64	.921875
3-16	.187500	7-16	.437500	11-16	.687500	15-16	.937500
13-64	.203125	29-64	.453125	45-64	.703125	61-64	.953125
7-32	.218750	15-32	.468750	23-32	.718750	31-32	.968750
15-64	.234375	31-64	.484375	47-64	.734375	63-64	.984375
1-4	.250000	1-2	.500000	3-4	.750000	1	1

DECIMALS OF A FOOT FOR EACH 1-32 OF AN INCH.

Inch.	0″	1″	2″	3″	4″	5″
0	0	.0833	.1667	.2500	.3333	.4167
	.0026	.0859	.1693	.2526	.3359	.4193
	.0052	.0885	.1719	.2552	.3385	.4219
	.0078	.0911	.1745	.2578	.3411	.4245
1/8	.0104	.0937	.1771	.2604	.3437	.4271
	.0130	.0964	.1797	.2630	.3464	.4297
	.0156	.0990	.1823	.2656	.3490	.4323
	.0182	.1016	.1849	.2682	.3516	.4349
1/4	.0208	.1042	.1875	.2708	.3542	.4375
	.0234	.1068	.1901	.2734	.3568	.4401
	.0260	.1094	.1927	.2760	.3594	.4427
	.0286	.1120	.1953	.2786	.3620	.4453
3/8	.0312	.1146	.1979	.2812	.3646	.4479
	.0339	.1172	.2005	.2839	.3672	.4505
	.0365	.1198	.2031	.2865	.3698	.4531
	.0691	.1224	.2057	.2891	.3724	.4557
1/2	.0417	.1250	.2083	.2917	.3750	.4583
	.0443	.1276	.2109	.2943	.3776	.4609
	.0469	.1302	.2135	.2969	.3802	.4635
	.0495	.1328	.2161	.2995	.3828	.4661
5/8	.0521	.1354	.2188	.3021	.3854	.4688
	.0547	.1380	.2214	.3047	.3880	.4714
	.0573	.1406	.2240	.3073	.3906	.4740
	.0599	.1432	.2266	.3099	.3932	.4766

Table—(Continued).

Inch.	0″	1″	2″	3″	4″	5″
	.0625	.1458	.2292	.3125	.3958	.4792
	.0651	.1484	.2318	.3151	.3984	.4818
	.0677	.1510	.2344	.3177	.4010	.4844
	.0703	.1536	.2370	.3203	.4036	.4870
	.0729	.1562	.2396	.3229	.4062	.4896
	.0755	.1589	.2422	.3255	.4089	.4922
	.0781	.1615	.2448	.3281	.4115	.4948
	.0807	.1641	.2474	.3307	.4141	.4974

DECIMALS OF A FOOT FOR EACH 1-32 OF AN INCH.

Inch.	6″	7″	8″	9″	10″	11″
0	.5000	.5833	.6667	.7500	.8333	.9167
	.5026	.5859	.6693	.7526	.8359	.9193
	.5052	.5885	.6719	.7552	.8385	.9219
	.5078	.5911	.6745	.7578	.8411	.9245
	.5104	.5937	.6771	.7604	.8437	.9271
	.5130	.5964	.6797	.7630	.8464	.9297
	.5156	.5990	.6823	.7656	.8490	.9323
	.5182	.6016	.6849	.7682	.8516	.9349
	.5208	.6042	.6875	.7708	.8542	.9375
	.5234	.6068	.6901	.7734	.8568	.9401
	.5260	.6094	.6927	.7760	.8594	.9427
	.5286	.6120	.6953	.7786	.8620	.9453
	.5312	.6146	.6979	.7812	.8646	.9479
	.5339	.6172	.7005	.7839	.8672	.9505
	.5365	.6198	.7031	.7865	.8698	.9531
	.5391	.6224	.7057	.7891	.8724	.9557
	.5417	.6250	.7083	.7917	.8750	.9583
	.5443	.6276	.7109	.7943	.8776	.9609
	.5469	.6302	.7135	.7969	.8802	.9635
	.5495	.6328	.7161	.7995	.8828	.9661
	.5521	.6354	.7188	.8021	.8854	.9688
	.5547	.6380	.7214	.8047	.8880	.9714
	.5573	.6406	.7240	.8073	.8906	.9740
	.5599	.6432	.7266	.8099	.8932	.9766
	.5625	.6458	.7292	.8125	.8958	.9792
	.5651	.6484	.7318	.8151	.8984	.9818
	.5677	.6510	.7344	.8177	.9010	.9844
	.5703	.6536	.7370	.8203	.9036	.9870
	.5729	.6562	.7396	.8229	.9062	.9896
	.5755	.6589	.7422	.8255	.9089	.9922
	.5781	.6615	.7448	.8281	.9115	.9948
	.5807	.6641	.7474	.8307	.9141	.9974

FORMULAS.

$$= \left\{ + \left[- : \left(\sqrt{\ \times / +} \right) : - \right] \right\} =$$

The term *formula*, as used in mathematics and in technical books, may be defined as *a rule in which symbols are used instead of words;* in fact, a formula may be regarded as a shorthand method of expressing a rule.

Most people having no knowledge of algebra regard formulas with distrust; they think that a person must be a good algebraic scholar in order to be able to use formulas. This idea, however, is erroneous. As a rule, no knowledge of any branch of mathematics except arithmetic is required to enable one to use a formula. Any formula can be expressed in words, and when so expressed it becomes a rule.

Formulas are much more convenient than rules; they show at a glance all the operations that are to be performed; they do not require to be read three or four times, as is the case with most rules, to enable one to understand their meaning; they take up much less space, both in the printed book and in one's note book, than rules; in short, whenever a rule can be expressed as a formula, the formula is to be preferred. In the following pages we purpose to show the reader how to use such formulas as he is likely to encounter in "pocketbooks," or other works of like nature.

The signs used in formulas are the ordinary signs indicative of operations and the signs of aggregation. All these signs are used in arithmetic, but, to refresh the reader's memory, we will explain their nature and uses before proceeding further.

The signs indicative of operations are six in number, viz.: $+, -, \times, \div, |, \sqrt{}$.

The sign $(+)$ indicates addition, and is called *plus;* when placed between two quantities, it indicates that the two quantities are to be added. Thus, in the expression $25 + 17$, the sign $(+)$ shows that 17 is to be added to 25.

The sign $(-)$ indicates subtraction, and is called *minus;* when placed between two quantities, it indicates that the quantity on the right is to be subtracted from that on the left. Thus, in the expression $25 - 17$, the sign $(-)$ shows that 17 is to be subtracted from 25.

The sign $(\times)$ indicates multiplication, and is read *times*, or *multiplied by;* when placed between two quantities, it indicates that the quantity on the left is to be multiplied by that on the right. Thus, in the expression 25×17, the sign $(\times)$ shows that 25 is to be multiplied by 17.

The sign $(\div)$ indicates division, and is read *divided by;* when placed between two quantities, it indicates that the quantity on the left is to be divided by that on the right. Thus, in the expression $25 \div 17$, the sign $(\div)$ shows that 25 is to be divided by 17.

Division is also indicated by placing a straight line between the two quantities. Thus, $25 \mid 17$, $25/17$, and $\frac{25}{17}$ all indicate that 25 is to be divided by 17. When both quantities are placed on the same horizontal line, the straight line indicates that the quantity on the left is to be divided by that on the right. When one quantity is below the other, the straight line between indicates that the quantity above the line is to be divided by the one below it.

The sign $(\sqrt{})$ indicates that some root of the quantity to the right is to be taken; it is called the *radical* sign. To indicate what root is to be taken, a small figure, called the *index*, is placed within the sign, this being always omitted when the square root is to be indicated. Thus, $\sqrt{25}$ indicates that the square root of 25 is to be taken; $\sqrt[3]{25}$ indicates that the cube root of 25 is to be taken, etc.

Note.—As the term "quantity" is a very convenient one to use, we will define it. In mathematics the word *quantity* is applied to anything that it is desired to subject to the ordinary operations of addition, subtraction, multiplication, etc., when we do not wish to be more specific and state exactly what the thing is. Thus, we can say "two or more numbers," or "two or more quantities." The word quantity is more general in its meaning than the word number.

The signs of aggregation are four in number, viz.: ——, (), [], and $\{\ \}$, respectively called the *vinculum*, the *parenthesis*, the *brackets*, and the *brace;* they are used when it is desired to indicate that all the quantities included by them are to be subjected to the same operation. Thus, if we desire to indicate that the sum of 5 and 8 is to be multiplied by 7, and we do not wish to actually add 5 and 8 before indicating the multiplication, we may employ any one of the four signs of aggregation as here shown: $\overline{5+8}\times7$, $(5+8)\times7$, $[5+8]\times7$, $\{5+8\}\times7$. The vinculum is placed above the quantities which are to be treated as one quantity and subjected to the same operations.

While any one of the four signs may be used as shown above, custom has restricted their use somewhat. The vinculum is rarely used except in connection with the radical sign. Thus, instead of writing $\sqrt[3]{}(5+8)$, $\sqrt[3]{}[5+8]$, or $\sqrt[3]{}\{5+8\}$ for the cube root of 5 plus 8, all of which would be correct, the vinculum is nearly always used, $\sqrt[3]{5+8}$.

In cases where but one sign of aggregation is needed (except, of course, when a root is to be indicated), the parenthesis is always used. Hence, $(5+8)\times7$ would be the usual way of expressing the product of 5 plus 8 and 7.

If two signs of aggregation are needed, the brackets and parenthesis are used, so as to avoid having a parenthesis within a parenthesis, the brackets being placed outside. For example, $[(20-5)\div3]\times9$ means that the difference between 20 and 5 is to be divided by 3, and this result multiplied by 9.

If three signs of aggregation are required, the brace, brackets, and parenthesis are used, the brace being placed outside, the brackets next, and the parenthesis inside. For example, $\{[(20-5)\div3]\times9-21\}\div8$ means that the quotient obtained by dividing the difference between 20 and 5 by 3 is to be multiplied by 9; and that 21 is to be subtracted from the product thus obtained, and the result divided by 8.

Should it be necessary to use all four signs of aggregation, the brace would be put outside, the brackets next, the parenthesis next, and the vinculum inside. For example, $\{[(\overline{20-5}+3)\times9-21]\div8\}\times12$. The reason for using the brace in this last instance will be explained, as it is not generally understood.

When several quantities are connected by the various signs indicating addition, subtraction, multiplication, and division, the operation indicated by the sign of multiplication

8

must always be performed first. Thus, $2 + 3 \times 4$ equals 14, 3 being multiplied by 4 before adding to 2. Similarly, $10 \div 2 \times 5$ equals 1, since 2×5 equals 10, and $10 \div 10$ equals 1. Hence, in the above case, if the brace were omitted, the result would be $\frac{1}{4}$; whereas, by inserting the brace, the result is 36.

Following the sign of multiplication comes the sign of division in its order of importance. For example, $5 - 9 \div 3$ equals 2, 9 being divided by 3 before subtracting from 5. The signs of addition and subtraction are of equal value; that is, if several quantities are connected by plus and minus signs, the indicated operations may be performed in the order in which the quantities are placed.

There is one other sign used, which is neither a sign of aggregation nor a sign indicative of an operation to be performed; it is $(=)$, and is called the sign of *equality;* it means that all on one side of it is exactly equal to all on the other side. For example, $2 = 2$, $5 - 3 = 2$, $5 \times (14 - 9) = 25$.

Having described the signs used in formulas, the formulas themselves will now be explained. First consider the well-known rule for finding the horsepower of a steam engine, which may be stated as follows:

Divide the continued product of the mean effective pressure in pounds per square inch, the length of the stroke in feet, the area of the piston in square inches, and the number of strokes per minute by 33,000; the result will be the horsepower.

This is a very simple rule, and very little, if anything, will be saved by expressing it as a formula, so far as clearness is concerned. The formula, however, will occupy a great deal less space, as we shall show.

An examination of the rule will show that four quantities (viz., the mean effective pressure, the length of the stroke, the area of the piston, and the number of strokes) are multiplied together, and the result is divided by 33,000. Hence, the rule might be expressed as follows:

$$\text{Horsepower} = \frac{\text{mean effective pressure}}{\text{(in pounds per square inch)}} \times \frac{\text{stroke}}{\text{(in feet)}}$$
$$\times \frac{\text{area of piston}}{\text{(in square inches)}} \times \frac{\text{number of strokes}}{\text{(per minute)}} \div 33,000.$$

This expression could be shortened by representing each quantity by a single letter, thus: representing horsepower by the letter "H," the mean effective pressure in pounds per square inch by "P," the length of the stroke in feet by "L," the area of the piston in square inches by "A," the number of strokes per minute by "N," and substituting these letters for the quantities that they represent, the above expression would reduce to

$$H = \frac{P \times L \times A \times N}{33,000},$$

a much simpler and shorter expression. This last expression is called a *formula*.

The formula just given shows, as we stated in the beginning, that a formula is really a shorthand method of expressing a rule. It is customary, however, to omit the sign of multiplication between two or more quantities when they are to be multiplied together, or between a number and a letter representing a quantity, it being always understood that when two letters are adjacent with no sign between them, the quantities represented by these letters are to be multiplied. Bearing this fact in mind, the formula just given can be further simplified to

$$H = \frac{P L A N}{33,000}.$$

The sign of multiplication, evidently, cannot be omitted between two or more numbers, as it would then be impossible to distinguish the numbers. A near approach to this, however, may be attained by placing a dot between the numbers that are to be multiplied together, and this is frequently done in works on mathematics when it is desired to economize space. In such cases it is usual to put the dot higher than the position occupied by the decimal point. Thus, $2 \cdot 3$ means the same as 2×3; $542 \cdot 749 \cdot 1,006$ indicates that the numbers 542, 749, and 1,006 are to be multiplied together.

It is also customary to omit the sign of multiplication in expressions similar to the following: $a \times \sqrt{b + c}$, $3 \times (b + c)$, $(b + c) \times a$, etc., writing them $a\sqrt{b + c}$, $3(b + c)$, $(b + c)a$, etc. The sign is not omitted when several quantities are included by a vinculum, and it is desired to indicate that the quantities so included are to be multiplied by another quantity. For example, $3 \times \overline{b+c}$, $\overline{b+c} \times a$, $\sqrt{\overline{b+c}} \times a$, etc., are always written as here printed.

Before proceeding further, we will explain one other device that is used by formula makers, and which is apt to puzzle one who encounters it for the first time. It is the use of what mathematicians call *primes* and *subs.*, and what printers call *superior* and *inferior* characters. As a rule, formula makers designate quantities by the initial letters of the names of the quantities. For example, they represent volume by v, pressure by p, height by h, etc. This practice is to be commended, as the letter itself serves in many cases to identify the quantity that it represents. Some authors carry the practice a little further and represent all quantities of the same nature by the same letter throughout the book, always having the same letter represent the same thing. Now, this practice necessitates the use of the primes and subs. above mentioned when two quantities have the same name, but represent different things. Thus, consider the word *pressure* as applied to steam at different stages between the boiler and the condenser. First, there is *absolute* pressure, which is equal to the gauge pressure in pounds per square inch plus the pressure indicated by the barometer reading (usually assumed in practice to be 14.7 pounds per square inch, when a barometer is not at hand). If this be represented by p, how shall we represent the gauge pressure? Since the absolute pressure is always greater than the gauge pressure, suppose we decide to represent it by a capital letter, and the gauge pressure by a small (lower-case) letter. Doing so, P represents absolute pressure, and p gauge pressure. Further, there is usually a "drop" in pressure between the boiler and the engine, so that the initial pressure, or pressure at the beginning of the stroke, is less than the pressure at the boiler. How shall we represent the initial pressure? We may do this in one of three ways, and still retain the letter p or P to represent the word pressure: First, by the use of the prime mark; thus, p' or P' (read p *prime* and p *major prime*) may be considered to represent the initial gauge pressure or the initial absolute pressure.

Second, by the use of sub. figures; thus, p_1 or P_1 (read p *sub. one* and p *major sub. one*). Third, by the use of sub. letters: thus, p_t or P_t (read p *sub. t* and P *major sub. t*). Likewise, p'' (read p *second*), p_2, or p_r might be used to represent the gauge pressure at release, etc. Sub. letters have the advantage of still further identifying the quantity represented; in many instances, however, it is not convenient to use them, in which case primes and subs. are used instead. The prime notation may be continued as follows: p''', p^{iv}, p^v, etc.; it is inadvisable to use superior figures, for example, p^1, p^2, p^3, p^a, etc., as they are liable to be mistaken for exponents.

The main thing to be remembered by the reader is that *when a formula is given in which the same letters occur several times, all like letters having the same primes or subs. represent the same quantities, while those that differ in any respect represent different quantities.* Thus, in the formula

$$t = \frac{w_1 s_1 t_1 + w_2 s_2 t_2 + w_3 s_3 t_3}{w_1 s_1 + w_2 s_2 + w_3 s_3},$$

w_1, w_2, and w_3 represent the weights of three different bodies; s_1, s_2, and s_3 their specific heats; and t_1, t_2, and t_3 their temperatures; while t represents the final temperature, after the bodies have been mixed together.

It is very easy to apply the above formula when the values of the quantities represented by the different letters are known. All that is required is to substitute the numerical values of the letters, and then perform the indicated operations. Thus, suppose that the values of w_1, s_1, and t_1 are, respectively, 2 pounds, .0951, and 80°; of w_2, s_2, and t_2, 7.8 pounds, 1, and 80°, and of w_3, s_3, and t_3, 3¼ pounds, .1138, and 780°; then, the final temperature t is, substituting these values for their respective letters in the formula,

$$t = \frac{2 \times .0951 \times 80 + 7.8 \times 1 \times 80 + 3\frac{1}{4} \times .1138 \times 780}{2 \times .0951 + 7.8 \times 1 + 3\frac{1}{4} \times .1138} =$$

$$\frac{15.216 + 624 + 288.483}{.1902 + 7.8 + .36985} = \frac{927.699}{8.36005} = 110.97°.$$

In substituting the numerical values, the signs of multiplication are, of course, written in their proper places; all the multiplications are performed before adding, according to the rule previously given.

The reader should now be able to apply any formula involving only algebraic expressions that he may meet with, not requiring the use of logarithms for their solution. We will, however, call his attention to one or two other facts which he may have forgotten.

Expressions similar to $\dfrac{160}{\frac{660}{25}}$ sometimes occur, the heavy line

indicating that 160 is to be divided by the quotient obtained by dividing 660 by 25. If both lines were light it would be

impossible to tell whether 160 was to be divided by $\dfrac{660}{25}$, or

whether $\dfrac{160}{660}$ was to be divided by 25. If this latter result

were desired, the expression would be written $\dfrac{\frac{160}{660}}{25}$. In every

case the heavy line indicates that all above it is to be divided by all below it.

In an expression like the following, $\dfrac{160}{7+\frac{660}{25}}$, the heavy line

is not necessary, since it is impossible to mistake the operation that is required to be performed. But, since $7+\dfrac{660}{25}$

$=\dfrac{175+660}{25}$, if we substitute $\dfrac{175+660}{25}$ for $7+\dfrac{660}{25}$, the heavy

line becomes necessary in order to make the resulting expression clear. Thus,

$$\frac{160}{7+\dfrac{660}{25}} = \frac{160}{\dfrac{175+660}{25}} = \frac{160}{\dfrac{835}{25}}.$$

Fractional exponents are sometimes used instead of the radical sign. That is, instead of indicating the square, cube, fourth root, etc. of some quantity, as 37 by $\sqrt{37}$, $\sqrt[3]{37}$, $\sqrt[4]{37}$, etc. these roots are indicated by $37^{\frac{1}{2}}$, $37^{\frac{1}{3}}$, $37^{\frac{1}{4}}$, etc. Should the numerator of the fractional exponent be some quantity other than 1, this quantity, whatever it may be, indicates that the quantity affected by the exponent is to be raised to the power indicated by the numerator; the denominator is *always* the index of the root. Hence, instead of expressing the cube root of the square of 37 as $\sqrt[3]{37^2}$, it may be expressed $37^{\frac{2}{3}}$, the denominator being the index of the root; in other words, $\sqrt[3]{37^2} = 37^{\frac{2}{3}}$. Likewise, $\sqrt[3]{(1 + a^2b)^2}$ may also be written $(1 + a^2b)^{\frac{2}{3}}$, a much simpler expression.

We will now give several examples showing how to apply some of the more difficult formulas that the reader may encounter.

The area of any segment of a circle that is less than (or equal to) a semicircle is expressed by the formula.

$$A = \frac{\pi r^2 E}{360} - \frac{c}{2}(r - h),$$

in which A = area of segment;

    π = 3.1416;

    r = radius;

    E = angle obtained by drawing lines from the center to the extremities of arc of segment;

    c = chord of segment;

    h = height of segment.

EXAMPLE.—What is the area of a segment whose chord is 10 in. long, angle subtended by chord is 83.46°, radius is 7.5 in., and height of segment is 1.91 in.?

SOLUTION.—Applying the formula just given,

$$A = \frac{\pi r^2 E}{360} - \frac{c}{2}(r - h) = \frac{3.1416 \times 7.5^2 \times 83.46}{360} - \frac{10}{2}(7.5 - 1.91)$$
$$= 40.968 - 27.95 = 13.018 \text{ sq. in., nearly.}$$

The area of any triangle may be found by means of the following formula, in which A = the area, and a, b, and c represent the lengths of the sides:

$$A = \frac{b}{2}\sqrt{a^2 - \left(\frac{a^2 + b^2 - c^2}{2b}\right)^2}.$$

EXAMPLE.—What is the area of a triangle whose sides are 21 ft., 46 ft., and 50 ft. long?

SOLUTION.—In order to apply the formula, suppose we let a represent the side that is 21 ft. long; b, the side that is 50 ft. long; and c, the side that is 46 ft. long. Then, substituting in the formula,

$$A = \frac{b}{2}\sqrt{a^2 - \left(\frac{a^2 + b^2 - c^2}{2b}\right)^2} = \frac{50}{2}\sqrt{21^2 - \left(\frac{21^2 + 50^2 - 46^2}{2 \times 50}\right)^2}$$

$$= \frac{50}{2}\sqrt{441 - \left(\frac{441 + 2,500 - 2,116}{100}\right)^2} = 25\sqrt{441 - \left(\frac{825}{100}\right)^2}$$

$$= 25\sqrt{441 - 8.25^2} = 25\sqrt{441 - 68.0625} = 25\sqrt{372.9375}$$

$$= 25 \times 19.312 = 482.8 \text{ sq. ft., nearly.}$$

The above operations have been extended much further than was necessary; this was done in order to show the reader every step of the process.

The Rankine-Gordon formula for determining the least load in pounds that will cause a long column to break is

$$P = \frac{SA}{1 + q\,\dfrac{l^2}{G^2}},$$

in which P = load (pressure) in lb.; S = ultimate strength (in lb. per sq. in.) of material composing column; A = area of cross-section of column in sq. in.; q = a factor (multiplier) whose value depends on the shape of the ends of the column and on the material composing the column; l = length of the column in in.; G = least radius of gyration of cross-section of column.

The values of S, q, and G^2 are all given in printed tables on pages 151, 153, and 156.

EXAMPLE.—What is the least load that will break a hollow steel column whose outside diameter is 14 in., inside diameter 11 in., length 20 ft., and whose ends are flat?

SOLUTION.—For steel, S = 150,000, and $q = \dfrac{1}{25,000}$ for flat-ended steel columns; A, the area of the cross-section, = .7854$(d_1^2 - d_2^2)$, d_1 and d_2 being the outside and inside diameters, respectively; $l = 20 \times 12 = 240$ in.; and $G^2 = \dfrac{d_1^2 + d_2^2}{16}$. Substituting these values in the formula,

$$P = \frac{SA}{1 + q\,\dfrac{l^2}{G^2}} = \frac{150,000 \times .7854(14^2 - 11^2)}{1 + \dfrac{1}{25,000} \times \dfrac{240^2}{\dfrac{14^2 + 11^2}{16}}} =$$

$$\frac{150,000 \times 58.905}{1 + .1163} = \frac{8,835.750}{1.1163} = 7,915,211 \text{ lb.}$$

INVOLUTION AND EVOLUTION.

By means of the following table the square, cube, square root, cube root, and reciprocal of any number may be obtained correct always to five significant figures, and in the majority of cases correct to six significant figures.

In any number, the figures beginning with the first digit* at the left and ending with the last digit at the right, are called the *significant figures* of the number. Thus, the number 405,800 has the four significant figures 4, 0, 5, 8; and the number .000090067 has the five significant figures 9, 0, 0, 6, and 7.

The part of a number consisting of its significant figures is called the *significant part* of the number. Thus, in the number 28,070, the significant part is 2807; in the number .00812, the significant part is 812; and in the number 170.3, the significant part is 1703.

In speaking of the significant figures or of the significant part of a number, the figures are considered, in their proper order, from the first digit at the left to the last digit at the right, but no attention is paid to the position of the decimal point. Hence, *all numbers that differ only in the position of the decimal point have the same significant part.* For example, .002103, 21.03, 21,080, and 210,300 have the same significant figures 2, 1, 0, and 3, and the same significant part 2103.

The *integral part* of a number is the part to the left of the decimal point.

It will be more convenient to explain first how to use the table for finding square and cube roots.

SQUARE ROOT.

First point off the given number into periods of two figures each, beginning with the decimal point and proceeding to the left and right. The following numbers are thus pointed off: 12703, 1'27'03; 12.703, 12.70'30; 220000, 22'00'00; .000442, .00'04'42.

---

* A cipher is not a digit.

Having pointed off the number, move the decimal point so that it will fall between the first and second periods of the significant part of the number. In the above numbers, the decimal point will be placed thus: 1.2703, 12.703, 22, 4.42.

If the number has but three (or less) significant figures, find the significant part of the number in the column headed n; the square root will be found in the column headed $\sqrt{n}$ or $\sqrt{10\,n}$, according to whether the part to the left of the decimal point contains one figure or two figures. Thus, $\sqrt{4.42} = 2.1024$, and $\sqrt{22} = \sqrt{10 \times 2.20} = 4.6904$. The decimal point is located in all cases by reference to the original number after pointing off into periods.

There will be as many figures in the root preceding the decimal point as there are periods preceding the decimal point in the given number; if the number is entirely decimal, the root is entirely decimal, and there will be as many ciphers following the decimal point in the root as there are cipher periods following the decimal point in the given number.

Applying this rule, $\sqrt{220000} = 469.04$ and $\sqrt{.000442} = .021024$.

The operation when the given number has more than three significant figures is best explained by an example.

EXAMPLE.—(a) $\sqrt{3.1416} = ?$ (b) $\sqrt{2342.9} = ?$

SOLUTION.—(a) Since the first period contains but one figure, there is no need of moving the decimal point. Look in the column headed n^2 and find two consecutive numbers, one a little greater and the other a little less than the given number; in the present case, $3.1684 = 1.78^2$ and $3.1329 = 1.77^2$. The first three figures of the root are therefore 177. Find the difference between the two numbers between which the given number falls, and the difference between the smaller number and the given number; divide the second difference by the first difference, carrying the quotient to three decimal places and increasing the second figure by 1 if the third is 5 or a greater digit. The two figures of the quotient thus determined will be the fourth and fifth figures of the root. In the present example, dropping decimal points in the remainders, $3.1684 - 3.1329 = 355$, the first difference:

8.1416 − 3 1329 = 87, the second difference; 87 + 355 = .245+, or .25. Hence, $\sqrt{3.1416}$ = 1.7725.

(b) $\sqrt{2342.9}$ = ? Pointing off into periods we get 23'42.90; moving the decimal point we get 23.4290; the first three figures of the root are 484; the first difference is 23.5225 − 23.4256 = 969; the second difference is 23.4290 − 23.4256 = 34; 34 + 969 = .035+, or .04. Hence, $\sqrt{2342.9}$ = 48.404.

CUBE ROOT.

The cube root of a number is found in the same manner as the square root, except the given number is pointed off into periods of three figures each. The following numbers would be pointed off thus: 3141.6, 3'141.6; 67296428, 67'296'428; 601426.314, 601'426.314; .0000000217, .000'000'021'700.

Having pointed off, move the decimal point so that it will fall between the first and second periods of the significant part of the number, as in square root. In the above numbers the decimal point will be placed thus: 3.1416, 67.296428, 601.426314, and 21.7.

If the given number has but three (or less) significant figures, find the significant part of the number in the column headed n; the cube root will be found in the column headed $\sqrt[3]{n}$, $\sqrt[3]{10n}$, or $\sqrt[3]{100n}$, according to whether one, two, or three figures precede the decimal point after it has been moved. Thus, the cube root of 21.7 will be found opposite 2.17, in column headed $\sqrt[3]{10n}$, while the cube root of 2.17 would be found in the column headed $\sqrt[3]{n}$, and the cube root of 217 in the column headed $\sqrt[3]{100n}$, all on the same line. If the given number contains more than three significant figures, proceed exactly as described for square root except that the column headed n^3 is used.

EXAMPLE.—(a) $\sqrt[3]{.0000062417}$ = ? (b) $\sqrt[3]{50932676}$ = ?

SOLUTION.—(a) Pointing off into periods, we get 000'006'241'700; moving the decimal point, we get 6.2417. The number falls between 6.22950 = 1.84^3 and 6.33163 = 1.85^3, the first difference = 10213: the second difference is

6.24170 — 6.22950 = 1220; 1220 + 10213 = .119+, or .12, the fourth and fifth figures of the root. The decimal point is located by the rule previously given; hence, $\sqrt[2]{.0000062417}$ = .018412.

(b) $\sqrt[2]{50932676}$ = ? As the number contains more than six significant figures, reduce it to six significant figures by replacing all after the sixth figure with ciphers, increasing the sixth figure by 1 when the seventh is 5 or a greater digit. In other words, the first five figures of $\sqrt[2]{50932700}$ and of $\sqrt[2]{50932676}$ are the same. Pointing off into periods, we get 50'932'700; moving the decimal point, we get 50.9327, which falls between 50.6580 = 3.70² and 51.0648 = 3.71²; the first difference is 4118; the second difference is 2797; 2797 + 4118 = .679+, or .68. The integral part of the root evidently contains three figures; hence, $\sqrt[2]{50932676}$ = 370.68, correct to five figures.

SQUARES AND CUBES.

If the given number contains but three (or less) significant figures, the square or cube is found in the column headed n^2 or n^3, opposite the given number in the column headed n. If the given number contains more than three significant figures, proceed in a manner similar to that described for extracting roots. To square a number, place the decimal point between the first and second significant figures and find in the column headed $\sqrt{n}$ or $\sqrt{10n}$ two consecutive numbers, one of which shall be a little greater and the other a little less than the given number. The remainder of the work is exactly as heretofore described. To locate the decimal point, employ the principle that the square of any number contains either twice as many figures as the number squared or twice as many less one. If the column headed $\sqrt{10n}$ is used, the square will contain twice as many figures, while if the column headed $\sqrt{n}$ is used, the square will contain twice as many figures as the number squared, less one. If the number contains an integral part, the principle is applied to the integral part only; if the number is wholly decimal, there will be twice as many ciphers following the decimal in the square or twice as many plus one as in the number squared, depending on whether $\sqrt{10n}$ or $\sqrt{n}$ column is used. For example, 273.42^2 will contain five figures in the integral part; 4516.2^2 will contain eight figures in the integral part, all after the fifth being denoted by ciphers; $.0029453^2$ will have five ciphers following the decimal point; $.052436^2$ will have two ciphers following the decimal point.

EXAMPLE.—(a) $273.42^2 = ?$ (b) $.052436^2 = ?$

SOLUTION.—(a) Placing the decimal point between the first and second significant figures, the result is 2.7342; this number occurs between $2.73313 = \sqrt{7.47}$ and $2.73496 = \sqrt{7.48}$ in the column headed $\sqrt{n}$. The first difference is $2.73496 - 2.73313 = 183$; the second difference is $2.73420 - 2.73313 = 107$; and $107 \div 183 = .584+$, or $.58$. Hence, $273.42^2 = 74,758$, correct to five significant figures.

(b) Shifting the decimal point to between the first and second significant figures, we get the number 5.2436, which falls between $5.23450 = \sqrt{27.4}$ and $5.24404 = \sqrt{27.5}$. The first difference is 954; the second difference is 910; $910 \div 954 = .953+$, or $.95$. Hence, $.052436^2 = .0027495$, to five significant figures.

A number is cubed in exactly the same manner, using the column headed $\sqrt[3]{n}$, $\sqrt[3]{10n}$, or $\sqrt[3]{100n}$, according to whether the first period of the significant part of the number contains one, two, or three figures, respectively. If the number contains an integral part, the number of figures in the integral part of the cube will be three times as many as in the given number if column headed $\sqrt[3]{100n}$ is used; it will be three times as many less 1 if the column headed $\sqrt[3]{10n}$ is used; and it will be three times as many less 2 if the column headed $\sqrt[3]{n}$ is used. If the given number is wholly decimal the cube will have either three times, three times plus one, or three times plus two, as many ciphers following the decimal as there are ciphers following the decimal point in the given number.

EXAMPLE.—(a) $129.684^3 = ?$ (b) $.76442^3 = ?$ (c) $.032425^3 = ?$

SOLUTION.—(a) Placing the decimal point between the

first and second significant figures, the number 1.29684 is found between 1.29664 = $\sqrt[3]{2.18}$ and 1.29662 = $\sqrt[3]{2.19}$. The first difference is 198; the second difference is 20; and 20 ÷ 198 = .101+, or .10. Hence, the first five significant figures are 21810; the number of figures in the integral part of the cube is 3 × 3 − 2 = 7; and 129.684^3 = 2,181,000, correct to five significant figures.

(b) 7.64420 occurs between 7.64032 = $\sqrt[3]{446}$ and 7.64603 = $\sqrt[3]{447}$. The first difference is 571; the second difference is 388; and 388 ÷ 571 = .679+, or .68. Hence, the first five significant figures are 44668; the number of ciphers following the decimal point is 3 × 0 = 0; and .76442^3 = .44668, correct to five significant figures.

(c) 3.2425 falls between 3.24278 = $\sqrt[3]{34.1}$ and 3.23961 = $\sqrt[3]{34.0}$. The first difference is 317; the second difference is 289; 289 ÷ 317 = .911+, or .91. Hence, the first five significant figures are 34091; the number of ciphers following the decimal point is 3 × 1 + 1 = 4; and .032425^3 = .000034091, correct to five significant figures.

RECIPROCALS.

The reciprocal of a number is 1 divided by the number. By using reciprocals, division is changed into multiplication, since $a \div b = \dfrac{a}{b} = a \times \dfrac{1}{b}$. The table gives the reciprocals of all numbers expressed with three significant figures to six significant figures. By proceeding in a manner similar to that just described for powers and roots, the reciprocal of any number correct to five significant figures may be obtained. The decimal point in the result may be located as follows: If the given number has an integral part, the number of ciphers following the decimal point in the reciprocal will be one less than the number of figures in the integral part of the given number; and if the given number is entirely decimal, the number of figures in the integral part of the reciprocal will be one greater than the number of ciphers following the decimal point in the given number. For example, the reciprocal of 3370 = .000296736 and of .00348 = 287.356.

When the number whose reciprocal is desired contains more than three significant figures, express the number to six significant figures (adding ciphers, if necessary, to make six figures) and find between what two numbers in the column headed $\frac{1}{n}$ the significant figures of the given number falls; then proceed exactly as previously described to determine the fourth and fifth figures.

EXAMPLE.— (a) The reciprocal of 379.426 =? (b) $\frac{1}{.0004692}$ = ?

SOLUTION. — (a) .379426 falls between .378788 = $\frac{1}{2.64}$ and .380228 = $\frac{1}{2.63}$. The first difference is 380228 — 378788 = 1440; the second difference is 380228 — 379426 = 802; 802 ÷ 1440 = .557, or .56. Hence, the first five significant figures are 26356, and the reciprocal of 379.426 is .0026356, to five significant figures.

(b) .469200 falls between .469484 = $\frac{1}{2.13}$ and .467290 = $\frac{1}{2.14}$. The first difference is 2194; the second difference is 284; 284 ÷ 2194 = .129+, or .13. Hence, $\frac{1}{.0004692}$ = 2131.3, correct to five significant figures.

n	n^2	n^3	$\sqrt{n}$	$\sqrt{10\,n}$	$\sqrt[3]{n}$	$\sqrt[3]{10\,n}$	$\sqrt[3]{100\,n}$	$\dfrac{1}{n}$
1.01	1.0201	1.03030	1.00499	3.17805	1.00332	2.16159	4.65701	.990099
1.02	1.0404	1.06121	1.00995	3.19374	1.00662	2.16870	4.67233	.980392
1.03	1.0609	1.09273	1.01489	3.20936	1.00990	2.17577	4.68755	.970874
1.04	1.0816	1.12486	1.01980	3.22490	1.01316	2.18278	4.70267	.961538
1.05	1.1025	1.15763	1.02470	3.24037	1.01640	2.18976	4.71769	.952381
1.06	1.1236	1.19102	1.02956	3.25576	1.01961	2.19669	4.73262	.943396
1.07	1 1449	1.22504	1.03441	3.27109	1.02281	2.20358	4.74746	.934579
1.08	1.1664	1.25971	1.03923	3.28634	1.02599	2.21042	4.76220	.925926
1.09	1.1881	1.29503	1.04403	3.30151	1.02914	2.21722	4.77686	.917431
1.10	1.2100	1.33100	1.04881	3.31662	1.03228	2.22398	4.79142	.909091
1.11	1.2321	1 36763	1.05357	3.33167	1.03540	2.23070	4.80590	.900901
1.12	1.2544	1.40493	1.05830	3.34664	1.03850	2.23738	4.82028	.892857
1.13	1.2769	1.44290	1.06301	3.36155	1.04158	2.24402	4.83459	.884956
1.14	1.2996	1.48154	1.06771	3.37639	1.04464	2.25062	4.84881	.877193
1.15	1.3225	1.52088	1.07238	3.39116	1.04769	2.25718	4.86294	.869565
1.16	1.3456	1.56090	1.07703	3.40588	1.05072	2.26370	4.87700	.862069
1.17	1.3689	1.60161	1.08167	3.42053	1.05373	2.27019	4.89097	.854701
1.18	1.3924	1.64303	1.08628	3.43511	1.05672	2.27664	4.90487	.847458
1.19	1.4161	1.68516	1.09087	3.44964	1.05970	2.28305	4.91868	.840336
1.20	1.4400	1.72800	1.09545	3.46410	1.06266	2.28943	4.93242	.833333
1.21	1.4641	1.77156	1.10000	3.47851	1.06560	2.29577	4.94609	.826446
1.22	1.4884	1.81585	1.10454	3.49285	1.06853	2.30208	4.95968	.819672
1.23	1.5129	1.86087	1.10905	3.50714	1.07144	2.30835	4.97319	.813008
1.24	1.5376	1.90662	1.11355	3.52136	1.07434	2.31459	4.98663	.806452
1.25	1.5625	1.95313	1.11803	3.53553	1.07722	2.32080	5.00000	.800000
1.26	1.5876	2.00038	1.12250	3.54965	1.08008	2.32697	5.01330	.793651
1.27	1.6129	2.04838	1.12694	3.56371	1.08293	2.33310	5.02653	.787402
1.28	1.6384	2.09715	1.13137	3.57771	1.08577	2.33921	5.03968	.781250
1.29	1.6641	2.14669	1.13578	3.59166	1.08859	2.34529	5.05277	.775194
1.30	1.6900	2.19700	1.14018	3.60555	1.09139	2.35134	5.06580	.769231
1.31	1.7161	2.24809	1.14455	3.61939	1.09418	2.35735	5.07875	.763359
1.32	1.7424	2.29997	1.14891	3.63318	1.09696	2.36333	5.09164	.757576
1.33	1.7689	2.35264	1.15326	3.64692	1.09972	2.36928	5.10447	.751880
1.34	1.7956	2.40610	1.15758	3.66060	1.10247	2.37521	5.11723	.746269
1.35	1.8225	2.46038	1.16190	3.67423	1.10521	2.38110	5.12993	.740741
1.36	1.8496	2.51546	1.16619	3.68782	1.10793	2.38696	5.14256	.735294
1.37	1.8769	2.57135	1.17047	3.70135	1.11064	2.39280	5.15514	.729927
1.38	1.9044	2.62807	1.17473	3.71484	1.11334	2.39861	5.16765	.724638
1.39	1.9321	2.68562	1.17898	3.72827	1.11602	2.40439	5.18010	.719425
1.40	1.9600	2.74400	1.18322	3.74166	1.11869	2.41014	5.19249	.714286
1.41	1.9881	2.80322	1.18743	3.75500	1.12135	2.41587	5.20483	.709220
1.42	2.0164	2.86329	1.19164	3.76829	1.12399	2.42156	5.21710	.704225
1.43	2.0449	2.92421	1.19583	3.78153	1.12662	2.42724	5.22932	.699301
1.44	2.0736	2.98598	1.20000	3.79473	1.12924	2.43288	5.24148	.694444
1.45	2.1025	3.04863	1.20416	3.80789	1.13185	2.43850	5.25359	.689655
1.46	2.1316	3.11214	1.20830	3.82099	1.13445	2.44409	5.26564	.684932
1.47	2.1609	3.17652	1.21244	3.83406	1.13703	2.44966	5.27763	.680272
1.48	2.1904	3.24179	1.21655	3.84708	1.13960	2.45520	5.28957	.675676
1.49	2.2201	3.30795	1.22066	3.86005	1.14216	2.46072	5.30146	.671141
1.50	2.2500	3.37500	1.22474	3.87298	1.14471	2.46621	5.31329	.666667

n	n^2	n^3	$\sqrt{n}$	$\sqrt{10\,n}$	$\sqrt[3]{n}$	$\sqrt[3]{10\,n}$	$\sqrt[3]{100\,n}$	$\dfrac{1}{n}$
1.51	2.2801	3.44295	1.22882	3.88587	1.14725	2.47168	5.32507	.662252
1.52	2.3104	3.51181	1.23288	3.89872	1.14978	2.47713	5.33680	.657895
1.53	2.3409	3.58158	1.23693	3.91152	1.15230	2.48255	5.34848	.653595
1.54	2.3716	3.65236	1.24097	3.92428	1.15480	2.48794	5.36011	.649351
1.55	2.4025	3.72388	1.24499	3.93700	1.15729	2.49332	5.37169	.645161
1.56	2.4336	3.79642	1.24900	3.94968	1.15978	2.49866	5.38321	.641026
1.57	2.4649	3.86989	1.25300	3.96232	1.16225	2.50399	5.39469	.636943
1.58	2.4964	3.94431	1.25698	3.97492	1.16471	2.50930	5.40612	.632911
1.59	2.5281	4.01968	1.26095	3.98748	1.16717	2.51458	5.41750	.628931
1.60	2.5600	4.09600	1.26491	4.00000	1.16961	2.51984	5.42884	.625000
1.61	2.5921	4.17328	1.26886	4.01248	1.17204	2.52508	5.44012	.621118
1.62	2.6244	4.25153	1.27279	4.02492	1.17446	2.53030	5.45136	.617284
1.63	2.6569	4.33075	1.27671	4.03733	1.17687	2.53549	5.46256	.613497
1.64	2.6896	4.41094	1.28062	4.04969	1.17927	2.54067	5.47370	.609756
1.65	2.7225	4.49213	1.28452	4.06202	1.18167	2.54582	5.48481	.606061
1.66	2.7556	4.57430	1.28841	4.07431	1.18405	2.55095	5.49586	.602410
1.67	2.7889	4.65746	1.29228	4.08656	1.18642	2.55607	5.50688	.598802
1.68	2.8224	4.74163	1.29615	4.09878	1.18878	2.56116	5.51785	.595238
1.69	2.8561	4.82681	1.30000	4.11096	1.19114	2.56623	5.52877	.591716
1.70	2.8900	4.91300	1.30384	4.12311	1.19348	2.57128	5.53966	.588235
1.71	2.9241	5.00021	1.30767	4.13521	1.19582	2.57631	5.55050	.584795
1.72	2.9584	5.08845	1.31149	4.14729	1.19815	2.58133	5.56130	.581395
1.73	2.9929	5.17772	1.31529	4.15933	1.20046	2.58632	5.57205	.578035
1.74	3.0276	5.26802	1.31909	4.17133	1.20277	2.59129	5.58277	.574713
1.75	3.0625	5.35938	1.32288	4.18330	1.20507	2.59625	5.59344	.571429
1.76	3.0976	5.45178	1.32665	4.19524	1.20736	2.60118	5.60408	.568182
1.77	3.1329	5.54523	1.33041	4.20714	1.20964	2.60610	5.61467	.564972
1.78	3.1684	5.63975	1.33417	4.21900	1.21192	2.61100	5.62523	.561798
1.79	3.2041	5.73534	1.33791	4.23084	1.21418	2.61588	5.63574	.558659
1.80	3.2400	5.83200	1.34164	4.24264	1.21644	2.62074	5.64622	.555556
1.81	3.2761	5.92974	1.34536	4.25441	1.21869	2.62558	5.65665	.552486
1.82	3.3124	6.02857	1.34907	4.26615	1.22093	2.63041	5.66705	.549451
1.83	3.3489	6.12849	1.35277	4.27785	1.22316	2.63522	5.67741	.546448
1.84	3.3856	6.22950	1.35647	4.28952	1.22539	2.64001	5.68773	.543478
1.85	3.4225	6.33163	1.36015	4.30116	1.22760	2.64479	5.69802	.540541
1.86	3.4596	6.43486	1.36382	4.31277	1.22981	2.64954	5.70827	.537634
1.87	3.4969	6.53920	1.36748	4.32435	1.23201	2.65428	5.71848	.534759
1.88	3.5344	6.64467	1.37113	4.33590	1.23420	2.65900	5.72865	.531915
1.89	3.5721	6.75127	1.37477	4.34741	1.23639	2.66371	5.73879	.529101
1.90	3.6100	6.85900	1.37840	4.35890	1.23856	2.66840	5.74890	.526316
1.91	3.6481	6.96787	1.38203	4.37035	1.24073	2.67307	5.75897	.523560
1.92	3.6864	7.07789	1.38564	4.38178	1.24289	2.67773	5.76900	.520833
1.93	3.7249	7.18906	1.38924	4.39318	1.24505	2.68237	5.77900	.518135
1.94	3.7636	7.30138	1.39284	4.40454	1.24719	2.68700	5.78896	.515464
1.95	3.8025	7.41488	1.39642	4.41588	1.24933	2.69161	5.79889	.512821
1.96	3.8416	7.52954	1.40000	4.42719	1.25146	2.69620	5.80879	.510204
1.97	3.8809	7.64537	1.40357	4.43847	1.25359	2.70078	5.81865	.507614
1.98	3.9204	7.76239	1.40712	4.44972	1.25571	2.70534	5.82848	.505051
1.99	3.9601	7.88060	1.41067	4.46094	1.25782	2.70989	5.83827	.502513
2.00	4.0000	8.00000	1.41421	4.47214	1.25992	2.71442	5.84804	.500000

9

n	n^2	n^3	$\sqrt{n}$	$\sqrt{10\,n}$	$\sqrt[3]{n}$	$\sqrt[3]{10\,n}$	$\sqrt[3]{100\,n}$	$\frac{1}{n}$
2.01	4.0401	8.12060	1.41774	4.48330	1.26202	2.71893	5.85777	.497512
2.02	4.0804	8.24241	1.42127	4.49444	1.26411	2.72343	5.86746	.495050
2.03	4.1209	8.36543	1.42478	4.50555	1.26619	2.72792	5.87713	.492611
2.04	4.1616	8.48966	1.42829	4.51664	1.26827	2.73239	5.88677	.490196
2.05	4.2025	8.61513	1.43178	4.52769	1.27033	2.73685	5.89637	.487805
2.06	4.2436	8.74182	1.43527	4.53872	1.27240	2.74129	5.90594	.485437
2.07	4.2849	8.86974	1.43875	4.54973	1.27445	2.74572	5.91548	.483092
2.08	4.3264	8.99891	1.44222	4.56070	1.27650	2.75014	5.92499	.480769
2.09	4.3681	9.12933	1.44568	4.57165	1.27854	2.75454	5.93447	.478469
2.10	4.4100	9.26100	1.44914	4.58256	1.28058	2.75893	5.94392	.476191
2.11	4.4521	9.39393	1.45258	4.59347	1.28261	2.76330	5.95334	.473934
2.12	4.4944	9.52813	1.45602	4.60435	1.28463	2.76766	5.96273	.471698
2.13	4.5369	9.66360	1.45945	4.61519	1.28665	2.77200	5.97209	.469484
2.14	4.5796	9.80034	1.46287	4.62601	1.28866	2.77633	5.98142	.467290
2.15	4.6225	9.93838	1.46629	4.63681	1.29066	2.78065	5.99073	.465116
2.16	4.6656	10.0777	1.46969	4.64758	1.29266	2.78495	6.00000	.462963
2.17	4.7089	10.2183	1.47309	4.65833	1.29465	2.78924	6.00925	.460829
2.18	4.7524	10.3602	1.47648	4.66905	1.29664	2.79352	6.01846	.458716
2.19	4.7961	10.5035	1.47986	4.67974	1.29862	2.79779	6.02765	.456621
2.20	4.8400	10.6480	1.48324	4.69042	1.30059	2.80204	6.03681	.454546
2.21	4.8841	10.7939	1.48661	4.70106	1.30256	2.80628	6.04594	.452489
2.22	4.9284	10.9410	1.48997	4.71169	1.30452	2.81051	6.05505	.450451
2.23	4.9729	11.0896	1.49332	4.72229	1.30648	2.81472	6.06413	.448431
2.24	5.0176	11.2394	1.49666	4.73286	1.30843	2.81892	6.07318	.446429
2.25	5.0625	11.3906	1.50000	4.74342	1.31037	2.82311	6.08220	.444444
2.26	5.1076	11.5432	1.50333	4.75395	1.31231	2.82728	6.09120	.442478
2.27	5.1529	11.6971	1.50665	4.76445	1.31424	2.83145	6.10017	.440529
2.28	5.1984	11.8524	1.50997	4.77493	1.31617	2.83560	6.10911	.438597
2.29	5.2441	12.0090	1.51327	4.78539	1.31809	2.83974	6.11803	.436681
2.30	5.2900	12.1670	1.51658	4.79583	1.32001	2.84387	6.12693	.434783
2.31	5.3361	12.3264	1.51987	4.80625	1.32192	2.84798	6.13579	.432900
2.32	5.3824	12.4872	1.52315	4.81664	1.32382	2.85209	6.14463	.431035
2.33	5.4289	12.6493	1.52643	4.82701	1.32572	2.85618	6.15345	.429185
2.34	5.4756	12.8129	1.52971	4.83735	1.32761	2.86026	6.16224	.427350
2.35	5.5225	12.9779	1.53297	4.84768	1.32950	2.86433	6.17101	.425532
2.36	5.5696	13.1443	1.53623	4.85798	1.33139	2.86838	6.17975	.423729
2.37	5.6169	13.3121	1.53948	4.86826	1.33326	2.87243	6.18846	.421941
2.38	5.6644	13.4813	1.54272	4.87852	1.33514	2.87646	6.19715	.420168
2.39	5.7121	13.6519	1.54596	4.88876	1.33700	2.88049	6.20582	.418410
2.40	5.7600	13.8240	1.54919	4.89898	1.33887	2.88450	6.21447	.416667
2.41	5.8081	13.9975	1.55242	4.90918	1.34072	2.88850	6.22308	.414938
2.42	5.8564	14.1725	1.55563	4.91935	1.34257	2.89249	6.23168	.413223
2.43	5.9049	14.3489	1.55885	4.92950	1.34442	2.89647	6.24025	.411523
2.44	5.9536	14.5268	1.56205	4.93964	1.34626	2.90044	6.24880	.409836
2.45	6.0025	14.7061	1.56525	4.94975	1.34810	2.90439	6.25732	.408163
2.46	6.0516	14.8869	1.56844	4.95984	1.34993	2.90834	6.26583	.406504
2.47	6.1009	15.0692	1.57162	4.96991	1.35176	2.91227	6.27431	.404858
2.48	6.1504	15.2530	1.57480	4.97996	1.35358	2.91620	6.28276	.403226
2.49	6.2001	15.4382	1.57797	4.98999	1.35540	2.92011	6.29119	.401606
2.50	6.2500	15.6250	1.58114	5.00000	1.35721	2.92402	6.29961	.400000

n	n^2	n^3	$\sqrt{n}$	$\sqrt{10\,n}$	$\sqrt[3]{n}$	$\sqrt[3]{10\,n}$	$\sqrt[3]{100\,n}$	$\dfrac{1}{n}$
2.51	6.3001	15.8133	1.58430	5.00999	1.35902	2.92791	6.30799	.398406
2.52	6.3504	16.0060	1.58745	5.01996	1.36062	2.93179	6.31636	.396825
2.53	6.4009	16.1943	1.59060	5.02991	1.36262	2.93567	6.32470	.395257
2.54	6.4516	16.3871	1.59374	5.03984	1.36441	2.93953	6.33303	.393701
2.55	6.5025	16.5814	1.59687	5.04975	1.36620	2.94338	6.34133	.392157
2.56	6.5536	16.7772	1.60000	5.05964	1.36798	2.94723	6.34960	.390625
2.57	6.6049	16.9746	1.60312	5.06952	1.36976	2.95106	6.35786	.389105
2.58	6.6564	17.1735	1.60624	5.07937	1.37153	2.95488	6.36610	.387597
2.59	6.7081	17.3740	1.60935	5.08920	1.37330	2.95869	6.37431	.386100
2.60	6.7600	17.5760	1.61245	5.09902	1.37507	2.96250	6.38250	.384615
2.61	6.8121	17.7796	1.61555	5.10882	1.37683	2.96629	6.39068	.383142
2.62	6.8644	17.9847	1.61864	5.11859	1.37859	2.97007	6.39883	.381679
2.63	6.9169	18.1914	1.62173	5.12835	1.38034	2.97385	6.40696	.380228
2.64	6.9696	18.3997	1.62481	5.13809	1.38208	2.97761	6.41507	.378788
2.65	7.0225	18.6096	1.62788	5.14782	1.38383	2.98137	6.42316	.377359
2.66	7.0756	18.8211	1.63095	5.15752	1.38557	2.98511	6.43123	.375940
2.67	7.1289	19.0342	1.63401	5.16720	1.38730	2.98885	6.43928	.374532
2.68	7.1824	19.2488	1.63707	5.17687	1.38903	2.99257	6.44731	.373134
2.69	7.2361	19.4651	1.64012	5.18652	1.39076	2.99629	6.45531	.371747
2.70	7.2900	19.6830	1.64317	5.19615	1.39248	3.00000	6.46330	.370370
2.71	7.3441	19.9025	1.64621	5.20577	1.39419	3.00370	6.47127	.369004
2.72	7.3984	20.1236	1.64924	5.21536	1.39591	3.00739	6.47922	.367647
2.73	7.4529	20.3464	1.65227	5.22494	1.39761	3.01107	6.48715	.366300
2.74	7.5076	20.5708	1.65529	5.23450	1.39932	3.01474	6.49507	.364964
2.75	7.5625	20.7969	1.65831	5.24404	1.40102	3.01841	6.50296	.363636
2.76	7.6176	21.0246	1.66132	5.25357	1.40272	3.02206	6.51083	.362319
2.77	7.6729	21.2539	1.66433	5.26308	1.40441	3.02571	6.51868	.361011
2.78	7.7284	21.4850	1.66733	5.27257	1.40610	3.02934	6.52652	.359712
2.79	7.7841	21.7176	1.67033	5.28205	1.40778	3.03297	6.53434	.358423
2.80	7.8400	21.9520	1.67332	5.29150	1.40946	3.03659	6.54213	.357143
2.81	7.8961	22.1880	1.67631	5.30094	1.41114	3.04020	6.54991	.355872
2.82	7.9524	22.4258	1.67929	5.31037	1.41281	3.04380	6.55767	.354610
2.83	8.0089	22.6652	1.68226	5.31977	1.41448	3.04740	6.56541	.353357
2.84	8.0656	22.9063	1.68523	5.32917	1.41614	3.05098	6.57314	.352113
2.85	8.1225	23.1491	1.68819	5.33854	1.41780	3.05456	6.58084	.350877
2.86	8.1796	23.3937	1.69115	5.34790	1.41946	3.05813	6.58853	.349650
2.87	8.2369	23.6399	1.69411	5.35724	1.42111	3.06169	6.59620	.348432
2.88	8.2944	23.8879	1.69706	5.36656	1.42276	3.06524	6.60385	.347222
2.89	8.3521	24.1376	1.70000	5.37587	1.42440	3.06878	6.61149	.346021
2.90	8.4100	24.3890	1.70294	5.38516	1.42604	3.07232	6.61911	.344828
2.91	8.4681	24.6422	1.70587	5.39444	1.42768	3.07585	6.62671	.343643
2.92	8.5264	24.8971	1.70880	5.40370	1.42931	3.07936	6.63429	.342466
2.93	8.5849	25.1538	1.71172	5.41295	1.43094	3.08287	6.64185	.341297
2.94	8.6436	25.4122	1.71464	5.42218	1.43257	3.08638	6.64940	.340136
2.95	8.7025	25.6724	1.71756	5.43139	1.43419	3.08987	6.65693	.338983
2.96	8.7616	25.9343	1.72047	5.44059	1.43581	3.09336	6.66444	.337838
2.97	8.8209	26.1961	1.72337	5.44977	1.43743	3.09684	6.67194	.336700
2.98	8.8804	26.4636	1.72627	5.45894	1.43904	3.10031	6.67942	.335571
2.99	8.9401	26.7309	1.72916	5.46809	1.44065	3.10378	6.68688	.334448
3.00	9.0000	27.0000	1.73205	5.47723	1.44225	3.10723	6.69433	.333333

n	n^2	n^3	$\sqrt{n}$	$\sqrt{10\,n}$	$\sqrt[3]{n}$	$\sqrt[3]{10\,n}$	$\sqrt[3]{100\,n}$	$\dfrac{1}{n}$
3.01	9.0601	27.2709	1.73494	5.48635	1.44385	3.11068	6.70176	.332226
3.02	9.1204	27.5436	1.73781	5.49645	1.44545	3.11412	6.70917	.331126
3.03	9.1809	27.8181	1.74069	5.50454	1.44704	3.11755	6.71657	.330033
3.04	9.2416	28.0945	1.74356	5.51362	1.44863	3.12096	6.72395	.328947
3.05	9.3025	28.3726	1.74642	5.52268	1.45022	3.12440	6.73132	.327869
3.06	9.3636	28.6526	1.74929	5.53173	1.45180	3.12781	6.73866	.326797
3.07	9.4249	28.9344	1.75214	5.54076	1.45338	3.13121	6.74600	.325733
3.08	9.4864	29.2181	1.75499	5.54977	1.45496	3.13461	6.75331	.324675
3.09	9.5481	29.5036	1.75784	5.55878	1.45653	3.13800	6.76061	.323625
3.10	9.6100	29.7910	1.76068	5.56776	1.45810	3.14138	6.76790	.322581
3.11	9.6721	30.0802	1.76352	5.57674	1.45967	3.14475	6.77517	.321543
3.12	9.7344	30.3713	1.76635	5.58570	1.46123	3.14812	6.78242	.320513
3.13	9.7969	30.6643	1.76918	5.59464	1.46279	3.15148	6.78966	.319489
3.14	9.8596	30.9591	1.77200	5.60357	1.46434	3.15484	6.79688	.318471
3.15	9.9225	31.2559	1.77482	5.61249	1.46590	3.15818	6.80409	.317460
3.16	9.9856	31.5545	1.77764	5.62139	1.46745	3.16152	6.81128	.316456
3.17	10.0489	31.8550	1.78045	5.63028	1.46899	3.16485	6.81846	.315457
3.18	10.1124	32.1574	1.78326	5.63915	1.47054	3.16817	6.82562	.314465
3.19	10.1761	32.4618	1.78606	5.64801	1.47208	3.17149	6.83277	.313480
3.20	10.2400	32.7680	1.78885	5.65685	1.47361	3.17480	6.83990	.312500
3.21	10.3041	33.0762	1.79165	5.66569	1.47515	3.17811	6.84702	.311527
3.22	10.3684	33.3862	1.79444	5.67450	1.47668	3.18140	6.85412	.310559
3.23	10.4329	33.6983	1.79722	5.68331	1.47820	3.18469	6.86121	.309598
3.24	10.4976	34.0122	1.80000	5.69210	1.47973	3.18798	6.86829	.308642
3.25	10.5625	34.3281	1.80278	5.70088	1.48125	3.19125	6.87534	.307692
3.26	10.6276	34.6460	1.80555	5.70964	1.48277	3.19452	6.88239	.306749
3.27	10.6929	34.9658	1.80831	5.71839	1.48428	3.19779	6.88942	.305810
3.28	10.7584	35.2876	1.81108	5.72713	1.48579	3.20104	6.89643	.304878
3.29	10.8241	35.6129	1.81384	5.73585	1.48730	3.20429	6.90344	.303951
3.30	10.8900	35.9370	1.81659	5.74456	1.48881	3.20753	6.91042	.303030
3.31	10.9561	36.2647	1.81934	5.75326	1.49031	3.21077	6.91740	.302115
3.32	11.0224	36.5944	1.82209	5.76194	1.49181	3.21400	6.92436	.301205
3.33	11.0889	36.9260	1.82483	5.77062	1.49330	3.21723	6.93130	.300300
3.34	11.1556	37.2597	1.82757	5.77927	1.49480	3.22044	6.93823	.299401
3.35	11.2225	37.5954	1.83030	5.78792	1.49629	3.22365	6.94515	.298508
3.36	11.2896	37.9331	1.83303	5.79655	1.49777	3.22686	6.95205	.297619
3.37	11.3569	38.2728	1.83576	5.80517	1.49926	3.23005	6.95894	.296736
3.38	11.4244	38.6145	1.83848	5.81378	1.50074	3.23325	6.96582	.295858
3.39	11.4921	38.9582	1.84120	5.82237	1.50222	3.23643	6.97268	.294985
3.40	11.5600	39.3040	1.84391	5.83095	1.50369	3.23961	6.97953	.294118
3.41	11.6281	39.6518	1.84662	5.83952	1.50517	3.24278	6.98637	.293255
3.42	11.6964	40.0017	1.84932	5.84808	1.50664	3.24595	6.99319	.292398
3.43	11.7649	40.3536	1.85203	5.85662	1.50810	3.24911	7.00000	.291545
3.44	11.8336	40.7076	1.85472	5.86515	1.50957	3.25227	7.00680	.290698
3.45	11.9025	41.0636	1.85742	5.87367	1.51103	3.25542	7.01358	.289855
3.46	11.9716	41.4217	1.86011	5.88218	1.51249	3.25856	7.02035	.289017
3.47	12.0409	41.7819	1.86279	5.89067	1.51394	3.26169	7.02711	.288184
3.48	12.1104	42.1442	1.86548	5.89915	1.51540	3.26482	7.03385	.287356
3.49	12.1801	42.5085	1.86815	5.90762	1.51685	3.26795	7.04058	.286533
3.50	12.2500	42.8750	1.87083	5.91608	1.51829	3.27107	7.04730	.285714

n	n^2	n^3	$\sqrt{n}$	$\sqrt{10\,n}$	$\sqrt[3]{n}$	$\sqrt[3]{10\,n}$	$\sqrt[3]{100\,n}$	$\dfrac{1}{n}$
3.51	12.3201	43.2436	1.87350	5.92453	1.51974	3.27418	7.05400	.284900
3.52	12.3904	43.6142	1.87617	5.93296	1.52118	3.27729	7.06070	.284091
3.53	12.4609	43.9870	1.87883	5.94138	1.52262	3.28039	7.06738	.283286
3.54	12.5316	44.3619	1.88149	5.94979	1.52406	3.28348	7.07404	.282486
3.55	12.6025	44.7389	1.88414	5.95819	1.52549	3.28657	7.08070	.281690
3.56	12.6736	45.1180	1.88680	5.96657	1.52692	3.28965	7.08734	.280899
3.57	12.7449	45.4993	1.88944	5.97495	1.52835	3.29273	7.09397	.280112
3.58	12.8164	45.8827	1.89209	5.98331	1.52978	3.29580	7.10059	.279330
3.59	12.8881	46.2683	1.89473	5.99166	1.53120	3.29887	7.10719	.278552
3.60	12.9600	46.6560	1.89737	6.00000	1.53262	3.30193	7.11379	.277778
3.61	13.0321	47.0459	1.90000	6.00833	1.53404	3.30498	7.12037	.277008
3.62	13.1044	47.4379	1.90263	6.01664	1.53545	3.30803	7.12694	.276243
3.63	13.1769	47.8321	1.90526	6.02495	1.53686	3.31107	7.13349	.275482
3.64	13.2496	48.2285	1.90788	6.03324	1.53827	3.31411	7.14004	.274725
3.65	13.3225	48.6271	1.91050	6.04152	1.53968	3.31714	7.14657	.273973
3.66	13.3956	49.0279	1.91311	6.04979	1.54109	3.32017	7.15309	.273224
3.67	13.4689	49.4309	1.91572	6.05805	1.54249	3.32319	7.15960	.272480
3.68	13.5424	49.8360	1.91833	6.06630	1.54389	3.32621	7.16610	.271739
3.69	13.6161	50.2434	1.92094	6.07454	1.54529	3.32922	7.17258	.271003
3.70	13.6900	50.6530	1.92354	6.08276	1.54668	3.33222	7.17905	.270270
3.71	13.7641	51.0648	1.92614	6.09096	1.54807	3.33522	7.18552	.269542
3.72	13.8384	51.4788	1.92873	6.09918	1.54946	3.33822	7.19197	.268817
3.73	13.9129	51.8951	1.93132	6.10737	1.55085	3.34120	7.19841	.268097
3.74	13.9876	52.3136	1.93391	6.11555	1.55223	3.34419	7.20483	.267380
3.75	14.0625	52.7344	1.93649	6.12372	1.55362	3.34716	7.21125	.266667
3.76	14.1376	53.1574	1.93907	6.13188	1.55500	3.35014	7.21765	.265957
3.77	14.2129	53.5826	1.94165	6.14003	1.55637	3.35310	7.22405	.265252
3.78	14.2884	54.0102	1.94422	6.14817	1.55775	3.35607	7.23043	.264550
3.79	14.3641	54.4399	1.94679	6.15630	1.55912	3.35902	7.23680	.263852
3.80	14.4400	54.8720	1.94936	6.16441	1.56049	3.36198	7.24316	.263158
3.81	14.5161	55.3063	1.95192	6.17252	1.56186	3.36492	7.24950	.262467
3.82	14.5924	55.7430	1.95448	6.18061	1.56322	3.36786	7.25584	.261780
3.83	14.6689	56.1819	1.95704	6.18870	1.56459	3.37080	7.26217	.261097
3.84	14.7456	56.6231	1.95959	6.19677	1.56595	3.37373	7.26848	.260417
3.85	14.8225	57.0666	1.96214	6.20484	1.56731	3.37666	7.27479	.259740
3.86	14.8996	57.5125	1.96469	6.21289	1.56866	3.37958	7.28108	.259067
3.87	14.9769	57.9606	1.96723	6.22093	1.57001	3.38249	7.28736	.258398
3.88	15.0544	58.4111	1.96977	6.22896	1.57137	3.38540	7.29363	.257732
3.89	15.1321	58.8639	1.97231	6.23699	1.57271	3.38831	7.29989	.257069
3.90	15.2100	59.3190	1.97484	6.24500	1.57406	3.39121	7.30614	.256410
3.91	15.2881	59.7765	1.97737	6.25300	1.57541	3.39411	7.31238	.255755
3.92	15.3664	60.2363	1.97990	6.26099	1.57675	3.39700	7.31861	.255102
3.93	15.4449	60.6985	1.98242	6.26897	1.57809	3.39988	7.32483	.254453
3.94	15.5236	61.1630	1.98494	6.27694	1.57942	3.40277	7.33104	.253807
3.95	15.6025	61.6299	1.98746	6.28490	1.58076	3.40564	7.33723	.253165
3.96	15.6816	62.0991	1.98997	6.29285	1.58209	3.40851	7.34342	.252525
3.97	15.7609	62.5706	1.99249	6.30079	1.58342	3.41138	7.34960	.251889
3.98	15.8404	63.0448	1.99499	6.30872	1.58475	3.41424	7.35576	.251256
3.99	15.9201	63.5212	1.99750	6.31664	1.58608	3.41710	7.36192	.250627
4.00	16.0000	64.0000	2.00000	6.32456	1.58740	3.41995	7.36806	.250000

n	n^2	n^3	$\sqrt{n}$	$\sqrt{10n}$	$\sqrt[3]{n}$	$\sqrt[3]{10n}$	$\sqrt[3]{100n}$	$\dfrac{1}{n}$
4.01	16.0801	64.4812	2.00250	6.33246	1.58872	3.42280	7.37420	.249377
4.02	16.1604	64.9648	2.00499	6.34035	1.59004	3.42564	7.38082	.248756
4.03	16.2409	65.4508	2.00749	6.34823	1.59136	3.42848	7.38644	.248139
4.04	16.3216	65.9393	2.00998	6.35610	1.59267	3.43131	7.39254	.247525
4.05	16.4025	66.4301	2.01246	6.36396	1.59399	3.43414	7.39864	.246914
4.06	16.4836	66.9234	2.01494	6.37181	1.59530	3.43697	7.40472	.246305
4.07	16.5649	67.4191	2.01742	6.37966	1.59661	3.43979	7.41080	.245700
4.08	16.6464	67.9173	2.01990	6.38749	1.59791	3.44260	7.41686	.245098
4.09	16.7281	68.4179	2.02237	6.39531	1.59922	3.44541	7.42291	.244499
4.10	16.8100	68.9210	2.02485	6.40312	1.60052	3.44822	7.42896	.243902
4.11	16.8921	69.4265	2.02731	6.41093	1.60182	3.45102	7.43499	.243309
4.12	16.9744	69.9345	2.02978	6.41872	1.60312	3.45382	7.44102	.242718
4.13	17.0569	70.4450	2.03224	6.42651	1.60441	3.45661	7.44703	.242131
4.14	17.1396	70.9579	2.03470	6.43428	1.60571	3.45939	7.45304	.241546
4.15	17.2225	71.4734	2.03715	6.44205	1.60700	3.46218	7.45904	.240964
4.16	17.3056	71.9913	2.03961	6.44981	1.60829	3.46496	7.46502	.240385
4.17	17.3889	72.5117	2.04206	6.45755	1.60958	3.46773	7.47100	.239808
4.18	17.4724	73.0346	2.04450	6.46529	1.61086	3.47050	7.47697	.239234
4.19	17.5561	73.5601	2.04695	6.47302	1.61215	3.47327	7.48292	.238664
4.20	17.6400	74.0880	2.04939	6.48074	1.61343	3.47603	7.48887	.238095
4.21	17.7241	74.6185	2.05183	6.48845	1.61471	3.47878	7.49481	.237530
4.22	17.8084	75.1514	2.05426	6.49615	1.61599	3.48154	7.50074	.236967
4.23	17.8929	75.6870	2.05670	6.50385	1.61726	3.48428	7.50666	.236407
4.24	17.9776	76.2250	2.05913	6.51153	1.61853	3.48703	7.51257	.235849
4.25	18.0625	76.7656	2.06155	6.51920	1.61981	3.48977	7.51847	.235294
4.26	18.1476	77.3088	2.06398	6.52687	1.62108	3.49250	7.52437	.234742
4.27	18.2329	77.8545	2.06640	6.53452	1.62234	3.49523	7.53025	.234192
4.28	18.3184	78.4028	2.06882	6.54217	1.62361	3.49796	7.53612	.233645
4.29	18.4041	78.9536	2.07123	6.54981	1.62487	3.50068	7.54199	.233100
4.30	18.4900	79.5070	2.07364	6.55744	1.62613	3.50340	7.54784	.232558
4.31	18.5761	80.0630	2.07605	6.56506	1.62739	3.50611	7.55369	.232019
4.32	18.6624	80.6216	2.07846	6.57267	1.62865	3.50882	7.55953	.231482
4.33	18.7489	81.1827	2.08087	6.58027	1.62991	3.51153	7.56535	.230947
4.34	18.8356	81.7465	2.08327	6.58787	1.63116	3.51423	7.57117	.230415
4.35	18.9225	82.3129	2.08567	6.59545	1.63241	3.51692	7.57698	.229885
4.36	19.0096	82.8819	2.08806	6.60303	1.63366	3.51962	7.58279	.229358
4.37	19.0969	83.4535	2.09045	6.61060	1.63491	3.52231	7.58858	.228833
4.38	19.1844	84.0277	2.09284	6.61816	1.63616	3.52499	7.59436	.228311
4.39	19.2721	84.6045	2.09523	6.62571	1.63740	3.52767	7.60014	.227790
4.40	19.3600	85.1840	2.09762	6.63325	1.63864	3.53035	7.60590	.227273
4.41	19.4481	85.7661	2.10000	6.64078	1.63988	3.53302	7.61166	.226757
4.42	19.5364	86.3509	2.10238	6.64831	1.64112	3.53569	7.61741	.226244
4.43	19.6249	86.9383	2.10476	6.65582	1.64236	3.53835	7.62315	.225734
4.44	19.7136	87.5284	2.10713	6.66333	1.64359	3.54101	7.62888	.225225
4.45	19.8025	88.1211	2.10950	6.67083	1.64483	3.54367	7.63461	.224719
4.46	19.8916	88.7165	2.11187	6.67832	1.64606	3.54632	7.64032	.224215
4.47	19.9809	89.3146	2.11424	6.68581	1.64729	3.54897	7.64603	.223714
4.48	20.0704	89.9154	2.11660	6.69328	1.64851	3.55162	7.65172	.223214
4.49	20.1601	90.5188	2.11896	6.70075	1.64974	3.55426	7.65741	.222717
4.50	20.2500	91.1250	2.12132	6.70820	1.65096	3.55689	7.66309	.222222

n	n^2	n^3	$\sqrt{n}$	$\sqrt{10\,n}$	$\sqrt[3]{n}$	$\sqrt[3]{10\,n}$	$\sqrt[3]{100\,n}$	$\frac{1}{n}$
4.51	20.3401	91.7339	2.12368	6.71565	1.65219	3.55953	7.66877	.221730
4.52	20.4304	92.3454	2.12603	6.72309	1.65341	3.56215	7.67443	.221239
4.53	20.5209	92.9597	2.12838	6.73053	1.65462	3.56478	7.68009	.220751
4.54	20.6116	93.5767	2.13073	6.73795	1.65584	3.56740	7.68573	.220264
4.55	20.7025	94.1964	2.13307	6.74537	1.65706	3.57002	7.69137	.219780
4.56	20.7936	94.8188	2.13542	6.75278	1.65827	3.57263	7.69700	.219298
4.57	20.8849	95.4440	2.13776	6.76018	1.65948	3.57524	7.70262	.218818
4.58	20.9764	96.0719	2.14009	6.76757	1.66069	3.57785	7.70824	.218341
4.59	21.0681	96.7026	2.14243	6.77495	1.66190	3.58045	7.71384	.217865
4.60	21.1600	97.3360	2.14476	6.78233	1.66310	3.58305	7.71944	.217391
4.61	21.2521	97.9722	2.14709	6.78970	1.66431	3.58564	7.72503	.216920
4.62	21.3444	98.6111	2.14942	6.79706	1.66551	3.58823	7.73061	.216450
4.63	21.4369	99.2528	2.15174	6.80441	1.66671	3.59082	7.73619	.215983
4.64	21.5296	99.8973	2.15407	6.81175	1.66791	3.59340	7.74175	.215517
4.65	21.6225	100.545	2.15639	6.81909	1.66911	3.59598	7.74731	.215054
4.66	21.7156	101.195	2.15870	6.82642	1.67030	3.59856	7.75286	.214592
4.67	21.8089	101.848	2.16102	6.83374	1.67150	3.60113	7.75840	.214133
4.68	21.9024	102.503	2.16333	6.84105	1.67269	3.60370	7.76394	.213675
4.69	21.9961	103.162	2.16564	6.84836	1.67388	3.60626	7.76946	.213220
4.70	22.0900	103.823	2.16795	6.85565	1.67507	3.60883	7.77498	.213766
4.71	22.1841	104.487	2.17025	6.86294	1.67626	3.61138	7.78049	.212314
4.72	22.2784	105.154	2.17256	6.87023	1.67744	3.61394	7.78599	.211864
4.73	22.3729	105.824	2.17486	6.87750	1.67863	3.61649	7.79149	.211417
4.74	22.4676	106.496	2.17715	6.88477	1.67981	3.61904	7.79697	.210971
4.75	22.5625	107.172	2.17945	6.89202	1.68099	3.62158	7.80245	.210526
4.76	22.6576	107.850	2.18174	6.89928	1.68217	3.62412	7.80793	.210084
4.77	22.7529	108.531	2.18403	6.90652	1.68334	3.62665	7.81339	.209644
4.78	22.8484	109.215	2.18632	6.91375	1.68452	3.62919	7.81885	.209205
4.79	22.9441	109.902	2.18861	6.92098	1.68569	3.63171	7.82429	.208768
4.80	23.0400	110.592	2.19089	6.92820	1.68687	3.63424	7.82974	.208333
4.81	23.1361	111.285	2.19317	6.93542	1.68804	3.63676	7.83517	.207900
4.82	23.2324	111.980	2.19545	6.94262	1.68920	3.63928	7.84059	.207469
4.83	23.3289	112.679	2.19773	6.94982	1.69037	3.64180	7.84601	.207039
4.84	23.4256	113.380	2.20000	6.95701	1.69154	3.64431	7.85142	.206612
4.85	23.5225	114.084	2.20227	6.96419	1.69270	3.64682	7.85683	.206186
4.86	23.6196	114.791	2.20454	6.97137	1.69386	3.64932	7.86222	.205761
4.87	23.7169	115.501	2.20681	6.97854	1.69503	3.65182	7.86761	.205339
4.88	23.8144	116.214	2.20907	6.98570	1.69619	3.65432	7.87299	.204918
4.89	23.9121	116.930	2.21133	6.99285	1.69734	3.65682	7.87837	.204499
4.90	24.0100	117.649	2.21359	7.00000	1.69850	3.65931	7.88374	.204082
4.91	24.1081	118.371	2.21585	7.00714	1.69965	3.66179	7.88909	.203666
4.92	24.2064	119.095	2.21811	7.01427	1.70081	3.66428	7.89445	.203252
4.93	24.3049	119.823	2.22036	7.02140	1.70196	3.66676	7.89979	.202840
4.94	24.4036	120.554	2.22261	7.02851	1.70311	3.66924	7.90513	.202429
4.95	24.5025	121.287	2.22486	7.03562	1.70426	3.67171	7.91046	.202020
4.96	24.6016	122.024	2.22711	7.04273	1.70540	3.67418	7.91578	.201613
4.97	24.7009	122.763	2.22935	7.04982	1.70655	3.67665	7.92110	.201207
4.98	24.8004	123.506	2.23159	7.05691	1.70769	3.67911	7.92641	.200803
4.99	24.9001	124.251	2.23383	7.06399	1.70884	3.68157	7.93171	.200401
5.00	25.0000	125.000	2.23607	7.07107	1.70998	3.68403	7.93701	.200000

n	n^2	n^3	$\sqrt{n}$	$\sqrt{10\,n}$	$\sqrt[3]{n}$	$\sqrt[3]{10\,n}$	$\sqrt[3]{100\,n}$	$\dfrac{1}{n}$
5.01	25.1001	125.752	2.23830	7.07814	1.71112	3.68649	7.94229	.199601
5.02	25.2004	126.506	2.24054	7.08520	1.71225	3.68894	7.94757	.199203
5.03	25.3009	127.264	2.24277	7.09225	1.71339	3.69138	7.95285	.198807
5.04	25.4016	128.024	2.24499	7.09930	1.71452	3.69383	7.95811	.198413
5.05	25.5025	128.788	2.24722	7.10634	1.71566	3.69627	7.96337	.198020
5.06	25.6036	129.554	2.24944	7.11337	1.71679	3.69871	7.96863	.197629
5.07	25.7049	130.324	2.25167	7.12039	1.71792	3.70114	7.97387	.197239
5.08	25.8064	131.097	2.25389	7.12741	1.71905	3.70358	7.97911	.196850
5.09	25.9081	131.872	2.25610	7.13442	1.72017	3.70600	7.98434	.196464
5.10	26.0100	132.651	2.25832	7.14143	1.72130	3.70843	7.98957	.196078
5.11	26.1121	133.433	2.26053	7.14843	1.72242	3.71085	7.99479	.195695
5.12	26.2144	134.218	2.26274	7.15542	1.72355	3.71327	8.00000	.195313
5.13	26.3169	135.006	2.26495	7.16240	1.72467	3.71566	8.00520	.194932
5.14	26.4196	135.797	2.26716	7.16938	1.72579	3.71810	8.01040	.194553
5.15	26.5225	136.591	2.26936	7.17635	1.72691	3.72051	8.01559	.194175
5.16	26.6256	137.388	2.27156	7.18331	1.72802	3.72292	8.02078	.193798
5.17	26.7289	138.188	2.27376	7.19027	1.72914	3.72532	8.02596	.193424
5.18	26.8324	138.992	2.27596	7.19722	1.73025	3.72772	8.03113	.193050
5.19	26.9361	139.798	2.27816	7.20417	1.73137	3.73012	8.03629	.192678
5.20	27.0400	140.608	2.28035	7.21110	1.73248	3.73251	8.04145	.192306
5.21	27.1441	141.421	2.28254	7.21803	1.73359	3.73490	8.04660	.191939
5.22	27.2484	142.237	2.28473	7.22496	1.73470	3.73729	8.05175	.191571
5.23	27.3529	143.056	2.28692	7.23187	1.73580	3.73968	8.05689	.191205
5.24	27.4576	143.878	2.28910	7.23878	1.73691	3.74206	8.06202	.190846
5.25	27.5625	144.703	2.29129	7.24569	1.73801	3.74443	8.06714	.190476
5.26	27.6676	145.532	2.29347	7.25259	1.73912	3.74681	8.07226	.190114
5.27	27.7729	146.363	2.29565	7.25948	1.74022	3.74918	8.07737	.189753
5.28	27.8784	147.198	2.29783	7.26636	1.74132	3.75158	8.08248	.189394
5.29	27.9841	148.036	2.30000	7.27324	1.74242	3.75392	8.08758	.189036
5.30	28.0900	148.877	2.30217	7.28011	1.74351	3.75629	8.09267	.188679
5.31	28.1961	149.721	2.30434	7.28697	1.74461	3.75865	8.09776	.188324
5.32	28.3024	150.569	2.30651	7.29383	1.74570	3.76100	8.10284	.187970
5.33	28.4089	151.419	2.30868	7.30068	1.74680	3.76336	8.10791	.187617
5.34	28.5156	152.273	2.31084	7.30753	1.74789	3.76571	8.11298	.187266
5.35	28.6225	153.130	2.31301	7.31437	1.74898	3.76806	8.11804	.186916
5.36	28.7296	153.991	2.31517	7.32120	1.75007	3.77041	8.12310	.186567
5.37	28.8369	154.854	2.31733	7.32803	1.75116	3.77275	8.12814	.186220
5.38	28.9444	155.721	2.31948	7.33485	1.75224	3.77509	8.13319	.185874
5.39	29.0521	156.591	2.32164	7.34166	1.75333	3.77740	8.13822	.185529
5.40	29.1600	157.464	2.32379	7.34847	1.75441	3.77976	8.14325	.185185
5.41	29.2681	158.340	2.32594	7.35527	1.75549	3.78210	8.14828	.184843
5.42	29.3764	159.220	2.32809	7.36206	1.75657	3.78442	8.15329	.184502
5.43	29.4849	160.108	2.33024	7.36885	1.75765	3.78675	8.15831	.184162
5.44	29.5936	160.989	2.33238	7.37564	1.75873	3.78907	8.16331	.183824
5.45	29.7025	161.879	2.33452	7.38241	1.75981	3.79139	8.16831	.183486
5.46	29.8116	162.771	2.33666	7.38918	1.76088	3.79371	8.17330	.183150
5.47	29.9209	163.667	2.33880	7.39594	1.76196	3.79603	8.17829	.182815
5.48	30.0304	164.567	2.34094	7.40270	1.76303	3.79834	8.18327	.182482
5.49	30.1401	165.469	2.34307	7.40945	1.76410	3.80065	8.18824	.182149
5.50	30.2500	166.375	2.34521	7.41620	1.76517	3.80295	8.19321	.181818

n	n^2	n^3	$\sqrt{n}$	$\sqrt{10\,n}$	$\sqrt[3]{n}$	$\sqrt[3]{10\,n}$	$\sqrt[3]{100\,n}$	$\dfrac{1}{n}$
5.51	30.3601	167.284	2.34734	7.42294	1.76624	3.80526	8.19818	.181488
5.52	30.4704	168.197	2.34947	7.42967	1.76731	3.80756	8.20313	.181159
5.53	30.5809	169.112	2.35160	7.43640	1.76838	3.80986	8.20808	.180832
5.54	30.6916	170.031	2.35372	7.44312	1.76944	3.80115	8.21303	.180505
5.55	30.8025	170.954	2.35584	7.44983	1.77051	3.81444	8.21797	.180180
5.56	30.9136	171.880	2.35797	7.45654	1.77157	3.81673	8.22290	.179856
5.57	31.0249	172.809	2.36008	7.46324	1.77263	3.81902	8.22783	.179533
5.58	31.1364	173.741	2.36220	7.46994	1.77369	3.82130	8.23275	.179212
5.59	31.2481	174.677	2.36432	7.47663	1.77475	3.82358	8.23766	.178891
5.60	31.3600	175.616	2.36643	7.48331	1.77581	3.82586	8.24257	.178571
5.61	31.4721	176.558	2.36854	7.48999	1.77686	3.82814	8.24747	.178253
5.62	31.5844	177.504	2.37065	7.49667	1.77792	3.83041	8.25237	.177936
5.63	31.6969	178.454	2.37276	7.50333	1.77897	3.83268	8.25726	.177620
5.64	31.8096	179.406	2.37487	7.50999	1.78003	3.83495	8.26215	.177305
5.65	31.9225	180.362	2.37697	7.51665	1.78108	3.83721	8.26703	.176991
5.66	32.0356	181.321	2.37908	7.52330	1.78213	3.83948	8.27190	.176678
5.67	32.1489	182.284	2.38118	7.52994	1.78318	3.84174	8.27677	.176367
5.68	32.2624	183.250	2.38328	7.53658	1.78422	3.84400	8.28164	.176056
5.69	32.3761	184.220	2.38537	7.54321	1.78527	3.84625	8.28649	.175747
5.70	32.4900	185.193	2.38747	7.54983	1.78632	3.84850	8.29134	.175439
5.71	32.6041	186.169	2.38956	7.55645	1.78736	3.85075	8.29619	.175131
5.72	32.7184	187.149	2.39165	7.56307	1.78840	3.85300	8.30103	.174825
5.73	32.8329	188.133	2.39374	7.56968	1.78944	3.85524	8.30587	.174520
5.74	32.9476	189.119	2.39583	7.57628	1.79048	3.85748	8.31069	.174216
5.75	33.0625	190.109	2.39792	7.58288	1.79152	3.85972	8.31552	.173913
5.76	33.1776	191.103	2.40000	7.58947	1.79256	3.86196	8.32034	.173611
5.77	33.2929	192.100	2.40208	7.59605	1.79360	3.86419	8.32515	.173310
5.78	33.4084	193.101	2.40416	7.60263	1.79463	3.86642	8.32995	.173010
5.79	33.5241	194.105	2.40624	7.60920	1.79567	3.86865	8.33476	.172712
5.80	33.6400	195.112	2.40832	7.61577	1.79670	3.87088	8.33955	.172414
5.81	33.7561	196.123	2.41039	7.62234	1.79773	3.87310	8.34434	.172117
5.82	33.8724	197.137	2.41247	7.62889	1.79876	3.87532	8.34913	.171821
5.83	33.9689	198.155	2.41454	7.63544	1.79979	3.87754	8.35390	.171527
5.84	34.1056	199.177	2.41661	7.64199	1.80082	3.87975	8.35868	.171233
5.85	34.2225	200.202	2.41868	7.64853	1.80185	3.88197	8.36345	.170940
5.86	34.3396	201.230	2.42074	7.65506	1.80288	3.88418	8.36821	.170649
5.87	34.4569	202.262	2.42281	7.66159	1.80390	3.88639	8.37297	.170358
5.88	34.5744	203.297	2.42487	7.66812	1.80492	3.88859	8.37772	.170068
5.89	34.6921	204.336	2.42693	7.67463	1.80595	3.89082	8.38247	.169779
5.90	34.8100	205.379	2.42899	7.68115	1.80697	3.89300	8.38721	.169492
5.91	34.9281	206.425	2.43105	7.68765	1.80799	3.89520	8.39194	.169205
5.92	35.0464	207.475	2.43311	7.69415	1.80901	3.89739	8.39667	.168919
5.93	35.1649	208.528	2.43516	7.70065	1.81003	3.89958	8.40140	.168634
5.94	35.2836	209.585	2.43721	7.70714	1.81104	3.90177	8.40612	.168350
5.95	35.4025	210.645	2.43926	7.71362	1.81206	3.90396	8.41083	.168067
5.96	35.5216	211.709	2.44131	7.72010	1.81307	3.90615	8.41554	.167785
5.97	35.6409	212.776	2.44336	7.72658	1.81409	3.90833	8.42025	.167504
5.98	35.7604	213.847	2.44540	7.73305	1.81510	3.91051	8.42494	.167224
5.99	35.8801	214.922	2.44745	7.73951	1.81611	3.91269	8.42964	.166945
6.00	36.0000	216.000	2.44949	7.74597	1.81712	3.91487	8.43433	.166667

112h POWERS, ROOTS, AND RECIPROCALS.

n	n^2	n^3	$\sqrt{n}$	$\sqrt{10\,n}$	$\sqrt[3]{n}$	$\sqrt[3]{10\,n}$	$\sqrt[3]{100\,n}$	$\dfrac{1}{n}$
6.01	36.1201	217.062	2.45153	7.75242	1.81818	3.91704	8.43901	.166389
6.02	36.2404	218.167	2.45357	7.75887	1.81914	3.91921	8.44390	.166113
6.03	36.3609	219.256	2.45561	7.76531	1.82014	3.92138	8.44836	.165838
6.04	36.4816	220.349	2.45764	7.77174	1.82115	3.92355	8.45308	.165563
6.05	36.6025	221.445	2.45967	7.77817	1.82215	3.92571	8.45769	.165289
6.06	36.7236	222.545	2.46171	7.78460	1.82316	3.92787	8.46235	.165017
6.07	36.8449	223.649	2.46374	7.79102	1.82416	3.93003	8.46700	.164745
6.08	36.9664	224.756	2.46577	7.79744	1.82516	3.93219	8.47165	.164474
6.09	37.0881	225.867	2.46779	7.80385	1.82616	3.93434	8.47629	.164204
6.10	37.2100	226.981	2.46982	7.81025	1.82716	3.93650	8.48093	.163934
6.11	37.3321	228.099	2.47184	7.81665	1.82816	3.93865	8.48556	.163666
6.12	37.4544	229.221	2.47386	7.82304	1.82915	3.94079	8.49018	.163399
6.13	37.5769	230.346	2.47588	7.82943	1.83015	3.94294	8.49481	.163132
6.14	37.6996	231.476	2.47790	7.83582	1.83115	3.94508	8.49942	.162866
6.15	37.8225	232.608	2.47992	7.84219	1.83214	3.94722	8.50404	.162602
6.16	37.9456	233.745	2.48193	7.84857	1.83313	3.94936	8.50864	.162338
6.17	38.0689	234.885	2.48395	7.85493	1.83412	3.95150	8.51324	.162075
6.18	38.1924	236.029	2.48596	7.86130	1.83511	3.95363	8.51784	.161812
6.19	38.3161	237.177	2.48797	7.86766	1.83610	3.95576	8.52243	.161551
6.20	38.4400	238.328	2.48998	7.87401	1.83709	3.95789	8.52702	.161290
6.21	38.5641	239.483	2.49199	7.88036	1.83808	3.96002	8.53160	.161031
6.22	38.6884	240.642	2.49399	7.88670	1.83906	3.96214	8.53618	.160772
6.23	38.8129	241.804	2.49600	7.89303	1.84005	3.96426	8.54075	.160514
6.24	38.9376	242.971	2.49800	7.89937	1.84103	3.96639	8.54532	.160256
6.25	39.0625	244.141	2.50000	7.90569	1.84202	3.96850	8.54988	.160000
6.26	39.1876	245.314	2.50200	7.91202	1.84300	3.97062	8.55444	.159744
6.27	39.3129	246.492	2.50400	7.91833	1.84398	3.97273	8.55899	.159490
6.28	39.4384	247.673	2.50599	7.92465	1.84496	3.97484	8.56354	.159236
6.29	39.5641	248.858	2.50799	7.93095	1.84594	3.97695	8.56808	.158983
6.30	39.6900	250.047	2.50998	7.93725	1.84691	3.97906	8.57262	.158730
6.31	39.8161	251.240	2.51197	7.94355	1.84789	3.98116	8.57715	.158479
6.32	39.9424	252.436	2.51396	7.94984	1.84887	3.98326	8.58168	.158228
6.33	40.0689	253.636	2.51595	7.95613	1.84984	3.98536	8.58620	.157978
6.34	40.1956	254.840	2.51794	7.96241	1.85082	3.98746	8.59072	.157729
6.35	40.3225	256.048	2.51992	7.96869	1.85179	3.98956	8.59524	.157480
6.36	40.4496	257.259	2.52190	7.97496	1.85276	3.99165	8.59975	.157233
6.37	40.5769	258.475	2.52389	7.98123	1.85373	3.99374	8.60425	.156986
6.38	40.7044	259.694	2.52587	7.98749	1.85470	3.99583	8.60875	.156740
6.39	40.8321	260.917	2.52784	7.99375	1.85567	3.99792	8.61325	.156495
6.40	40.9600	262.144	2.52982	8.00000	1.85664	4.00000	8.61774	.156250
6.41	41.0881	263.375	2.53180	8.00625	1.85760	4.00208	8.62222	.156006
6.42	41.2164	264.609	2.53377	8.01249	1.85857	4.00416	8.62671	.155763
6.43	41.3449	265.848	2.53574	8.01873	1.85953	4.00624	8.63118	.155521
6.44	41.4736	267.090	2.53772	8.02496	1.86050	4.00832	8.63566	.155280
6.45	41.6025	268.336	2.53969	8.03119	1.86146	4.01039	8.64012	.155039
6.46	41.7316	269.586	2.54165	8.03741	1.86242	4.01246	8.64459	.154799
6.47	41.8609	270.840	2.54362	8.04363	1.86338	4.01453	8.64904	.154560
6.48	41.9904	272.098	2.54558	8.04984	1.86434	4.01660	8.65350	.154321
6.49	42.1201	273.359	2.54755	8.05605	1.86530	4.01866	8.65795	.154083
6.50	42.2500	274.625	2.54951	8.06226	1.86626	4.02073	8.66239	.153846

n	n^2	n^3	$\sqrt{n}$	$\sqrt{10\,n}$	$\sqrt[3]{n}$	$\sqrt[3]{10\,n}$	$\sqrt[3]{100\,n}$	$\dfrac{1}{n}$
6.51	42.3801	275.894	2.55147	8.06846	1.86721	4.02279	8.66663	.153610
6.52	42.5104	277.168	2.55343	8.07465	1.86817	4.02465	8.67127	.153374
6.53	42.6409	278.445	2.55539	8.08064	1.86912	4.02690	8.67570	.153139
6.54	42.7716	279.726	2.55734	8.08703	1.87006	4.02896	8.68012	.152905
6.55	42.9025	281.011	2.55930	8.09321	1.87103	4.03101	8.68455	.152672
6.56	43.0336	282.300	2.56125	8.09938	1.87198	4.03306	8.68896	.152439
6.57	43.1649	283.593	2.56320	8.10555	1.87293	4.03511	8.69338	.152207
6.58	43.2964	284.890	2.56515	8.11172	1.87388	4.03715	8.69778	.151976
6.59	43.4281	286.191	2.56710	8.11788	1.87483	4.03920	8.70219	.151745
6.60	43.5600	287.496	2.56905	8.12404	1.87578	4.04124	8.70659	.151515
6.61	43.6921	288.805	2.57099	8.13019	1.87672	4.04328	8.71098	.151286
6.62	43.8244	290.118	2.57294	8.13634	1.87767	4.04532	8.71537	.151057
6.63	43.9569	291.434	2.57488	8.14248	1.87862	4.04735	8.71976	.150830
6.64	44.0896	292.755	2.57682	8.14862	1.87956	4.04939	8.72414	.150602
6.65	44.2225	294.080	2.57876	8.15475	1.88050	4.05142	8.72852	.150376
6.66	44.3556	295.408	2.58070	8.16088	1.88144	4.05345	8.73289	.150150
6.67	44.4889	296.741	2.58263	8.16701	1.88239	4.05548	8.73726	.149925
6.68	44.6224	298.078	2.58457	8.17313	1.88333	4.05750	8.74162	.149701
6.69	44.7561	299.418	2.58650	8.17924	1.88427	4.05953	8.74598	.149477
6.70	44.8900	300.763	2.58844	8.18535	1.88520	4.06155	8.75034	.149254
6.71	45.0241	302.112	2.59037	8.19146	1.88614	4.06357	8.75469	.149031
6.72	45.1584	303.464	2.59230	8.19756	1.88708	4.06558	8.75904	.148810
6.73	45.2929	304.821	2.59422	8.20366	1.88801	4.06760	8.76338	.148588
6.74	45.4276	306.182	2.59615	8.20975	1.88895	4.06961	8.76772	.148368
6.75	45.5625	307.547	2.59808	8.21584	1.88988	4.07163	8.77205	.148148
6.76	45.6976	308.916	2.60000	8.22192	1.89081	4.07364	8.77638	.147929
6.77	45.8329	310.289	2.60192	8.22800	1.89175	4.07564	8.78071	.147711
6.78	45.9684	311.666	2.60384	8.23408	1.89268	4.07765	8.78503	.147493
6.79	46.1041	313.047	2.60576	8.24015	1.89361	4.07965	8.78935	.147275
6.80	46.2400	314.432	2.60768	8.24621	1.89454	4.08166	8.79366	.147059
6.81	46.3761	315.821	2.60960	8.25227	1.89546	4.08365	8.79797	.146843
6.82	46.5124	317.215	2.61151	8.25833	1.89639	4.08565	8.80227	.146628
6.83	46.6489	318.612	2.61343	8.26438	1.89732	4.08765	8.80657	.146413
6.84	46.7856	320.014	2.61534	8.27043	1.89824	4.08964	8.81087	.146199
6.85	46.9225	321.419	2.61725	8.27647	1.89917	4.09164	8.81516	.145985
6.86	47.0596	322.829	2.61916	8.28251	1.90009	4.09362	8.81945	.145773
6.87	47.1969	324.243	2.62107	8.28855	1.90102	4.09561	8.82373	.145560
6.88	47.3344	325.661	2.62298	8.29458	1.90194	4.09760	8.82801	.145349
6.89	47.4721	327.083	2.62488	8.30060	1.90286	4.09958	8.83229	.145138
6.90	47.6100	328.509	2.62679	8.30662	1.90378	4.10157	8.83656	.144928
6.91	47.7481	329.939	2.62869	8.31264	1.90470	4.10355	8.84082	.144718
6.92	47.8864	331.374	2.63059	8.31865	1.90562	4.10552	8.84509	.144509
6.93	48.0249	332.813	2.63249	8.32466	1.90653	4.10750	8.84934	.144300
6.94	48.1636	334.255	2.63439	8.33067	1.90745	4.10948	8.85360	.144092
6.95	48.3025	335.702	2.63629	8.33667	1.90837	4.11145	8.85785	.143885
6.96	48.4416	337.154	2.63818	8.34266	1.90928	4.11342	8.86210	.143678
6.97	48.5809	338.609	2.64008	8.34865	1.91019	4.11539	8.86634	.143472
6.98	48.7204	340.068	2.64197	8.35464	1.91111	4.11736	8.87058	.143267
6.99	48.8601	341.532	2.64386	8.36062	1.91202	4.11932	8.87481	.143062
7.00	49.0000	343.000	2.64575	8.36660	1.91293	4.12129	8.87904	.142857

n	n^2	n^3	$\sqrt{n}$	$\sqrt{10\,n}$	$\sqrt[3]{n}$	$\sqrt[3]{10\,n}$	$\sqrt[3]{100\,n}$	$\dfrac{1}{n}$
7.01	49.1401	344.472	2.64764	8.37257	1.91384	4.12325	8.88327	.142653
7.02	49.2804	345.948	2.64953	8.37854	1.91475	4.12521	8.88749	.142450
7.03	49.4209	347.429	2.65141	8.38451	1.91566	4.12716	8.89171	.142248
7.04	49.5616	348.914	2.65330	8.39047	1.91657	4.12912	8.89592	.142046
7.05	49.7025	350.403	2.65518	8.39643	1.91747	4.13107	8.90013	.141844
7.06	49.8436	351.896	2.65707	8.40238	1.91838	4.13303	8.90434	.141643
7.07	49.9849	353.393	2.65895	8.40833	1.91929	4.13498	8.90854	.141443
7.08	50.1264	354.895	2.66083	8.41427	1.92019	4.13695	8.91274	.141243
7.09	50.2681	356.401	2.66271	8.42021	1.92109	4.13887	8.91693	.141044
7.10	50.4100	357.911	2.66458	8.42615	1.92200	4.14082	8.92112	.140845
7.11	50.5521	359.425	2.66646	8.43208	1.92290	4.14276	8.92531	.140647
7.12	50.6944	360.944	2.66833	8.43801	1.92380	4.14470	8.92949	.140449
7.13	50.8369	362.467	2.67021	8.44393	1.92470	4.14664	8.93367	.140253
7.14	50.9796	363.994	2.67208	8.44965	1.92560	4.14858	8.93784	.140056
7.15	51.1225	365.526	2.67395	8.45577	1.92650	4.15051	8.94201	.139860
7.16	51.2656	367.062	2.67582	8.46168	1.92740	4.15245	8.94618	.139665
7.17	51.4089	368.602	2.67769	8.46759	1.92829	4.15438	8.95034	.139470
7.18	51.5524	370.146	2.67955	8.47349	1.92919	4.15631	8.95450	.139276
7.19	51.6961	371.695	2.68142	8.47939	1.93008	4.15824	8.95866	.139082
7.20	51.8400	373.248	2.68328	8.48528	1.93098	4.16017	8.96281	.138889
7.21	51.9841	374.805	2.68514	8.49117	1.93187	4.16209	8.96696	.138696
7.22	52.1284	376.367	2.68701	8.49706	1.93277	4.16402	8.97110	.138504
7.23	52.2729	377.933	2.68887	8.50294	1.93366	4.16594	8.97524	.138313
7.24	52.4176	379.503	2.69072	8.50882	1.93455	4.16786	8.97938	.138122
7.25	52.5625	381.078	2.69258	8.51469	1.93544	4.16978	8.98351	.137931
7.26	52.7076	382.657	2.69444	8.52056	1.93633	4.17169	8.98764	.137741
7.27	52.8529	384.241	2.69629	8.52643	1.93722	4.17361	8.99176	.137552
7.28	52.9984	385.828	2.69815	8.53229	1.93810	4.17552	8.99588	.137363
7.29	53.1441	387.420	2.70000	8.53815	1.93899	4.17743	9.00000	.137174
7.30	53.2900	389.017	2.70185	8.54400	1.93988	4.17934	9.00411	.136986
7.31	53.4361	390.618	2.70370	8.54985	1.94076	4.18125	9.00822	.136799
7.32	53.5824	392.223	2.70555	8.55570	1.94165	4.18315	9.01233	.136612
7.33	53.7289	393.833	2.70740	8.56154	1.94253	4.18506	9.01643	.136426
7.34	53.8756	395.447	2.70924	8.56738	1.94341	4.18696	9.02053	.136240
7.35	54.0225	397.065	2.71109	8.57321	1.94430	4.18886	9.02462	.136054
7.36	54.1696	398.688	2.71293	8.57904	1.94518	4.19076	9.02871	.135870
7.37	54.3169	400.316	2.71477	8.58487	1.94606	4.19266	9.03280	.135685
7.38	54.4644	401.947	2.71662	8.59069	1.94694	4.19455	9.03689	.135501
7.39	54.6121	403.583	2.71846	8.59651	1.94782	4.19644	9.04097	.135318
7.40	54.7600	405.224	2.72029	8.60233	1.94870	4.19834	9.04504	.135135
7.41	54.9081	406.869	2.72213	8.60814	1.94957	4.20023	9.04911	.134953
7.42	55.0564	408.518	2.72397	8.61394	1.95045	4.20212	9.05318	.134771
7.43	55.2049	410.172	2.72580	8.61974	1.95132	4.20400	9.05725	.134590
7.44	55.3536	411.831	2.72764	8.62554	1.95220	4.20589	9.06131	.134409
7.45	55.5025	413.494	2.72947	8.63134	1.95307	4.20777	9.06537	.134228
7.46	55.6516	415.161	2.73130	8.63713	1.95395	4.20965	9.06942	.134048
7.47	55.8009	416.833	2.73313	8.64292	1.95482	4.21153	9.07347	.133869
7.48	55.9504	418.509	2.73496	8.64870	1.95569	4.21341	9.07752	.133690
7.49	56.1001	420.190	2.73679	8.65448	1.95656	4.21529	9.08156	.133511
7.50	56.2500	421.875	2.73861	8.66025	1.95743	4.21716	9.08560	.133333

n	n^2	n^3	$\sqrt{n}$	$\sqrt{10\,n}$	$\sqrt[3]{n}$	$\sqrt[3]{10\,n}$	$\sqrt[3]{100\,n}$	$\dfrac{1}{n}$
7.51	56.4001	423.565	2.74044	8.66608	1.95830	4.21904	9.08964	.133156
7.52	56.5504	425.259	2.74226	8.67179	1.95917	4.22091	9.09867	.132979
7.53	56.7009	426.958	2.74408	8.67756	1.96004	4.22278	9.09770	.132802
7.54	56.8516	428.661	2.74591	8.68332	1.96091	4.22465	9.10173	.132626
7.55	57.0025	430.369	2.74773	8.68907	1.96177	4.22651	9.10575	.132450
7.56	57.1536	432.081	2.74955	8.69488	1.96264	4.22838	9.10977	.132275
7.57	57.3049	433.798	2.75136	8.70057	1.96350	4.23024	9.11378	.132100
7.58	57.4564	435.520	2.75318	8.70632	1.96437	4.23210	9.11779	.131926
7.59	57.6081	437.245	2.75500	8.71206	1.96523	4.23396	9.12180	.131752
7.60	57.7600	438.976	2.75681	8.71780	1.96610	4.23582	9.12581	.131579
7.61	57.9121	440.711	2.75862	8.72353	1.96696	4.23768	9.12981	.131406
7.62	58.0644	442.451	2.76043	8.72926	1.96782	4.23954	9.13380	.131234
7.63	58.2169	444.195	2.76225	8.73499	1.96868	4.24139	9.13780	.131062
7.64	58.3696	445.994	2.76405	8.74071	1.96954	4.24324	9.14179	.130890
7.65	58.5225	447.697	2.76586	8.74643	1.97040	4.24509	9.14577	.130719
7.66	58.6756	449.455	2.76767	8.75214	1.97126	4.24694	9.14976	.130548
7.67	58.8289	451.218	2.76948	8.75785	1.97211	4.24879	9.15374	.130378
7.68	58.9824	452.985	2.77128	8.76356	1.97297	4.25063	9.15771	.130208
7.69	59.1361	454.757	2.77308	8.76926	1.97383	4.25248	9.16169	.130039
7.70	59.2900	456.533	2.77489	8.77496	1.97468	4.25432	9.16566	.129870
7.71	59.4441	458.314	2.77669	8.78066	1.97554	4.25616	9.16963	.129702
7.72	59.5984	460.100	2.77849	8.78635	1.97639	4.25800	9.17359	.129534
7.73	59.7529	461.890	2.78029	8.79204	1.97724	4.25984	9.17754	.129366
7.74	59.9076	463.685	2.78209	8.79773	1.97809	4.26168	9.18150	.129199
7.75	60.0625	465.484	2.78388	8.80341	1.97895	4.26351	9.18545	.129032
7.76	60.2176	467.289	2.78568	8.80909	1.97980	4.26534	9.18940	.128866
7.77	60.3729	469.097	2.78747	8.81476	1.98065	4.26717	9.19335	.128700
7.78	60.5284	470.911	2.78927	8.82043	1.98150	4.26900	9.19729	.128535
7.79	60.6841	472.729	2.79106	8.82610	1.98234	4.27083	9.20123	.128370
7.80	60.8400	474.552	2.79285	8.83176	1.98319	4.27266	9.20516	.128205
7.81	60.9961	476.380	2.79464	8.83742	1.98404	4.27448	9.20910	.128041
7.82	61.1524	478.212	2.79643	8.84308	1.98489	4.27631	9.21303	.127877
7.83	61.3089	480.049	2.79821	8.84873	1.98573	4.27813	9.21695	.127714
7.84	61.4656	481.890	2.80000	8.85438	1.98658	4.27995	9.22087	.127551
7.85	61.6225	483.737	2.80179	8.86002	1.98742	4.28177	9.22479	.127389
7.86	61.7796	485.588	2.80357	8.86566	1.98826	4.28359	9.22871	.127227
7.87	61.9369	487.443	2.80535	8.87130	1.98911	4.28540	9.23262	.127065
7.88	62.0944	489.304	2.80713	8.87694	1.98995	4.28722	9.23653	.126904
7.89	62.2521	491.169	2.80891	8.88257	1.99079	4.28903	9.24043	.126743
7.90	62.4100	493.039	2.81069	8.88819	1.99163	4.29084	9.24433	.126582
7.91	62.5681	494.914	2.81247	8.89382	1.99247	4.29265	9.24823	.126422
7.92	62.7264	496.793	2.81425	8.89944	1.99331	4.29446	9.25213	.126263
7.93	62.8849	498.677	2.81603	8.90505	1.99415	4.29627	9.25602	.126103
7.94	63.0436	500.566	2.81780	8.91067	1.99499	4.29807	9.25991	.125945
7.95	63.2025	502.460	2.81957	8.91628	1.99582	4.29987	9.26380	.125786
7.96	63.3616	504.358	2.82135	8.92188	1.99666	4.30168	9.26768	.125628
7.97	63.5209	506.262	2.82312	8.92749	1.99750	4.30348	9.27156	.125471
7.98	63.6804	508.170	2.82489	8.93308	1.99833	4.30528	9.27544	.125313
7.99	63.8401	510.082	2.82666	8.93868	1.99917	4.30707	9.27931	.125156
8.00	64.0000	512.000	2.82843	8.94427	2.00000	4.30887	9.28318	.125000

n	n^2	n^3	$\sqrt{n}$	$\sqrt{10\,n}$	$\sqrt[3]{n}$	$\sqrt[3]{10\,n}$	$\sqrt[3]{100\,n}$	$\dfrac{1}{n}$
8.01	64.1601	513.922	2.83019	8.94986	2.00083	4.31066	9.28704	.124844
8.02	64.3204	515.850	2.83196	8.95545	2.00167	4.31246	9.29091	.124688
8.03	64.4809	517.782	2.83373	8.96103	2.00250	4.31425	9.29477	.124533
8.04	64.6416	519.718	2.83549	8.96660	2.00333	4.31604	9.29862	.124378
8.05	64.8025	521.660	2.83725	8.97218	2.00416	4.31783	9.30248	.124224
8.06	64.9636	523.607	2.83901	8.97775	2.00499	4.31961	9.30633	.124070
8.07	65.1249	525.558	2.84077	8.98332	2.00582	4.32140	9.31018	.123916
8.08	65.2864	527.514	2.84253	8.98888	2.00664	4.32318	9.31402	.123762
8.09	65.4481	529.475	2.84429	8.99444	2.00747	4.32497	9.31786	.123609
8.10	65.6100	531.441	2.84605	9.00000	2.00830	4.32675	9.32170	.123457
8.11	65.7721	533.412	2.84781	9.00555	2.00912	4.32853	9.32553	.123305
8.12	65.9344	535.387	2.84956	9.01110	2.00995	4.33031	9.32936	.123153
8.13	66.0969	537.368	2.85132	9.01665	2.01078	4.33208	9.33319	.123001
8.14	66.2596	539.353	2.85307	9.02219	2.01160	4.33386	9.33702	.122850
8.15	66.4225	541.343	2.85482	9.02774	2.01242	4.33563	9.34084	.122699
8.16	66.5856	543.338	2.85657	9.03327	2.01325	4.33741	9.34466	.122549
8.17	66.7489	545.339	2.85832	9.03881	2.01407	4.33918	9.34847	.122399
8.18	66.9124	547.343	2.86007	9.04434	2.01489	4.34095	9.35229	.122249
8.19	67.0761	549.353	2.86182	9.04986	2.01571	4.34272	9.35610	.122100
8.20	67.2400	551.368	2.86356	9.05539	2.01653	4.34448	9.35990	.121951
8.21	67.4041	553.388	2.86531	9.06091	2.01735	4.34625	9.36370	.121803
8.22	67.5684	555.412	2.86705	9.06642	2.01817	4.34801	9.36751	.121655
8.23	67.7329	557.442	2.86880	9.07193	2.01899	4.34977	9.37130	.121507
8.24	67.8976	559.476	2.87054	9.07744	2.01980	4.35153	9.37510	.121359
8.25	68.0625	561.516	2.87228	9.08295	2.02062	4.35329	9.37889	.121212
8.26	68.2276	563.560	2.87402	9.08845	2.02144	4.35505	9.38268	.121065
8.27	68.3929	565.609	2.87576	9.09395	2.02225	4.35681	9.38646	.120919
8.28	68.5584	567.664	2.87750	9.09945	2.02307	4.35856	9.39024	.120773
8.29	68.7241	569.723	2.87924	9.10494	2.02388	4.36032	9.39402	.120627
8.30	68.8900	571.787	2.88097	9.11043	2.02469	4.36207	9.39780	.120482
8.31	69.0561	573.856	2.88271	9.11592	2.02551	4.36382	9.40157	.120337
8.32	69.2224	575.930	2.88444	9.12140	2.02632	4.36557	9.40534	.120192
8.33	69.3889	578.010	2.88617	9.12688	2.02713	4.36732	9.40911	.120048
8.34	69.5556	580.094	2.88791	9.13236	2.02794	4.36907	9.41287	.119904
8.35	69.7225	582.183	2.88964	9.13783	2.02875	4.37081	9.41663	.119761
8.36	69.8896	584.277	2.89137	9.14330	2.02956	4.37255	9.42039	.119617
8.37	70.0569	586.376	2.89310	9.14877	2.03037	4.37430	9.42414	.119474
8.38	70.2244	588.480	2.89482	9.15423	2.03118	4.37604	9.42789	.119332
8.39	70.3921	590.590	2.89655	9.15969	2.03199	4.37778	9.43164	.119190
8.40	70.5600	592.704	2.89828	9.16515	2.03279	4.37952	9.43589	.119048
8.41	70.7281	594.823	2.90000	9.17061	2.03360	4.38126	9.43913	.118906
8.42	70.8964	596.948	2.90172	9.17606	2.03440	4.38299	9.44287	.118765
8.43	71.0649	599.077	2.90345	9.18150	2.03521	4.38473	9.44661	.118624
8.44	71.2336	601.212	2.90517	9.18695	2.03601	4.38646	9.45034	.118483
8.45	71.4025	603.351	2.90689	9.19239	2.03682	4.38819	9.45407	.118343
8.46	71.5716	605.496	2.90861	9.19783	2.03762	4.38992	9.45780	.118203
8.47	71.7409	607.645	2.91033	9.20326	2.03842	4.39165	9.46152	.118064
8.48	71.9104	609.800	2.91204	9.20869	2.03923	4.39338	9.46525	.117925
8.49	72.0801	611.960	2.91376	9.21412	2.04003	4.39511	9.46897	.117786
8.50	72.2500	614.125	2.91548	9.21954	2.04083	4.39683	9.47268	.117647

n	n^2	n^3	$\sqrt{n}$	$\sqrt{10\,n}$	$\sqrt[3]{n}$	$\sqrt[3]{10\,n}$	$\sqrt[3]{100\,n}$	$\dfrac{1}{n}$
8.51	72.4201	616.295	2.91719	9.22497	2.04168	4.39855	9.47640	.117509
8.52	72.5904	618.470	2.91890	9.23088	2.04243	4.40028	9.48011	.117371
8.53	72.7609	620.650	2.92062	9.23580	2.04328	4.40200	9.48381	.117233
8.54	72.9316	622.836	2.92233	9.24121	2.04402	4.40372	9.48752	.117096
8.55	73.1025	625.026	2.92404	9.24662	2.04482	4.40543	9.49122	.116959
8.56	73.2736	627.222	2.92575	9.25203	2.04562	4.40715	9.49492	.116822
8.57	73.4449	629.423	2.92746	9.25743	2.04641	4.40887	9.49861	.116686
8.58	73.6164	631.629	2.92916	9.26283	2.04721	4.41058	9.50231	.116550
8.59	73.7881	633.840	2.93087	9.26823	2.04801	4.41229	9.50600	.116414
8.60	73.9600	636.056	2.93258	9.27362	2.04880	4.41400	9.50969	.116279
8.61	74.1321	638.277	2.93428	9.27901	2.04959	4.41571	9.51337	.116144
8.62	74.3044	640.504	2.93598	9.28440	2.05039	4.41742	9.51705	.116009
8.63	74.4769	642.736	2.93769	9.28978	2.05118	4.41913	9.52073	.115875
8.64	74.6496	644.973	2.93939	9.29516	2.05197	4.42084	9.52441	.115741
8.65	74.8225	647.215	2.94109	9.30054	2.05276	4.42254	9.52808	.115607
8.66	74.9956	649.462	2.94279	9.30591	2.05355	4.42425	9.53175	.115473
8.67	75.1689	651.714	2.94449	9.31128	2.05434	4.42595	9.53542	.115340
8.68	75.3424	653.972	2.94618	9.31665	2.05513	4.42765	9.53908	.115207
8.69	75.5161	656.235	2.94788	9.32202	2.05592	4.42935	9.54274	.115075
8.70	75.6900	658.503	2.94958	9.32738	2.05671	4.43105	9.54640	.114943
8.71	75.8641	660.776	2.95127	9.33274	2.05750	4.43274	9.55006	.114811
8.72	76.0384	663.055	2.95296	9.33809	2.05828	4.43444	9.55371	.114679
8.73	76.2129	665.339	2.95466	9.34345	2.05907	4.43614	9.55736	.114548
8.74	76.3876	667.628	2.95635	9.34880	2.05986	4.43783	9.56101	.114417
8.75	76.5625	669.922	2.95804	9.35414	2.06064	4.43952	9.56466	.114286
8.76	76.7376	672.221	2.95973	9.35949	2.06143	4.44121	9.56830	.114155
8.77	76.9129	674.526	2.96142	9.36483	2.06221	4.44290	9.57194	.114025
8.78	77.0884	676.836	2.96311	9.37017	2.06299	4.44459	9.57557	.113895
8.79	77.2641	679.151	2.96479	9.37550	2.06378	4.44627	9.57921	.113766
8.80	77.4400	681.472	2.96648	9.38083	2.06456	4.44796	9.58284	.113636
8.81	77.6161	683.798	2.96816	9.38616	2.06534	4.44964	9.58647	.113507
8.82	77.7924	686.129	2.96985	9.39149	2.06612	4.45133	9.59009	.113379
8.83	77.9689	688.465	2.97153	9.39681	2.06690	4.45301	9.59372	.113250
8.84	78.1456	690.807	2.97321	9.40213	2.06768	4.45469	9.59734	.113122
8.85	78.3225	693.154	2.97489	9.40744	2.06846	4.45637	9.60095	.112994
8.86	78.4996	695.506	2.97658	9.41276	2.06924	4.45805	9.60457	.112867
8.87	78.6769	697.864	2.97825	9.41807	2.07002	4.45972	9.60818	.112740
8.88	78.8544	700.227	2.97993	9.42338	2.07080	4.46140	9.61179	.112613
8.89	79.0321	702.595	2.98161	9.42868	2.07157	4.46307	9.61540	.112486
8.90	79.2100	704.969	2.98329	9.43398	2.07235	4.46474	9.61900	.112360
8.91	79.3881	707.348	2.98496	9.43928	2.07313	4.46642	9.62260	.112233
8.92	79.5664	709.732	2.98664	9.44458	2.07390	4.46809	9.62620	.112108
8.93	79.7449	712.122	2.98831	9.44987	2.07468	4.46976	9.62980	.111982
8.94	79.9236	714.517	2.98998	9.45516	2.07545	4.47142	9.63339	.111857
8.95	80.1025	716.917	2.99166	9.46044	2.07622	4.47309	9.63698	.111732
8.96	80.2816	719.323	2.99333	9.46573	2.07700	4.47476	9.64057	.111607
8.97	80.4609	721.734	2.99500	9.47101	2.07777	4.47642	9.64415	.111483
8.98	80.6404	724.151	2.99666	9.47629	2.07854	4.47808	9.64774	.111359
8.99	80.8201	726.573	2.99833	9.48156	2.07931	4.47974	9.65132	.111235
9.00	81.0000	729.000	3.00000	9.48683	2.08008	4.48140	9.65489	.111111

n	n^2	n^3	$\sqrt{n}$	$\sqrt{10\,n}$	$\sqrt[3]{n}$	$\sqrt[3]{10\,n}$	$\sqrt[3]{100\,n}$	$\dfrac{1}{n}$
9.01	81.1801	731.433	3.00167	9.49210	2.08085	4.48306	9.65847	.110988
9.02	81.3604	733.871	8.00333	9.49737	2.08162	4.48472	9.66204	.110865
9.03	81.5409	736.314	3.00500	9.50263	2.08239	4.48638	9.66561	.110742
9.04	81.7216	738.763	3.00666	9.50789	2.08316	4.48803	9.66918	.110620
9.05	81.9025	741.218	3.00832	9.51315	2.08393	4.48968	9.67274	.110497
9.06	82.0836	743.677	3.00998	9.51840	2.08470	4.49134	9.67630	.110375
9.07	82.2649	746.143	3.01164	9.52365	2.08546	4.49299	9.67986	.110254
9.08	82.4464	748.613	3.01330	9.52890	2.08623	4.49464	9.68342	.110132
9.09	82.6281	751.089	3.01496	9.53415	2.08699	4.49629	9.68697	.110011
9.10	82.8100	753.571	3.01662	9.53939	2.08776	4.49794	9.69052	.109890
9.11	82.9921	756.058	3.01828	9.54463	2.08852	4.49959	9.69407	.109770
9.12	83.1744	758.551	3.01993	9.54987	2.08929	4.50123	9.69762	.109649
9.13	83.3569	761.048	3.02159	9.55510	2.09005	4.50288	9.70116	.109529
9.14	83.5396	763.552	3.02324	9.56033	2.09081	4.50452	9.70470	.109409
9.15	83.7225	766.061	3.02490	9.56556	2.09158	4.50616	9.70824	.109290
9.16	83.9056	768.575	3.02655	9.57079	2.09234	4.50780	9.71177	.109170
9.17	84.0889	771.095	3.02820	9.57601	2.09310	4.50945	9.71531	.109051
9.18	84.2724	773.621	3.02985	9.58123	2.09386	4.51108	9.71884	.108933
9.19	84.4561	776.152	3.03150	9.58645	2.09462	4.51272	9.72236	.108814
9.20	84.6400	778.688	3.03315	9.59166	2.09538	4.51436	9.72589	.108696
9.21	84.8241	781.230	3.03480	9.59687	2.09614	4.51599	9.72941	.108578
9.22	85.0084	783.777	3.03645	9.60208	2.09690	4.51763	9.73293	.108460
9.23	85.1929	786.330	3.03809	9.60729	2.09765	4.51926	9.73645	.108342
9.24	85.3776	788.889	3.03974	9.61249	2.09841	4.52089	9.73996	.108225
9.25	85.5625	791.453	3.04138	9.61769	2.09917	4.52252	9.74348	.108108
9.26	85.7476	794.023	3.04302	9.62289	2.09992	4.52415	9.74699	.107991
9.27	85.9329	796.598	3.04467	9.62808	2.10068	4.52578	9.75049	.107875
9.28	86.1184	799.179	3.04631	9.63328	2.10144	4.52740	9.75400	.107759
9.29	86.3041	801.765	3.04795	9.63846	2.10219	4.52903	9.75750	.107643
9.30	86.4900	804.357	3.04959	9.64365	2.10294	4.53065	9.76100	.107527
9.31	86.6761	806.954	3.05123	9.64883	2.10370	4.53228	9.76450	.107411
9.32	86.8624	809.558	3.05287	9.65401	2.10445	4.53390	9.76799	.107296
9.33	87.0489	812.166	3.05450	9.65919	2.10520	4.53552	9.77148	.107181
9.34	87.2356	814.781	3.05614	9.66437	2.10595	4.53714	9.77497	.107066
9.35	87.4225	817.400	3.05778	9.66954	2.10671	4.53876	9.77846	.106952
9.36	87.6096	820.026	3.05941	9.67471	2.10746	4.54038	9.78195	.106838
9.37	87.7969	822.657	3.06105	9.67988	2.10821	4.54199	9.78543	.106724
9.38	87.9844	825.294	3.06268	9.68504	2.10896	4.54361	9.78891	.106610
9.39	88.1721	827.936	3.06431	9.69020	2.10971	4.54522	9.79239	.106496
9.40	88.3600	830.584	3.06594	9.69536	2.11045	4.54684	9.79586	.106383
9.41	88.5481	833.238	3.06757	9.70052	2.11120	4.54845	9.79933	.106270
9.42	88.7364	835.897	3.06920	9.70567	2.11195	4.55006	9.80280	.106157
9.43	88.9249	838.562	3.07083	9.71082	2.11270	4.55167	9.80627	.106045
9.44	89.1136	841.232	3.07246	9.71597	2.11344	4.55328	9.80974	.105932
9.45	89.3025	843.909	3.07409	9.72111	2.11419	4.55488	9.81320	.105820
9.46	89.4916	846.591	3.07571	9.72625	2.11494	4.55649	9.81666	.105708
9.47	89.6809	849.278	3.07734	9.73139	2.11568	4.55809	9.82012	.105597
9.48	89.8704	851.971	3.07896	9.73653	2.11642	4.55970	9.82357	.105485
9.49	90.0601	854.670	3.08058	9.74166	2.11717	4.56130	9.82703	.105374
9.50	90.2500	857.375	3.08221	9.74679	2.11791	4.56290	9.83048	.105263

n	n^2	n^3	$\sqrt{n}$	$\sqrt{10\,n}$	$\sqrt[3]{n}$	$\sqrt[3]{10\,n}$	$\sqrt[3]{100\,n}$	$\dfrac{1}{n}$
9.51	90.4401	860.085	3.08383	9.75192	2.11865	4.56450	9.83392	.105153
9.52	90.6304	862.801	3.08545	9.75705	2.11940	4.56610	9.83737	.105042
9.53	90.8209	865.523	3.08707	9.76217	2.12014	4.56770	9.84081	.104932
9.54	91.0116	868.251	3.08869	9.76729	2.12088	4.56930	9.84425	.104822
9.55	91.2025	870.984	3.09031	9.77241	2.12162	4.57089	9.84769	.104712
9.56	91.3936	873.723	3.09192	9.77753	2.12236	4.57249	9.85113	.104603
9.57	91.5849	876.467	3.09354	9.78264	2.12310	4.57408	9.85456	.104493
9.58	91.7764	879.218	3.09516	9.78775	2.12384	4.57568	9.85799	.104384
9.59	91.9681	881.974	3.09677	9.79285	2.12458	4.57727	9.86142	.104275
9.60	92.1600	884.736	3.09839	9.79796	2.12532	4.57886	9.86485	.104167
9.61	92.3521	887.504	3.10000	9.80306	2.12605	4.58045	9.86827	.104058
9.62	92.5444	890.277	3.10161	9.80816	2.12679	4.58203	9.87169	.103950
9.63	92.7369	893.056	3.10322	9.81326	2.12753	4.58362	9.87511	.103842
9.64	92.9296	895.841	3.10483	9.81835	2.12826	4.58521	9.87853	.103734
9.65	93.1225	898.632	3.10644	9.82344	2.12900	4.58679	9.88195	.103627
9.66	93.3156	901.429	3.10805	9.82853	2.12974	4.58838	9.88536	.103520
9.67	93.5089	904.231	3.10966	9.83362	2.13047	4.58996	9.88877	.103413
9.68	93.7024	907.039	3.11127	9.83870	2.13120	4.59154	9.89217	.103306
9.69	93.8961	909.853	3.11288	9.84378	2.13194	4.59312	9.89558	.103199
9.70	94.0900	912.673	3.11448	9.84886	2.13267	4.59470	9.89898	.103093
9.71	94.2841	915.499	3.11609	9.85393	2.13340	4.59628	9.90238	.102987
9.72	94.4784	918.330	3.11769	9.85901	2.13414	4.59786	9.90578	.102881
9.73	94.6729	921.167	3.11929	9.86408	2.13487	4.59943	9.90918	.102775
9.74	94.8676	924.010	3.12090	9.86914	2.13560	4.60101	9.91257	.102669
9.75	95.0625	926.859	3.12250	9.87421	2.13633	4.60258	9.91596	.102564
9.76	95.2576	929.714	3.12410	9.87927	2.13706	4.60416	9.91935	.102459
9.77	95.4529	932.575	3.12570	9.88433	2.13779	4.60573	9.92274	.102354
9.78	95.6484	935.441	3.12730	9.88939	2.13852	4.60730	9.92612	.102250
9.79	95.8441	938.314	3.12890	9.89444	2.13925	4.60887	9.92950	.102145
9.80	96.0400	941.192	3.13050	9.89949	2.13997	4.61044	9.93288	.102041
9.81	96.2361	944.076	3.13209	9.90454	2.14070	4.61200	9.93626	.101937
9.82	96.4324	946.966	3.13369	9.90959	2.14143	4.61357	9.93964	.101833
9.83	96.6289	949.862	3.13528	9.91464	2.14216	4.61513	9.94301	.101729
9.84	96.8256	952.764	3.13688	9.91968	2.14288	4.61670	9.94638	.101626
9.85	97.0225	955.672	3.13847	9.92472	2.14361	4.61826	9.94975	.101523
9.86	97.2196	958.585	3.14006	9.92975	2.14433	4.61983	9.95311	.101420
9.87	97.4169	961.505	3.14166	9.93479	2.14506	4.62139	9.95648	.101317
9.88	97.6144	964.430	3.14325	9.93982	2.14578	4.62296	9.95984	.101215
9.89	97.8121	967.362	3.14484	9.94485	2.14651	4.62451	9.96320	.101112
9.90	98.0100	970.299	3.14643	9.94987	2.14723	4.62607	9.96655	.101010
9.91	98.2081	973.242	3.14802	9.95490	2.14795	4.62762	9.96991	.100908
9.92	98.4064	976.191	3.14960	9.95992	2.14867	4.62918	9.97326	.100807
9.93	98.6049	979.147	3.15119	9.96494	2.14940	4.63073	9.97661	.100705
9.94	98.8036	982.108	3.15278	9.96995	2.15012	4.63229	9.97996	.100604
9.95	99.0025	985.075	3.15436	9.97497	2.15084	4.63384	9.98331	.100503
9.96	99.2016	988.048	3.15595	9.97998	2.15156	4.63539	9.98665	.100402
9.97	99.4009	991.027	3.15753	9.98499	2.15228	4.63694	9.98999	.100301
9.98	99.6004	994.012	3.15911	9.98999	2.15300	4.63849	9.99333	.100200
9.99	99.8001	997.003	3.16070	9.99500	2.15372	4.64004	9.99667	.100100
10.00	100.000	1000.00	3.16228	10.0000	2.15443	4.64159	10.0000	.100000

DECIMAL EQUIVALENTS OF 64ths.

The decimal fractions printed in large type give the exact value of the corresponding fraction to the fourth decimal place. A given decimal fraction is rarely exactly equal to any of these values, and the numbers in small type show which common fraction is nearest to the given decimal. Thus, lay off the fraction .1330 in 64ths. The nearest decimal fractions are .1250 and .1406. The value of any fraction in small type is the mean of the two adjacent fractions. In this instance the mean fraction is .1328, and as .1330 is greater than this, .1406 or 9/64 will be chosen. In the same manner the nearest 64ths corresponding to the decimal fractions .8670 and .8979 are found to be 55/64 and 57/64, respectively.

Fraction	Decimal	Fraction	Decimal	Fraction	Decimal	Fraction	Decimal
	.0078		.2578		.5078		.7578
1/64	.0156	17/64	.2656	33/64	.5156	49/64	.7656
	.0235		.2735		.5235		.7735
1/32	.0313	9/32	.2813	17/32	.5313	25/32	.7813
	.0391		.2891		.5391		.7891
3/64	.0469	19/64	.2969	35/64	.5469	51/64	.7969
	.0547		.3047		.5547		.8047
1/16	.0625	5/16	.3125	9/16	.5625	13/16	.8125
	.0703		.3203		.5703		.8203
5/64	.0781	21/64	.3281	37/64	.5781	53/64	.8281
	.0860		.3360		.5860		.8360
3/32	.0938	11/32	.3438	19/32	.5938	27/32	.8438
	.1016		.3516		.6016		.8516
7/64	.1094	23/64	.3594	39/64	.6094	55/64	.8594
	.1172		.3672		.6172		.8672
1/8	.1250	3/8	.3750	5/8	.6250	7/8	.8750
	.1328		.3828		.6328		.8828
9/64	.1406	25/64	.3906	41/64	.6406	57/64	.8906
	.1485		.3985		.6485		.8985
5/32	.1563	13/32	.4063	21/32	.6563	29/32	.9063
	.1641		.4141		.6641		.9141
11/64	.1719	27/64	.4219	43/64	.6719	59/64	.9219
	.1797		.4297		.6797		.9297
3/16	.1875	7/16	.4375	11/16	.6875	15/16	.9375
	.1953		.4453		.6953		.9453
13/64	.2031	29/64	.4531	45/64	.7031	61/64	.9531
	.2110		.4610		.7110		.9610
7/32	.2188	15/32	.4688	23/32	.7188	31/32	.9688
	.2266		.4766		.7266		.9766
15/64	.2344	31/64	.4844	47/64	.7344	63/64	.9844
	.2422		.4922		.7422		.9922
1/4	.2500	1/2	.5000	3/4	.7500	1	1.0000
	.2578		.5078		.7578		1.0078

MENSURATION.

In the following formulas, the letters have the meanings here given, unless otherwise stated.

D = larger diameter;

d = smaller diameter;

R = radius corresponding to D;

r = radius corresponding to d;

p = perimeter or circumference;

C = area of convex surface = area of flat surface which can be rolled into the shape shown;

S = area of entire surface = C + area of the end or ends;

A = area of plane figure;

π = 3.1416, nearly = ratio of any circumference to its diameter;

V = volume of solid.

The other letters used will be found on the cuts.

CIRCLE.

$$p = \pi d = 3.1416\, d.$$

$$p = 2\pi r = 6.2832\, r.$$

$$p = 2\sqrt{\pi A} = 3.5449 \sqrt{A}.$$

$$p = \frac{2A}{r} = \frac{4A}{d}.$$

$$d = \frac{p}{\pi} = \frac{p}{3.1416} = .3183\, p.$$

$$d = 2\sqrt{\frac{A}{\pi}} = 1.1284 \sqrt{A}.$$

$$r = \frac{p}{2\pi} = \frac{p}{6.2832} = .1592\, p.$$

$$r = \sqrt{\frac{A}{\pi}} = .5642 \sqrt{A}.$$

$$A = \frac{\pi d^2}{4} = .7854\, d^2.$$

$$A = \pi r^2 = 3.1416\, r^2.$$

$$A = \frac{pr}{2} = \frac{pd}{4}.$$

TRIANGLES.

$D = B + C.$ $E + B + C = 180°.$

$B = D - C.$ $E' + B + C = 180°.$

$E' = E.$ $B' = B.$

The above letters refer to angles.

For a right-angled triangle, c being the hypotenuse,

$$c = \sqrt{a^2 + b^2}.$$

$$a = \sqrt{c^2 - b^2}.$$

$$b = \sqrt{c^2 - a^2}.$$

$c = $ length of side opposite an acute angle of an oblique-angled triangle.

$$c = \sqrt{a^2 + b^2 - 2be}.$$

$$h = \sqrt{a^2 - e^2}.$$

$c = $ length of side opposite an obtuse angle of an oblique-angled triangle.

$$c = \sqrt{a^2 + b^2 + 2be}.$$

$$h = \sqrt{a^2 - e^2}.$$

For a triangle inscribed in a semicircle; i. e., any right-angled triangle,

$$c : b :: a : h.$$

$$h = \frac{ab}{c} = \frac{ce}{a}.$$

$$a : b + e = e : a = h : c.$$

For any triangle,

$$A = \frac{bh}{2} = \tfrac{1}{2}bh.$$

$$A = \frac{b}{2} \sqrt{a^2 - \left(\frac{a^2 + b^2 - c^2}{2b}\right)^2}.$$

RECTANGLE AND PARALLELOGRAM.

$$A = ab.$$

TRAPEZOID.

$$A = \tfrac{1}{2}h(a+b).$$

TRAPEZIUM.

Divide into two triangles and a trapezoid.

$$A = \tfrac{1}{2}b\,h' + \tfrac{1}{2}a(h'+h) + \tfrac{1}{2}c\,h;$$
$$\text{or,} \quad A = \tfrac{1}{2}[b\,h' + c\,h + a(h'+h)].$$

Or, divide into two triangles by drawing a diagonal. Consider the diagonal as the base of both triangles, call its length l; call the altitudes of the triangles h_1 and h_2; then

$$A = \tfrac{1}{2}l\,(h_1 + h_2).$$

ELLIPSE.

$$p* = \pi\sqrt{\frac{D^2 + d^2}{2} - \frac{(D-d)^2}{8.8}}.$$

$$A = \frac{\pi}{4}D\,d = .7854\,D\,d.$$

SECTOR.

$$A = \tfrac{1}{2}lr.$$

$$A = \frac{\pi r^2 E}{360} = .008727\,r^2\,E.$$

$$l = \text{length of arc.}$$

SEGMENT.

$$A = \tfrac{1}{2}\,[lr - c\,(r-h)].$$

$$A = \frac{\pi r^2 E}{360} - \frac{c}{2}(r-h).$$

$$l = \frac{\pi r E}{180} = .0175\,r\,E.$$

$$E = \frac{180\,l}{\pi r} = 57.2956\,\frac{l}{r}.$$

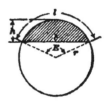

* The perimeter of an ellipse cannot be exactly determined without a very elaborate calculation, and this formula is merely an approximation giving fairly close results.

RING.

$$A = \frac{\pi}{4}(D^2 - d^2).$$

CHORD.

c = length of chord.

$$r = \frac{c^2 + 4h^2}{8h} = \frac{c^2}{2h}.$$

$$c = 2\sqrt{2hr - h^2}.$$

$$l = \frac{8e - c}{3}, \text{ approximately.}$$

HELIX.

To construct a helix.

l = length of helix;
n = number of turns;
t = pitch.

$$t = \sqrt{\frac{l^2}{n^2} - \pi^2 d^2}.$$

$$l = n\sqrt{\pi^2 d^2 + t^2}.$$

$$n = \frac{l}{\sqrt{\pi^2 a^2 + t^2}}.$$

CYLINDER.

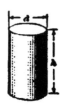

$$C = \pi d h.$$

$$S = 2\pi r h + 2\pi r^2$$
$$= \pi d h + \frac{\pi}{2} d^2.$$

$$V = \pi r^2 h = \frac{\pi}{4} d^2 h.$$

$$V = \frac{p^2 h}{4\pi} = .0796 p^2 h.$$

FRUSTUM OF CYLINDER.

h = ½ sum of greatest and least heights.

$$C = p h = \pi d h.$$

$$S = \pi d h + \frac{\pi}{4} d^2 + \text{area of elliptical top.}$$

$$V = A h = \frac{\pi}{4} d^2 h.$$

CONE.

$$C = \tfrac{1}{2}\pi\, dl = \pi\, rl.$$

$$S = \pi\, rl + \pi\, r^2 = \pi\, r\sqrt{r^2 + h^2} + \pi\, r^2.$$

$$V = \frac{\pi\, d^2}{4} \times \frac{h}{3} = \frac{.7854\, d^2 h}{3} = \frac{p^2 h}{12\pi}.$$

FRUSTUM OF CONE.

$$C = \tfrac{1}{2}l(P + p) = \frac{\pi}{2} l\,(D + d).$$

$$S = \frac{\pi}{2}\left[l\,(D + d) + \tfrac{1}{2}\,(D^2 + d^2) \right].$$

$$V = \frac{\pi}{4}\,(D^2 + D d + d^2) \times \tfrac{1}{3} h$$

$$= .2618\, h\,(D^2 + D d + d^2).$$

SPHERE.

$$S = \pi\, d^2 = 4\,\pi\, r^2 = 12.5664\, r^2.$$

$$V = \tfrac{1}{6}\pi\, d^3 = \tfrac{4}{3}\pi\, r^3 = .5236\, d^3 = 4.1888\, r^3.$$

CIRCULAR RING.

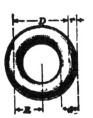

$D =$ mean diameter;

$R =$ mean radius.

$$S = 4\pi^2 R\, r = 9.8696\, D\, d.$$

$$V = 2\pi^2 R\, r^2 = 2.4674\, D\, d^2.$$

WEDGE.

$$V = \tfrac{1}{6} w\, h(a + b + c).$$

PRISMOID.

A prismoid is a solid having two parallel plane ends, the edges of which are connected by plane triangular or quadrilateral surfaces.

A = area one end;
a = area of other end;
m = area of section midway between ends;
l = perpendicular distance between ends.

$$V = \tfrac{1}{6} l(A + a + 4m).$$

The area m is not in general a mean between the areas of the two ends, but its sides are means between the corresponding lengths of the ends.

Approximately, $V = \dfrac{A + a}{2} l.$

REGULAR PYRAMID.

P = perimeter of base;
A = area of base.

$$C = \tfrac{1}{2} Pl.$$
$$S = \tfrac{1}{2} Pl + A.$$
$$V = \frac{Ah}{3}.$$

To obtain area of base, divide it into triangles, and find their sum.

The formula for V applies to any pyramid whose base is A and altitude h.

FRUSTUM OF REGULAR PYRAMID.

a = area of upper base;
A = area of lower base;
p = perimeter of upper base;
P = perimeter of lower base.

$$C = \tfrac{1}{2} l(P + p).$$
$$S = \tfrac{1}{2} l(P + p) + A + a.$$
$$V = \tfrac{1}{3} h \left(A + a + \sqrt{Aa} \right).$$

The formula for V applies to the frustum of any pyramid.

LENGTH OF SPIRAL.

$l = \pi n \left(\dfrac{D + d}{2} \right).$ n = number of coil;
l = length of spiral;
$l = \dfrac{\pi}{t} (R^2 - r^2).$ t = pitch.

PRISM OR PARALLELOPIPED.

$$C = Ph.$$
$$S = Ph + 2A.$$
$$V = Ah.$$

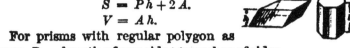

For prisms with regular polygon as bases, $P =$ length of one side $\times$ number of sides.

To obtain area of base, if it is a polygon, divide it into triangles, and find sum of partial areas.

FRUSTUM OF PRISM.

If a section perpendicular to the edges is a triangle, square, parallelogram, or *regular* polygon,

$$V = \frac{\text{sum of lengths of edges}}{\text{number of edges}} \times \text{ area of right section.}$$

REGULAR POLYGONS.

Divide the polygon into equal triangles and find the sum of the partial areas. Otherwise, square the length of one side and multiply by proper number from the following table:

Name.	No. Sides.	Multiplier.
Triangle	3	.433
Square	4	1.000
Pentagon	5	1.720
Hexagon	6	2.598
Heptagon	7	3.634
Octagon	8	4.828
Nonagon	9	6.182
Decagon	10	7.694

IRREGULAR AREAS.

Divide the area into trapezoids, triangles, parts of circles, etc., and find the sum of the partial areas.

If the figure is very irregular, the approximate area may be found as follows: Divide the figure into trapezoids by equidistant parallel lines $b, c, d,$ etc. The lengths of these lines being measured, then, calling a the first and n the last length, and y the width of strips,

$$\text{Area} = y\left(\frac{a+n}{2} + b + c + \text{etc.} + m\right).$$

MECHANICS.

FALLING BODIES.

Let $g = 32.16 =$ constant acceleration due to the attraction of the earth;

$t =$ number of seconds that the body falls;

$v =$ velocity in feet per second at the end of the time t;

$h =$ distance that the body falls during the time t.

Then, $v = g t = \dfrac{2h}{t} = \sqrt{2gh} = 8.02 \sqrt{h}$.

$$h = \frac{vt}{2} = \frac{g t^2}{2} = \frac{v^2}{2g} = .015547\, v^2.$$

$$t = \frac{v}{g} = \frac{2h}{v} = \sqrt{\frac{2h}{g}} = .24938 \sqrt{h}.$$

PROJECTILES.

The formulas under this and the preceding heading are rigidly true only for bodies moving in a vacuum or in space (as the stars and planets); they are approximately true for bodies moving in air, provided they are dense and the velocity is not very great. Fairly good results may be obtained by applying the formulas for projectiles in calculating the range of a jet of water issuing from a small orifice in the side of a vessel.

Let $g = 32.16 =$ acceleration due to gravity;

$v =$ initial velocity in feet per second;

$r =$ range;

$y =$ vertical height of starting point above ground;

$A =$ elevation in degrees $=$ angle that the direction of the projectile at the start makes with the horizontal.

Then the range, or distance from the starting point to the point where the projectile crosses a horizontal line through the starting point, is

$$r = \frac{v^2}{g} \sin 2A.$$

If the body is projected in a horizontal direction, the range is the distance from the starting point to the point where the projectile strikes the ground, and

$$r = v\sqrt{\frac{2y}{g}} = .24938 \, v\sqrt{y}.$$

The range of a projectile fired in a horizontal direction, 80 ft. above the ground, with a velocity of 300 ft. per second, equals $r = .24938 \times 300 \times \sqrt{30} = 409.77$ ft.

CENTRIFUGAL FORCE.

$F =$ centrifugal force in pounds;

$W =$ weight of revolving body in pounds;

$r =$ distance from the axis of motion to the center of gravity of the body in feet;

$N =$ number of revolutions per minute;

$v =$ velocity in feet per second.

$$F = \frac{W v^2}{g \, r} = .00034 \, W r \, N^2.$$

In calculating the centrifugal force of flywheels, it is customary to neglect the arms and take r equal to the mean radius of the rim; in such cases W is taken as one-half the weight of the rim. The result thus obtained, divided by π, is approximately the force tending to burst the flywheel rim.

EXAMPLE.—What is the force tending to burst a flywheel rim weighing 7 tons, making 150 rev. per min., and having a mean radius of 5 ft.?

SOLUTION.—

$$F = \frac{.00034 \times (\frac{1}{2} \times 7 \times 2,000) \, 5 \times 150^2}{3.1416} = 85,227 \text{ lb.}$$

CENTER OF GRAVITY.

The center of gravity of a body, or of a system of bodies, is that point from which, if the body or system were suspended, it would be in equilibrium.

If a line or a surface has two axes, or a solid has three axes of symmetry, the center of gravity lies at their point of intersection, and corresponds with the geometrical center of the figure.

An axis of symmetry is any line so drawn that, if part of the figure on one side of the line is folded on this line, it will coincide exactly with the other part, point for point and line for line. Thus, in Fig. 1, if the part $a\,b$ is folded on the line $A\,B$, the upper half will coincide exactly with the lower half; also, if $b\,c$ is folded on the line $C\,D$, the right-hand half will coincide exactly with the left-hand half. Hence, the point O where $A\,B$ and $C\,D$ intersect is the center of gravity of the rectangle $a\,b\,c\,d$. If the figure has one axis of symmetry, the center of gravity may be found as follows: Let

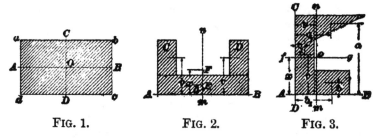

FIG. 1.   FIG. 2.   FIG. 3.

$m\,n$ be an axis of symmetry of the area in Fig. 2. The center of gravity will lie somewhere on this line. Draw any line $A\,B$ perpendicular to $m\,n$. Divide the area into squares, rectangles, triangles, parallelograms, circles, etc., whose centers of gravity are easily found, and measure the perpendicular distances of these centers of gravity from, the line $A\,B$. Add the sum of the products obtained by multiplying each area by the distance of its center of gravity from the line $A\,B$, and divide by the area of the entire figure; the result is the distance x of the center of gravity from $A\,B$ measured on $m\,n$, or the point F.

If the figure has no axis of symmetry, as in Fig. 3, draw any line, as $A\,B$, and find the distance x of the center of gravity from $A\,B$, and through x draw $f\,g$ parallel to $A\,B$. Choose any other line, $C\,D$, and find the distance y of the center of gravity from $C\,D$ by the same method, and through y draw $m\,n$ parallel to $C\,D$. The point of intersection o of $f\,g$ and $m\,n$ is the center of gravity.

Thus, suppose that the area of the triangle, Fig. 3, is A sq. in., and the distance of its center of gravity from $A\,B$ is

a in., and from $C D$, a_1 in.; that the area of the small rectangle is B sq. in., and the distance of its center of gravity from $A B$ is b in., and from $C D$ is b_1 in.; that the area of the large rectangle is C sq. in., and the distance of its center of gravity from $A B$ is c in., and from $C D$ is c_1 in.; then,

$$x = \frac{(A \times a) + (B \times b) + (C \times c)}{A + B + C},$$

and

$$y = \frac{(A \times a_1) + (B \times b_1) + (C \times c_1)}{A + B + C}.$$

To find the center of gravity mechanically, suspend the object from a point near its edge and mark on it the direction of a plumb-line from that point; then suspend it from another point and again mark the direction of a plumb-line. The intersection of these two lines will be directly over the center of gravity.

The center of gravity of a body having parallel sides may be found by drawing the outline of one of the sides upon heavy paper, and cutting out the exact shape of the figure. Then suspend the paper from the two points and find the center of gravity, as in the last case.

The center of gravity of a triangle lies on a line drawn from a vertex to the middle point of the opposite side, and at a distance from that side equal to one-third of the length of the line. Or, draw a line from another vertex to the middle point of the side opposite, and the intersection of the two lines will be the center of gravity.

For a parallelogram, the center of gravity is at the intersection of the two diagonals.

For an irregular four-sided figure, draw a diagonal, dividing it into two triangles. Draw a line joining these centers of gravity. Draw the other diagonal, dividing the figure into two other triangles, and join the centers of gravity by a straight line. The intersection of these lines is the center of gravity of the figure.

For a figure having more than four sides, find the center of gravity by the general method explained in connection with Fig. 3.

For an arc of a circle, the center of gravity lies on the radius drawn to the middle point of the arc (an **axis of** symmetry) and at a distance from the center equal to the length of the chord multiplied by the radius and divided by the length of the arc.

For a semicircle, the distance from the center $= \dfrac{2\,r}{\pi}$ $= .6366\,r$, when $r =$ the radius.

For the area included in a half circle, the distance of the center of gravity from the center $= \dfrac{4\,r}{3\,\pi} = .4244\,r$.

For circular sector, the distance of the center of gravity from the center equals two-thirds of the length of the chord multiplied by the radius and divided by the length of the arc.

For a circular segment, let A be its area and C the length of its chord; then the distance of the center of gravity from the center of the circle is equal to $\dfrac{C^3}{12\,A}$.

For a solid having three axes of symmetry, all perpendicular to each other, like a sphere, cube, right parallelopiped, etc., their point of intersection is the center of gravity.

For a cone or pyramid, draw a line from the apex to the center of gravity of the base; the required center of gravity is one-fourth the length of this line from the base, measured on the line.

For two bodies, the larger weighing W lb., and the smaller P lb., the center of gravity will lie on the line joining the centers of gravity of the two bodies and at a distance from the larger body equal to $\dfrac{Pa}{P+W}$, where a is the distance between the centers of gravity of the two bodies.

For any number of bodies, first find the center of gravity of two of them as above, and consider them as one weight whose center of gravity is at the point just found. Find the center of gravity of this combined weight and a third body. So continue for the rest of the bodies, and the last center of gravity will be the center of gravity of the whole system of bodies.

MOMENT OF INERTIA.

The *moment of inertia* of a body or section is a mathematical expression that is much used in computations relating to rotating bodies and to the strength of materials.

It may be defined as follows:

The moment of inertia of a body, rotating about a given axis, is the sum of the products obtained by multiplying the weights of the elementary particles of which it is composed by the square of their distances from the axis.

It is often desirable to use the moment of inertia for a plane section; but as a plane surface has no weight, it is apparent that the above definition does not correctly apply. The following definition applies to plane surfaces:

The moment of inertia of a plane surface about a given axis is the sum of the products obtained by multiplying each elementary areas into which the surface may be conceived to be divided by the square of its distance from the axis.

The axis about which the body or surface rotates, or is assumed to rotate, i. e., the axis from which the distance to each area or particle is measured, is called the *axis of rotation*. The least moment of inertia is that value of the moment of inertia of a body or section when the axis of rotation passes through the center of gravity, since its value is less for that position of the axis than for any other.

To find the moment of inertia of a body about a given axis:

Divide the body or section into many small parts and multiply the weight or area of each part by the square of the distance from its center of gravity to the axis of rotation; the sum of these products will be the moment of inertia.

NOTE.—The results obtained by the above rules are really only approximate; for practically it is impossible to divide a body or surface into parts sufficiently small for absolute accuracy. The smaller the parts the more accurate will be the result; but the results obtained by these rules will always be *slightly too small*.

The moment of inertia is usually designated by the letter I.

Formulas for the values of I about an axis of rotation passing through the center of gravity of the section are given for various forms of sections in Table V, page 153.

The moment of inertia about an axis of rotation not passing through the center of gravity is *equal to the moment of inertia about a parallel axis through the center of gravity plus the product of the entire weight of the body (or area of the section) multiplied by the square of the distance between the two axes.*

EXAMPLE.—It is desired to find the moment of inertia of a 6″ I-beam of the dimensions shown in Fig. 1 about an axis $x\,y$ perpendicular to the web of the beam at the center.

SOLUTION.—Since the axis about which the moment of inertia is to be found is an *axis of symmetry* of the beam, it is necessary to make the computations only for the half section of the beam lying at one side of the axis, and multiply the result by 2. As stated before, the smaller the parts into which the area is divided, the more accurate will be the result.

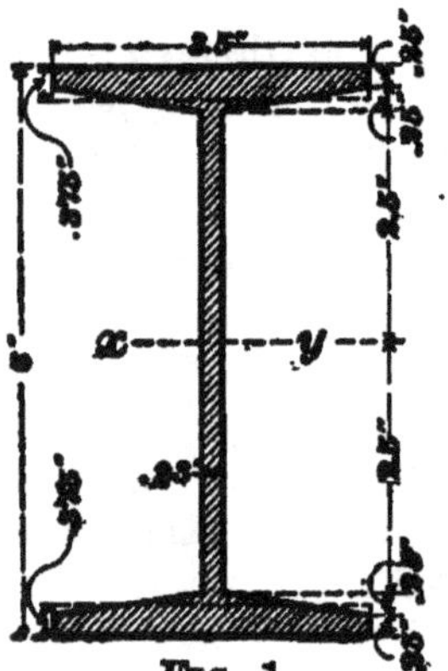

FIG. 1.

It will be sufficiently accurate for present purposes to divide the section in the manner shown in Fig. 2.

The operations are given at the side of the figure, and will be readily understood. The sum of the products is the approximate value of the moment of inertia of this half of the section about the axis $x\,y$, and when multiplied by 2 is the approximate value of I for the entire section. It is found to equal 23.444.

Area.	Square of Distance.
$3.50 \times .25 = .875$	$.875 \times 2.875^2 = 7.232$
$3.27 \times .125 = .409$	$.409 \times 2.667^2 = 2.907$
$.23 \times .50 = .115$	$.115 \times 2.50^2 = .719$
$.23 \times .50 = .115$	$.115 \times 2.00^2 = .460$
$.23 \times .50 = .115$	$.115 \times 1.50^2 = .259$
$.23 \times .50 = .115$	$.115 \times 1.00^2 = .115$
$.23 \times .50 = .115$	$.115 \times 0.50^2 = .029$
$.23 \times .25 = .058$	$.058 \times 0.125^2 = .001$
$\overline{1.917}$	$\overline{11.722}$
2	2
$A = \overline{3.834}$	$I = \overline{23.444}$

If the web of the beam is divided into areas ¼ in. in height (instead of ½ in.), the value of I obtained will be 23.46 in. If the section is considered to be of the form indicated by the dotted lines in Fig. 1, and to have the same area as the original section, then, by the formula for the moment of inertia of an I-beam given in Table V, page 153, the value of

$$I = \frac{3.50 \times 6^3 - 8.27 \times 5.25^3}{12} = 23.57.$$

The true value is almost exactly 23.48 in. Any one of these values would be sufficiently correct for most practical purposes.

If it is desired to find the moment of inertia of a body about a given axis with reference to the *weight* of the body, the process is substantially the same as in the example given for the plane section, except that the *weight* of each small part of the body is taken instead of the *area* of each small part of the section.

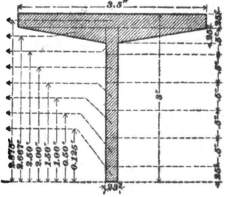

FIG. 2.

CENTER OF OSCILLATION.

The *center of oscillation* of a pendulum or other body vibrating or rotating about a fixed axis or center is that point at which, if the entire weight of the body were concentrated, the body would continue to vibrate in the same intervals of time.

When a pendulum, or other suspended body, is oscillating backward and forward, it is plain that those particles that are farther from the point of suspension travel through greater distances, and therefore move with greater velocities than those particles that are nearer the point of suspension.

11

But there is evidently some point on the pendulum that travels through the same distance and has the same velocity as the average distance and average velocity of all the particles. This point is called the *center of oscillation;* it is *not* situated at the center of gravity. It always exists in the ball of a revolving governor or other rotating body. The axis or center around which the body rotates (corresponding to the point of suspension in pendulum) is the *axis of rotation.*

The distance from the axis, or center of rotation, to the center of oscillation is sometimes called the *true length of the pendulum;* it is also called the *radius of oscillation;* the latter name is preferable. To find the radius of oscillation:

Divide the moment of inertia of the body about the given axis of rotation by the product of the total weight of the body, multiplied by the distance from the given axis to the center of gravity of the body.

The centers of oscillation and of rotation (point of suspension) are *interchangeable.* If the position of a pendulum is reversed, and suspended from its center of oscillation, the pendulum will vibrate in the same intervals of time.

EXAMPLE.—It is desired to find the position of the center of oscillation of a wrought-iron bar 1 in. square and 12 in. long, axis of rotation perpendicular to the bar at one end:

Weight of Each Cu. In.		Sq. of Dist.	
$.281 \times 0.5^2$	=	0.070
$.281 \times 1.5^2$	=	0.632
$.281 \times 2.5^2$	=	1.756
$.281 \times 3.5^2$	=	3.442
$.281 \times 4.5^2$	=	5.690
$.281 \times 5.5^2$	=	8.500
$.281 \times 6.5^2$	=	11.872
$.281 \times 7.5^2$	=	15.806
$.281 \times 8.5^2$	=	20.302
$.281 \times 9.5^2$	=	25.360
$.281 \times 10.5^2$	=	30.980
$.281 \times 11.5^2$	=	37.162
3.372		$161.572 = I$

SOLUTION.—For the purposes of the example it will be sufficiently accurate to find the moment of inertia by considering the bar to be divided into 12 equal cubes, each containing 1 cu. in. of metal, as indicated in the figure, and the weight of each cube to be concentrated at its center of gravity.

The weight of 1 cu. in. of wrought iron is .281 lb., and of a bar 1 in. square and 1 ft. long it is .281 × 12 = 3.372 lb. Hence, $I = .281 × .5^2 + .281 × 1.5^2 +$ etc. = 161.572. (See page 128.) The exact value of I is 161.856; this shows that the approximate method is very close.

According to the rule previously given, if the moment of inertia is divided by the product of the weight of the body, by the distance from the axis of rotation to the center of gravity, the quotient will be the radius of oscillation.

Therefore, the distance from the exact center of oscillation of a wrought-iron bar, 1 in. square and 12 in. long, to an axis of rotation perpendicular to the end of the bar, is

$$\frac{161.856}{3.372 × 6} = 8 \text{ in.,}$$

or two-thirds of the length of the bar.

The value of I for a bar of any cross-section, provided it is uniform throughout its length, revolving about an axis perpendicular to it and passing through its end, is

$$\frac{W l^2}{3},$$

in which W is the weight of the bar, and l is its length.

Hence, $$I = \frac{W l^2}{3} = \frac{3.372 × 12^2}{3} = 161.856.$$

If the axis passes through the center of gravity of the bar,

$$I = \frac{W l^2}{12}.$$

CENTER OF PERCUSSION.

The *center of percussion* with respect to a given axis of rotation may be defined as the point of application of the resultant of the forces that cause the body to rotate. It is that point at which if a force is applied, the force will have no effect at the axis of rotation.

Strike anything solid, as an anvil, with a stick. If the end of the stick hits the anvil, the opposite end will sting your hand and will jerk in the direction in which the blow is struck; if the center of the stick hits the anvil it will again sting your hand, but you will jerk it in a direction opposite to the movement of the blow. But somewhere between the end and the center of the stick will be a point where it may hit the anvil and not sting your hand at all. This point is the center of percussion.

Level off the surface of some wet sand and lay a strip of board upon it (say 18 in. long and 3 in. wide). Strike or press the board near the center and the entire length of the board will be imprinted in the sand; but press it near one end and the opposite end will be raised up from the sand and will make no imprint. Between the center and the end of the board is a point that if pressed upon will cause no movement in the opposite end, i. e., the end of the board will neither press into the sand nor be lifted from it, but the imprint in the sand will diminish to zero at the end of the board. The point pressed or struck will be the center of percussion. If the board is of uniform width, the center of percussion will be at one-third of the distance from one end of the board.

Similarly in the preceding illustration, if the stick is of uniform size and weight, and your hand grasps it at one end, the point at which it can strike the anvil without affecting your hand will be at one-third the distance from the opposite end.

In all cases the center of percussion is identical with the center of oscillation, and its position is found in the same manner.

EXAMPLE.—It is desired to find the position of the center of oscillation or percussion of two balls fastened upon a rod. The first, weighing 2 lb., is at a distance of 18 in. from the axis of rotation, and the second, weighing 1 lb., is at a distance of 36 in. from the axis. (See figure.)

SOLUTION.—For simplicity, the rod will be assumed to have no weight. Consider the weight of each ball to be concentrated at its center of gravity.

The moment of inertia is found as follows.

$$
\begin{array}{rcl}
\text{Wt.} & \text{Sq. of Dist.} & \\
2 \times 18^2 & = & 648 \\
1 \times 36^2 & = & 1,296 \\
\hline
& & 1,944 = I.
\end{array}
$$

The center of gravity of the two balls is found to be at a distance of 6 in. from the larger, or 24 in. from the axis of rotation (see page 124), and the combined weight of the two balls is $2 + 1 = 3$ lb. Therefore, the center of percussion is found to be at a distance of $\dfrac{1,944}{3 \times 24} = 27$ in. from the axis of rotation.

But, in an actual case, the rod would have weight, and its moment of inertia must be considered as well as the moment of inertia of the balls.

If we assume that the rod is of steel, $\frac{3}{8}$ in. in diameter and 36 in. long, it will weigh $\left(\dfrac{3}{8}\right)^2 \times .7854 \times 36 \times .283 = 1.125$ lb. .283 lb. is the weight of 1 cu. in. of steel.

Using the formula given on page 129,

$$
I = \frac{W l^2}{3} = \frac{1.125 \times 36^2}{3} = 486.
$$

Adding this result to the former, $1,944 + 486 = 2,430 =$ moment of inertia of rods and balls. The center of gravity of the combination is found by the formula (see page 124) $\dfrac{P\,a}{P + W}$. Substituting, $\dfrac{1.125 \times 6}{1.125 + 3} = 1\frac{7}{11}$. $24 - 1\frac{7}{11} = 22\frac{4}{11}$ in. $=$ distance from end of rod to center of gravity.

Applying the rule given for finding the center of oscillation, the distance of the center of percussion from the end of the bar is $\dfrac{2,430}{(1 + 2 + 1.125) \times 22\frac{4}{11}} = 26.34$ in., very nearly.

RADIUS OF GYRATION.

The *center of gyration* is that point in a revolving body at which, if the entire mass of the body were concentrated, the moment of inertia with respect to a given axis would be the same as in the body.

An ounce of cork occupies about 94 times as much space as

an ounce of platinum; but the ounce of platinum can have the same moment of inertia as the ounce of cork, if its center of gyration has the same position with respect to the axis of rotation.

The center of gyration is not at the center of gravity, nor at the center of oscillation, but at some point in a straight line between those centers.

The *radius of gyration* is the distance from the axis of rotation to the center of gyration.

The square of the radius of gyration is the average of the squares of the distances from the axis of rotation to each elementary particle of the body, or to each elementary area of the section, as the case may be. But the sum of these squares of distances, multiplied by the weight or area of each elementary part, equals the moment of inertia; therefore, the moment of inertia divided by the weight of the body or area of the section equals the square of the radius of gyration; the square root of this quotient is the radius of gyration.

But, according to the rule for finding the radius of oscillation, the quotient obtained by dividing the moment of inertia by the weight or area equals the product of the distance from the axis of rotation to the center of gravity, multiplied by the radius of oscillation; and, therefore, *the radius of gyration is a mean proportional between these distances.*

If the distance from the axis of rotation to the center of gravity is known, and the radius of oscillation is known, the radius of gyration may be found by multiplying these two known distances together and extracting the square root of the product.

In the example of the I-beam, Fig. 2, page 126, the sum of the areas of the half section of the beam is 1.917, and the area of the entire section is 3.834 sq. in. Therefore, the radius of gyration of this beam about an axis through the center of gravity perpendicular to the web $= \sqrt{\frac{23.44}{3.834}} = 2.47$ in.

In the example of the iron bar 12 in. long (see figure, page 128), the distance from the axis of rotation to the center of gravity is 6 in., and the radius of oscillation was found to equal 8 in. Therefore, the radius of gyration about an

axis perpendicular to the bar at one end $= \sqrt{6 \times 8} = 6.93$ in.
Or, the moment of inertia of the bar $= 161.586$, and the
weight of the bar $= 3.872$ lb. Therefore, the radius of gyration $= \sqrt{\dfrac{161.586}{3.872}} = 6.93$ in., very nearly.

The radius of gyration is used in determining the strength
of columns. The axis must be taken in such a direction that
the result will be the *least* radius of gyration of the column;
this condition is usually obtained when the axis is perpendicular to the least diameter or side of the column.

The various relations between these quantities may be
concisely expressed by the following formulas, in which
$A =$ area of section (or weight of body if the weight is used);
$g =$ distance from axis of rotation to center of gravity;
$G =$ radius of gyration;
$r_o =$ radius of oscillation;
$I =$ moment of inertia.

Then,

$$I = A G^2. \qquad\qquad I = A g r_o. \qquad\qquad G^2 = g r_o.$$

$$G = \sqrt{\frac{I}{A}}. \qquad\qquad g = \frac{I}{A r_o}. \qquad\qquad r_o = \frac{I}{A g}$$

$$G = \sqrt{g r_o}. \qquad\qquad g = \frac{G^2}{r_o}. \qquad\qquad r_o = \frac{G^2}{g}.$$

$$g : G = G : r_o.$$

To find the radius of oscillation, radius of gyration, and
moment of inertia, experimentally.

The connecting-rod of an engine is represented in the

figure. It is desired to find the moment of inertia of the rod
about an axis of rotation through the center of the crosshead
pin A.

This may be accomplished, experimentally, as follows:
Suspend the rod from the crosshead pin in such a manner

that it will swing freely; cause it to swing, or oscillate, and note the exact time of the vibrations. Remove the crosshead pin and reverse the rod, but, instead of suspending it by the crankpin, suspend it by a movable pin B, that can be clamped at any desired point upon the rod. C is another view of this pin. There will be a point on the rod from which it may be suspended by means of the movable pin, so that it will vibrate in exactly the same intervals of time as when suspended from the crosshead pin. This point is the *center of oscillation*, for the center of oscillation and the center of rotation are interchangeable; the point will be found at about one-third the length of the rod from the crankpin. Find this center of oscillation, experimentally, and carefully measure the distance from the center of the movable pin to the center of the crosshead-pin hole. This distance is the *radius of oscillation* $= r_o$. Next remove the movable pin, and find the center of gravity (lengthwise) of the rod by balancing it across a knife edge, and measure the distance from the center of gravity thus found to the center of the crosshead-pin hole; this distance $= g$. Finally, weigh the rod.

The product of the weight ($= A$), the radius of oscillation ($= r_o$), and the distance from the center of crosshead pin (axis of rotation) to the center of gravity ($= g$) will be the moment of inertia. For, by the formula, $I = A g r_o$. The radius of gyration G may be found by the formula

$$G = \sqrt{\frac{I}{A}}, \text{ or } G = \sqrt{g r_o.}$$

MOMENT OF RESISTANCE.

If the moment of inertia of the cross-section of a beam is divided by the distance from the neutral axis (see definition on next page) to the extreme fiber, i. e., the fiber that is farthest from the axis, the quotient will be the quantity known as the *moment of resistance*.

It is evident that, if a beam is strained by a vertical load, the greatest stress will be in the extreme upper and lower fibers of the beam.

The intensity of the stress that can be borne by the extreme fibers is the limit of the strength of the beam.

The upper fibers are compressed and the lower fibers are stretched, but somewhere along or near the center of a vertical section of the beam, the fibers are neither extended nor compressed; the position of these fibers is called the *neutral surface*, and the line where this neutral surface intersects a right section of the beam is the *neutral axis* of the section.

The neutral axis passes through the center of gravity of the section.

If the moment of resistance is multiplied by the amount of stress that may be allowed per square inch upon the extreme fiber, the product will represent the efficiency of the beam to resist bending moment.

EXAMPLE.—Referring to the 6″ I-beam, Figs. 1 and 2, pages 126 and 127, for which the moment of inertia of the section has been found, it is desired to ascertain the load that a wrought-iron beam of the same dimensions as Fig. 1 will carry at the center of a span 8 ft. between supports.

SOLUTION.—The moment of resistance for the section = $\frac{23.48}{3} = 7.83$. In Table II, page 151, the ultimate strength or fiber stress for wrought iron is given as 50,000 lb. per sq. in., and in Table I, page 151, the factor of safety given for wrought iron under a steady stress is 4; therefore, the safe fiber stress for wrought iron $= \frac{S}{f} = \frac{50,000}{4} = 12,500$ lb. per sq. in., and the moment of resistance multiplied by the safe fiber stress, or $\frac{SR}{f} = 7.83 \times 12,500 = 97,875$ in.-lb. But $l = 8$ ft., or 96 in.; equating the bending moment for a load at the center of a beam $\left(= \frac{Wl}{4} \right)$ with the moment of resistance, or putting $M = \frac{SR}{4} = \frac{Wl}{4}$; then $\frac{96\,W}{4} = 97,875$; therefore, $W = 4,078$ lb., the load that can be safely supported at the center of the beam.

MECHANICAL POWERS.

$$F:W=l:L. \quad FL=Wl.$$
$$F=\frac{Wl}{L}. \qquad W=\frac{FL}{l}.$$
$$l=\frac{Fa}{W+F} \qquad L=\frac{Wa}{W+F}$$

$$F:W=l:L. \quad FL=Wl.$$
$$F=\frac{Wl}{L}. \qquad W=\frac{FL}{l}.$$
$$L=\frac{Wa}{W-F} \qquad l=\frac{Fa}{W-F}$$

$$F:W=l:L. \quad FL=Wl.$$
$$F=\frac{Wl}{L}. \qquad W=\frac{FL}{l}.$$
$$L=\frac{Wa}{F-W}. \qquad l=\frac{Fa}{F-W}.$$

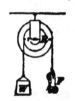

$$F:W=r:R. \quad FR=Wr.$$
$$F=\frac{Wr}{R}. \qquad R=\frac{Wr}{F}.$$
$$W=\frac{RF}{r}. \qquad r=\frac{RF}{W}.$$

$$F=\frac{Wrr'}{RR'}. \qquad W=\frac{FRR'}{rr'}.$$

$n=$ number of revolutions of large gear.
$$n:n'=r':R.$$
$$v:v'=rr':RR'.$$
$v=$ velocity of W; $v'=$ velocity of F.

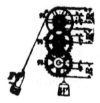

$$F=\frac{Wrr'r''}{RR'R''}. \qquad W=\frac{FRR'R''}{rr'r''}.$$
$$n:n''=r'r'':RR'.$$
$$v:v'=rr'r'':RR'R''.$$
r, r', r'', etc. $=$ radii of the pinions;
R, R', R''. etc. $=$ radii of the wheels.

Let db and qb represent the magnitudes and directions of two forces that act to move the body b. By completing the parallelogram there will be obtained a diagonal force fb, whose magnitude and direction are equal to the effect produced by db and qb. fb is called the resultant of db and qb.

If three or more forces act in different directions to move a body b, find the resultant of any two of them, and consider it as a single force. Between this and the next force find a second resultant. Thus, pb, qb, and rb are magnitudes and directions of the forces. $pb + qb + rb = gb + rb = fb$, the magnitude and direction of the three forces, pb, qb, and rb.

A SINGLE FIXED PULLEY.

$$F = W.$$
$$v = v'.$$

$v = $ velocity of W; $v' = $ velocity of F.

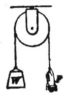

A SINGLE MOVABLE PULLEY.

$F : W = 1 : 2$, or $F = \frac{1}{2} W$.

If the force F be applied at a and act upwards, the result will be the same.

$$v' = 2 v.$$

$v = $ velocity of W; $v' = $ velocity of F.

A DOUBLE MOVABLE PULLEY.

$F : W = 1 : 4$, or $F = \frac{1}{4} W$.

Let $u = $ number of parts of rope, not counting the free end.

$$F = W \div u. \quad v : v' = 1 : u.$$

$v = $ velocity of W; $v' = $ velocity of F.

QUADRUPLE MOVABLE PULLEY.

$$F = \tfrac{1}{8} W. \quad F : W = 1 : 8.$$

Let u = number of parts of rope, not counting the free end; then,

$$F = W \div u. \quad v : v' = 1 : u.$$

v = velocity of W; v' = velocity of F.

COMPOUND PULLEY.

u = number of movable pulleys.

$$F = \frac{W}{2^u}. \qquad W = 2^u F.$$

$$v : v' = 1 : 2^u.$$

v = velocity of W; v' = velocity of F.

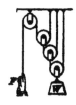

DIFFERENTIAL PULLEY.

$$W = \frac{2 P R}{R - r}.$$

AN OBLIQUE FIXED PULLEY.

$$F : W = 1 : 2 \cos z.$$

$$W = 2 F \cos z. \quad F = \frac{W}{2 \cos z}.$$

INCLINED PLANE.

$$F = \frac{W h}{l} = W \sin a.$$

$$W = \frac{F l}{h} = \frac{F}{\sin a}.$$

WEDGE.

F = force required to drive the wedge;
R = resistance.

$$F = \frac{R a}{l}. \qquad R = \frac{F l}{a}.$$

SCREW.

P = pitch of the screw;
r = radius on which the force F acts.

$$F : W :: P : 2\pi r.$$

$$F = \frac{WP}{2\pi r}. \qquad W = \frac{2\pi r F}{P}.$$

WORK.

Work is the overcoming of resistance through a distance. The unit of work is the *foot-pound;* that is, it equals 1 pound raised vertically 1 foot. The amount of work done is equal to the resistance in pounds multiplied by the distance in feet through which it is overcome. If a body is lifted, the resistance is the weight or the overcoming of the attraction of gravity, the work done being the weight in pounds multiplied by the height of the lift in feet. If a body moves in a horizontal direction, the work done is the friction overcome, or the force needed to move a resistant body or combination of bodies, multiplied by the distance moved. In order to compare the different amounts of work done by different systems of forces, time is also considered.

One *horsepower* is 550 ft.-lb. of work in 1 second, or 33,000 ft.-lb. in 1 minute, or 1,980,000 ft.-lb. in 1 hour.

The work necessary to be done in raising a body weighing W lb. through a height of h ft. equals Wh ft.-lb. The total work that any moving body is capable of doing in being brought to rest equals its kinetic energy, or $\dfrac{Wv^2}{2g}$, when v is the velocity in feet per second.

Thus, the work that a cannon ball weighing 800 lb. and traveling with a velocity of 1,200 ft. per sec. could do, is

$$\frac{800 \times 1,200^2}{2 \times 32.16} = 17,910,447 \text{ ft.-lb.}$$

If stopped in 1 min., the horsepower would be 17,910,447 ÷ 33,000 = 542.8, nearly.

FORCE OF A BLOW.

In order to determine the force of a blow, the velocity of the object at the instant of striking must be known, and also the time required to bring the body to rest. It is a very difficult matter to determine the exact time, but a close approximation to the striking force may be obtained by dividing the kinetic energy of the body at the instant of striking by the average amount of penetration or compression produced by the striking body.

Let F = striking force in pounds;

    W = weight of striking body in pounds;

    v = velocity of striking body in feet per second;

    R = distance penetrated or amount of compression = the distance through which the resistance acts, in feet;

    t = time required to bring the body to rest;

    h = height in feet which would produce the velocity v.

Then, $F = \dfrac{Wv}{gt}$, or $F = \dfrac{Wv^2}{2gR} = \dfrac{Wh}{R}$.

EXAMPLE.—A steam hammer weighing 1,000 lb. (with its piston) falls from a height of 8 ft., and compresses a piece of iron $\frac{1}{8}$ in.; what is its striking force?

SOLUTION.—If gravity be considered as the only force acting, the steam on top of the piston being used to prevent a rebound of the hammer,

$$F = \frac{Wh}{R} = \frac{1,000 \times 8}{(\frac{1}{8} \div 12)} = 1,000 \times 8 \times 8 \times 12 = 768,000 \text{ lb.}$$

Divide $\frac{1}{8}$ in. by 12, to obtain the amount of compression in feet or parts of a foot.

BELTING.

D = diameter of larger pulley in inches;

d = diameter of smaller pulley in inches;

N = revolutions per minute of larger pulley;

n = revolutions per minute of smaller pulley;

W = width of double belt in inches;

w = width of single belt in inches;

H = horsepower that can be transmitted by the belt.

Then,
$$H = \frac{D N w}{2,750} \text{ for single belts.}$$

$$H = \frac{D N W}{1,925} \text{ for double belts.}$$

$$w = \frac{2,750\ H}{D N} = \frac{2,750\ H}{d\ n}.$$

$$W = \frac{1,925\ H}{D N} = \frac{1,925\ H}{d\ n}.$$

$$D = \frac{2,750\ H}{w N} \text{ for single belt.}$$

$$D = \frac{1,925\ H}{W N} \text{ for double belt.}$$

$$N = \frac{2,750\ H}{w D} \text{ for single belt.}$$

$$N = \frac{1,925\ H}{W D} \text{ for double belt.}$$

The above rules are for open belts and pulleys having the same diameter, the arc of contact being, in this case, half the circumference, or 180°. For open belts and pulleys of different diameters, the arc of contact is less than 180° on the smaller pulley, and a different constant, to be taken from the following table, must be substituted in the formulas. To find the arc of contact, let l be the distance in inches between the centers of the pulleys. Then, $\frac{D - d}{2\ l} = $ cosine of half the angle Find this half angle from a table of natural cosines, and

Degrees.	Fraction of Circumference.	Single Belt Constant.	Double Belt Constant.
90	¼ = .25	6,080	4,250
112½	⁵⁄₁₆ = .3125	4,730	3,310
120	⅓ = .3333	4,400	3,080
135	⅜ = .375	3,850	2,700
150	⁵⁄₁₂ = .4167	3,410	2,390
157½	⁷⁄₁₆ = .4375	3,220	2,250
180 to 270	½ to ¾ = .5 to .75	2,750	1,925

multiply by 2. The result is the arc of contact in degrees. Find the number in the first column of the table, which is nearest to this result, and use the constant corresponding to that number. If a table of natural cosines is not at hand, measure the length of the arc of contact on the smaller pulley and divide it by the circumference of the pulley. Find the fraction in the second column that corresponds nearest to this result, and opposite this its corresponding constant.

EXAMPLE.—What must be the width of a single belt to transmit 12 horsepower, when the diameter of the larger pulley is 42 in., of the smaller pulley 20 in., distance between their centers 14 ft. = 168 in., and R, P. M. of smaller pulley 150?

SOLUTION.— $\dfrac{42-20}{2\times 168}$ = .06548 = cosine of half the arc of contact, which thus = 86° 15′, nearly; 86° 15′ × 2 = 172¼° = arc of contact; the nearest number in the table is 180°, and the corresponding constant is 2,750; hence, $w = \dfrac{2,750\times 12}{20\times 150} = 11$ in.

Oak-tanned leather makes the best belts. When belts are run with the hair side over the pulley, they have greater adhesion.

The ordinary thickness of leather belts is $\frac{3}{16}$ in., and their weight is about 60 lb. per cu. ft.

Ordinarily, four-ply cotton belting is considered equivalent to single-leather belting.

RULES FOR CALCULATING THE SPEED OF GEARS OR PULLEYS.

In calculating for gears, multiply or divide by the diameter or the number of teeth, as may be required. In calculating for pulleys, multiply or divide by their diameters in inches.

The driving wheel is called the *driver*, and the driven wheel the *driven* or *follower*.

PROBLEM I.

The revolutions of driver and driven, and the diameter of the driven, being given, required the diameter of the driver.

Rule.—*Multiply the diameter of the driven by its number of revolutions, and divide by the number of revolutions of the driver.*

PROBLEM II.

The diameter and revolutions of the driver being given, required the diameter of the driven to make a given number of revolutions in the same time.

Rule.—Multiply the diameter of the driver by its number of revolutions, and divide the product by the required number of revolutions.

PROBLEM III.

The diameter or number of teeth, and number of revolutions of the driver, with the diameter or number of teeth of the driven, being given, required the revolutions of the driven.

Rule.—Multiply the diameter or number of teeth of the driver by its number of revolutions, and divide by the diameter or number of teeth of the driven.

PROBLEM IV.

The diameter of driver and driven, and the number of revolutions of the driven, being given, required the number of revolutions of the driver.

Rule.—Multiply the diameter of the driven by its number of revolutions, and divide by the diameter of the driver.

PUMPS.

In all pumps, whether lifting, force, steam, single-acting, double-acting, or centrifugal, the number of foot-pounds of work performed by the pump is equal to the weight of the water discharged in pounds, multiplied by the vertical distance in feet between the level of the water in the well or source and the point of discharge, plus the work done in overcoming the friction and other resistances. (It is assumed that the water is delivered with practically no velocity.)

To find the discharge of a pump in gallons per minute:

Let T = piston travel in feet per minute;

d = diameter of cylinder in inches;

G = number of gallons discharged per minute.

Then, $G = .03264\ T d^2.$

To find the horsepower of a pump, use the following formula, in which T and d are the same as above, and h is the vertical distance in feet between the level of the water at the source and the point of discharge:

H. P. $= .00033724\ G h = .00001238\ T d^2 h.$

In both the above formulas, allowance has been made for friction, leakage, etc.

12

x

DUTY.

The duty of a pump is the number of foot-pounds of work actually done for 100 lb. of coal burned.

$$\text{Duty} = 835.53 \frac{G h}{W},$$

where W = weight of coal burned, in pounds.

HYDROMECHANICS.

HYDROSTATICS.

Hydrostatics treats of liquids at rest under the action of forces. If a liquid is acted on by a pressure, the pressure per unit of area exerted anywhere on the mass of liquid is transmitted undiminished in all directions, and acts with the same force on all surfaces, in a direction at right angles to those surfaces.

General Law for the Downward Pressure on the Bottom of Any Vessel.—The pressure on the bottom of a vessel containing a liquid is independent of the shape of the vessel, and is equal to the weight of a prism of the liquid whose base is the same as the bottom of the vessel, and whose altitude is the distance between the bottom and the upper surface of the liquid, plus the pressure per unit of area upon the upper surface of the liquid multiplied by the area of the bottom of the vessel.

General Law for Upward Pressure.—The upward pressure on any submerged horizontal surface equals the weight of a prism of the liquid whose base has an area equal to the area of the submerged surface, and whose altitude is the distance between the submerged surface and the upper surface of the liquid, plus the pressure per unit of area on the upper surface of the liquid multiplied by the area of the submerged surface.

General Law for Lateral Pressure.—The pressure on any vertical surface due to the weight of the liquid is equal to the weight of a prism of the liquid whose base has the same area as the vertical surface, and whose altitude is the depth of the center of gravity of the vertical surface below the level of the liquid. Any additional pressure is to be added, as in the previous cases.

Pressure on Oblique Surfaces.—The pressure exerted by a liquid in any direction on a plane surface is equal to the weight of a prism of the liquid whose base is the projection of the surface at right angles to the given direction, and whose height is the depth of the center of gravity of the surface below the level of the liquid.

If a cylinder is filled with water, and a pressure applied, the total pressure on any half section of the cylinder is equal to the projected area of the half cylinder (or the diameter multiplied by the length of the cylinder) multiplied by the depth of the center of gravity of the half cylinder, multiplied by the weight of a cubic inch of water, plus the diameter of the shell, multiplied by the pressure per square inch, multiplied by the length of the cylinder.

If d = the diameter, and l = the length of the cylinder, the pressure due to the weight of the water when the cylinder is vertical upon the half cylinder = $d \times l \times \dfrac{l}{2} \times$ the weight of a cubic inch of water = $d \times \dfrac{p}{2} \times$ the weight of a cubic inch of water; d and l are to be measured in inches.

The pressure in pounds per square inch due to a head of water is equal to the head in feet multiplied by .434.

The head equals the pressure in pounds per square inch multiplied by 2.304.

EXAMPLE.—(a) What is the pressure per square inch corresponding to a head of water of 175 ft.? (b) If the pressure had been 90 lb. per sq. in., what would the head have been?

SOLUTION.—(a) 175 × .434 = 75.95 lb. per sq. in.

(b) 90 × 2.304 = 207.36 ft.

HYDROKINETICS.

Hydrokinetics, also called hydrodynamics and hydraulics, treats of water in motion. When water flows in a pipe, conduit, or channel of any kind, the velocity is not the same at all points of the flow, unless all cross-sections of the pipe or channel are equal. That velocity which, being multiplied by the area of the cross-section of the stream, will equal the total quantity discharged, is called the *mean velocity*.

Let Q = quantity that passes any section in 1 second;

A = area of the section;

v = mean velocity in feet per second.

Then, $Q = A v$, and $v = \dfrac{Q}{A}$.

The vertical distance between the level surface of the water and the center of the aperture through which it flows, is called the *head*.

Let V = mean velocity of efflux through a small aperture;

h = head in feet at the center of the aperture;

w = weight of water flowing through the aperture per second.

Then, $V = \sqrt{2gh}$; that is, the velocity of efflux is the same as if the water had fallen through a height equal to the head.

Let Q = theoretical number of cubic feet discharged per second;

V_m = mean velocity through orifice in feet per second;

A = area of orifice;

h = theoretical head necessary to give a mean velocity V_m;

Q_a = actual quantity discharged in cubic feet per second.

Then, for an orifice in a thin plate, or a square-edged orifice (the hole itself may be of any shape, triangular, square, circular, etc., but the edges must not be rounded), the actual quantity discharged is

$$Q_a = .615\,Q = .615\,A\,V_m.$$

The *weir* is a device used for measuring the discharge of water. It is a retangular orifice through which the water flows.

If d = the depth of the opening in feet, and b its breadth in feet, the area of the opening is $A = d \times b$, and the theoretical discharge is $Q = d \times b \times V_m = db \times \tfrac{2}{3}\sqrt{2gd}$, the head for this case being taken as d.

The actual discharge when the top of the weir lies at the surface of the water is

$$Q_a = .615\,Q = .615 \times db \times \tfrac{2}{3}\sqrt{2gd} = .615 \times \tfrac{2}{3}\,b\,\sqrt{2gd^3} =$$
$$\tfrac{3}{2}\,b\,\sqrt{d^3}.$$

If h_1 is the depth in feet of the top of a weir below the surface of the water, and h is the depth in feet of the bottom of the weir below the surface of the water, the actual discharge Q_a, in cubic feet per second, is

$$Q_a = .615 \times \tfrac{2}{3}\, b\, \sqrt{2\,g}\,(\sqrt{h^3} - \sqrt{h_1^3}) = 3.288\, b\, (\sqrt{h^3} - \sqrt{h_1^3}).$$

FLOW OF WATER IN PIPES.

Let V_m = mean velocity of discharge in feet per second;

   h = total head in feet = vertical distance between the level of water in reservoir and the point of discharge;

   l = length of pipe in feet;

   d = diameter of pipe in inches;

   f = coefficient of friction.

Then, for straight cylindrical pipes of uniform diameter, the mean velocity of efflux may be calculated by the formula,

$$V_m = 2.315 \sqrt{\frac{h\,d}{f\,l + .125\,d}}. \qquad (a)$$

NOTE.—The head is always taken as the vertical distance between the point of discharge and the level of the water at the source, or point from which it is taken, and is always measured in feet. It matters not how long the pipe is—whether vertical or inclined, whether straight or curved, nor whether any part of the pipe goes below the level of the point of discharge or not—the head is always measured as stated above.

EXAMPLE.—What is the mean velocity of efflux from a 6″ pipe, 5,780 ft. long, if the head is 170 ft.? Take f = .021.

SOLUTION —

$$V_m = 2.315 \sqrt{\frac{h\,d}{f\,l + .125\,d}} = 2.315 \sqrt{\frac{170 \times 6}{.021 \times 5,780 + (.125 \times 6)}}$$

= 6.69 ft. per sec.

When the pipe is very long compared with the diameter, as in the above example, the following formula may be used:

$$V_m = 2.315 \sqrt{\frac{h\,d}{f\,l}}, \qquad (b)$$

in which the letters have the same meaning as in the preceding formula. This formula may be used when the length of the pipe exceeds 10,000 times its diameter.

The actual head necessary to produce a certain velocity V_m may be calculated by the formula

$$h = \frac{f\,l\,V_m^2}{5.36\,d} + .0233\,V_m^2. \qquad (c)$$

If the head, the length of the pipe, and the diameter of the pipe are given, to find the discharge, use the formula

$$Q = .09445\,d^2\sqrt{\frac{h\,d}{f\,l + .125\,d}}; \qquad (d)$$

that is, the discharge in gallons per second equals .09445 times the square of the diameter of the pipe in inches, multiplied by the square root of the head in feet, multiplied by the diameter of the pipe in inches, divided by the coefficient of friction times the length of the pipe in feet, plus .125 times the diameter of the pipe in inches.

To find the value of f, calculate V_m by formula (b) assuming that $f = .025$, and get the final value of f from the following table:

V_m	f	V_m	f	V_m	f
.1	.0686	.7	.0349	2	.0265
.2	.0527	.8	.0336	3	.0243
.3	.0457	.9	.0325	4	.0230
.4	.0415	1	.0315	6	.0214
.5	.0387	1¼	.0297	8	.0205
.6	.0365	1½	.0284	12	.0193

EXAMPLE.—The length of a pipe is 6,270 ft., its diameter is 8 in., and the total head at the point of discharge is 215 ft. How many gallons are discharged per minute?

SOLUTION.—

$$V_m = 2.315\sqrt{\frac{215 \times 8}{.025 \times 6,270}} = 7.67 \text{ ft. per sec., nearly.}$$

Using the value of $f = .0205$ for $V_m = 8$ (see table), $Q =$

$$.09455 \times 8^2\sqrt{\frac{215 \times 8}{.0205 \times 6,270 + (.125 \times 8)}} = 22.03 \text{ gal. per sec.} =$$

$22.03 \times 60 = 1,321.8$ gal. per min.

If it is desired to find the head necessary to give a discharge of a certain number of gallons per second through a pipe whose length and diameter are known, calculate the mean velocity of efflux by using the formula

$$V_m = \frac{24.51\ Q}{d^2}; \qquad (e)$$

find the value of f from the table, corresponding to this value of V_m, and substitute these values of f and V_m in the formula for the head.

EXAMPLE.—A 4" pipe, 2,000 ft. long, is to discharge 24,000 gal. of water per hr.; what head is necessary?

SOLUTION.— $\dfrac{24,000}{60 \times 60} = 6\frac{2}{3}$ gal. per sec. $V_m = \dfrac{24.51 \times 6\frac{2}{3}}{4^2}$

$= 10.2$ ft. per sec.

From the table, $f = .0205$ for $V_m = 8$, and $.0193$ for $V_m = 12$; assume that $f = .02$ for $V_m = 10.2$.

Then, $h = \dfrac{.02 \times 2,000 \times 10.2^2}{5.36 \times 4} + .0233 \times 10.2^2 = 196.53$ ft.

To find the diameter of a pipe that will give any required discharge in gallons per second, the total length of the pipe and the head being known, *find the value of d by formula (f); substitute this value in formula (e), and find the value of V_m. Then find from the table the value of f corresponding to this value of V_m. Substitute the values of d and f just found in the right-hand member of formula (g) and solve for d; the result will be the diameter of the pipe, accurate enough for all practical purposes.*

$$d = 1.229 \sqrt[5]{\frac{l\ Q^2}{h}}. \quad (f) \qquad d = 2.57 \sqrt[5]{\frac{(fl + \frac{1}{2}d)\ Q^2}{h}}. \quad (g)$$

EXAMPLE.—A pipe 2,000 ft. long is required to discharge 24,000 gal. of water per hr. The head being 195 ft., what should be the diameter of the pipe?

SOLUTION. — $Q = \dfrac{24,000}{60 \times 60} = 6\frac{2}{3}$ gal. per sec. Substitu-

ting in formula (f), $d = 1.229 \sqrt[5]{\dfrac{2,000 \times (6\frac{2}{3})^2}{195}} = 4.18 +$ in.

Substituting this value in formula (e), $V_m = \dfrac{24.51 \times 6\frac{2}{3}}{4.18^2} = 9.352$ ft. per sec. From the table, the value of f for $V_m = 9.352$ is $.0201$. Substituting this value of f and the value of d, found above, in formula (g),

$$d = 2.57 \sqrt[5]{\frac{(.0201 \times 2,000 + \frac{1}{2} \times 4.18) \times (6\frac{2}{3})^2}{195}} = 4.01 +; \text{ say, 4 in.}$$

STRENGTH OF MATERIALS.

The ultimate strengths of different materials vary greatly from the average values given in the following tables. In actual practice, the safest proc dure would be to make a test of the materi.l for its ultimate strength and coefficient of elasticity, or else specify in the contract that it shall not fall below certain prescribed limits. In the following formulas,

A = area of cross-section of material in square inches;

E = coefficient of elasticity in pounds per square inch;

G^2 = square of least radius of gyration;

I = moment of inertia about an axis passing through the center of gravity of the cross-section;

M = maximum bending moment in inch-pounds;

P = total stress in pounds;

R = moment of resistance;

S = ultimate stress in lb. per sq. in. of area of section;

W = weight placed on a beam in pounds;

b = breadth of cross-section of beam in inches;

d = depth of beam (in.) = diam. of circ. section = altitude of triangular section = length of vertical side;

e = amount of elongation or shortening in inches;

f = factor of safety;

l = length in inches;

p = pressure in pounds per square inch;

π = ratio of circumference to diameter = 3.1416, nearly;

q = a constant used in formula for columns;

r = radius of a circular section;

s = elastic set or deflection in inches of a beam under a transverse (bending) stress;

t = thickness of a shell or hollow section.

For tension, compression (where the piece does not exceed 10 times its least diameter), and shear,

$$P = \frac{A S}{f}. \qquad (1)$$

To find the breaking stress (P), make f = 1. For safe load, take f from Table I, and S from Table II, according to the nature and character of stress.

TABLE I.

FACTORS OF SAFETY (f).

Name of Material.	Steady Stress.	Varying Stress.	Shocks (Machines).
Cast iron	6	15	20
Wrought iron	4	6	10
Steel	5	7	15
Wood	8	10	15
Brick and stone	15	25	30

TABLE II.

ULTIMATE STRENGTHS (S).

Name of Material.	Tension.	Compression.	Shear.	Flexure.
Cast iron	20,000	90,000	20,000	36,000
Wrought iron	50,000	50,000	47,000	50,000
Steel	100,000	150,000	70,000	120,000
Wood	10,000	8,000	600 to 3,000	9,000
Stone		6,000		2,000
Brick	200	2,500		

EXAMPLE.—A square cast-iron pillar 18 in. long is required to sustain a steady load of 75,000 lb.; what must be the length of a side?

SOLUTION.—From the table, $f = 6$, and $S = 90,000$. By formula (1),

$$P = \frac{AS}{f}, \text{ or } A = \frac{Pf}{S} = \frac{75,000 \times 6}{90,000} = 5 \text{ sq. in.}$$

Length of side $= \sqrt{5} = 2.236$ in., say $2\frac{1}{4}$ in.

The amount of elongation or of shortening of a piece under a stress is given by the formula

$$e = \frac{Pl}{AE} \quad (2)$$

The coefficient of elasticity (E) must be taken from the following table:

TABLE III.

Name of Material.	Coefficient of Elasticity.	Elastic Limit for Tension.
Cast iron	15,000,000	6,000
Wrought iron....................	25,000,000	25,000
Steel	30,000,000	50,000
Wood.................................	1,500,000	3,000

A wrought-iron bar 24 ft. long, $1\frac{1}{4}$ in. in diameter, would elongate, under a tensile stress of 15 tons,

$$e = \frac{(15 \times 2{,}000) \times (24 \times 12)}{\frac{1}{4}\pi\,(1\frac{1}{4})^2 \times 25{,}000{,}000} = .196 \text{ in.}$$

To find the breaking strength of a beam, use the formula

$$M = SR. \quad (8)$$

Obtain M and R from the two following tables, according to the kind of beam and nature of cross-section. A simple beam is one merely supported at its ends. In the expression for R, d is always understood to be the vertical side or depth; hence, that beam is the stronger which always has its greatest depth or longest side vertical. The moment of inertia I is taken about an axis perpendicular to d, and lying in the same plane.

TABLE IV.

Kind of Beam and Manner of Loading.	Bending Moment. M	Deflection. s
Cantilever, load at end	Wl	$\frac{1}{3}\dfrac{Wl^3}{EI}$
Cantilever, uniformly loaded	$\frac{1}{2}Wl$	$\frac{1}{8}\dfrac{Wl^3}{EI}$
Simple beam, load at middle	$\frac{1}{4}Wl$	$\frac{1}{48}\dfrac{Wl^3}{EI}$
Simple beam, uniformly loaded	$\frac{1}{8}Wl$	$\frac{5}{384}\dfrac{Wl^3}{EI}$
Beam fixed at both ends, load at middle...................	$\frac{1}{8}Wl$	$\frac{1}{192}\dfrac{Wl^3}{EI}$
Beam fixed at both ends, uniformly loaded	$\frac{1}{12}Wl$	$\frac{1}{384}\dfrac{Wl^3}{EI}$

TABLE V.

Name of Section.		I	R	G^2
Solid circular		$\dfrac{\pi d^4}{64}$	$\dfrac{\pi d^3}{32}$	$\dfrac{d^2}{16}$
Hollow circular		$\dfrac{\pi(d^4 - d_1^4)}{64}$	$\dfrac{\pi(d^4 - d_1^4)}{32d}$	$\dfrac{d^2 + d_1^2}{16}$
Solid square		$\dfrac{d^4}{12}$	$\dfrac{d^3}{6}$	$\dfrac{d^2}{12}$
Hollow square		$\dfrac{d^4 - d_1^4}{12}$	$\dfrac{d^4 - d_1^4}{6d}$	$\dfrac{d^2 + d_1^2}{12}$
Solid rectangular		$\dfrac{bd^3}{12}$	$\dfrac{bd^2}{6}$	$\dfrac{b^2}{12}$
Hollow rectangular		$\dfrac{bd^3 - b_1 d_1^3}{12}$	$\dfrac{bd^3 - b_1 d_1^3}{6d}$	$\dfrac{b^3 d - b_1^3 d_1}{12(bd - b_1 d_1)}$
Solid triangular		$\dfrac{bd^3}{36}$	$\dfrac{bd^2}{24}$	$\dfrac{d^2}{18}$
Solid elliptical		$\dfrac{\pi bd^3}{64}$	$\dfrac{\pi bd^2}{32}$	$\dfrac{b^2}{16}$
Hollow elliptical		$\dfrac{\pi}{64}(bd^3 - b_1 d_1^3)$	$\dfrac{\pi(bd^3 - b_1 d_1^3)}{32d}$	$\dfrac{b^3 d - b_1^3 d_1}{16(bd - b_1 d_1)}$
I-beam		$\dfrac{bd^3 - b_1 d_1^3}{12}$	$\dfrac{bd^3 - b_1 d_1^3}{6d}$	$\dfrac{b^3 d - b_1^3 d_1}{12(bd - b_1 d_1)}$
Cross with equal arms (approximate)				$\dfrac{d^2}{22.5}$
Angle with equal arms (approximate)				$\dfrac{d^2}{25}$

Thus, the breaking strength of a cast-iron simple beam uniformly loaded and 20 ft. long between the supports, having a hollow rectangular cross-section 8 in. by 6 in. outside and 6 in. by 4 in. inside, is given by the formula

$$M = S R, \text{ or } \tfrac{1}{4} W l = 36,000 \times \frac{b\,d^3 - b_1\,d_1^3}{6\,d};$$

whence, $\quad W = \dfrac{36,000 \times 8 \times (6 \times 8^3 - 4 \times 6^3)}{(20 \times 12) \times (6 \times 8)} = 55,200.$

Using a factor of safety of 6, the beam should support

$$\frac{55,200}{6} = 9,200 \text{ lb.}$$

with perfect safety. The value of S for beams should be taken from the flexure column of Table II.

To find the amount of deflection in a beam due to a load, substitute the values of W, l, E, and I in the different expressions for the deflection s in Table IV.

The value of I is to be taken from Table V.

EXAMPLE.—What is the deflection of a wrought-iron beam fixed at both ends, 7 ft. long between the supports, having a solid rectangular cross-section 6 in. wide and 2¾ in. deep, carrying a load of 21,000 lb. in the middle?

SOLUTION.—From the table,

$$s = \frac{W l^3}{192\,E I} = \frac{W l^3}{192\,E \times \dfrac{b\,d^3}{12}} = \frac{21,000 \times (7 \times 12)^3 \times 12}{192 \times 25,000,000 \times 6 \times (2\frac{3}{4})^3} = .249''.$$

EXAMPLE.—It is desired to calculate the depth (d) of a cast-iron cantilever 36 in. in length ($= l$) that will sustain at its end a weight of 4,000 lb. ($= W$), the lever to be of rectangular section and 2 in. in width.

SOLUTION.—The ultimate stress per square inch for cast iron in flexure is given in Table II as 36,000 lb. ($= S$). The weight will be a steady load, and therefore, according to Table I, a factor of safety of 6 should be used. By formula (3), $M = S R$. For a cantilever beam carrying a load at the end, $M = W l$ (Table IV); and for a rectangular section, $R = \dfrac{b\,d^2}{6}$ (Table V).

Then, as $W = 4,000$, $l = 36$, $b = 2$, $f = 6$, we have

$$\frac{S R}{f} = M, \text{ or } \frac{S b\,d^2}{6 f} = W l.$$

The value of d is found by substituting in this equation the known values of S, b, W, l, and f, as follows:

$$\frac{36,000 \times 2 \times d^2}{6 \times 6} = 4,000 \times 36; \text{ whence, } d = 8.49 \text{ in.}$$

At the point where the beam is supported, the required depth is found to be 8.49, or, practically, 8¼ in. At a point 6 in. from the support, the depth may again be calculated by substituting in the equation the value of l (the overhanging length beyond this point); $l = 30$, and the equation becomes

$$\frac{36,000 \times 2 \times d^2}{6 \times 6} = 4,000 \times 30.$$
$$d = 7.75 \text{ in.}$$

At a point 12 in. from the support, $l = 24$, and

$$\frac{36,000 \times 2 \times d^2}{6 \times 6} = 4,000 \times 24; \text{ whence, } d = 6.93 \text{ in.}$$

At a point 18 in. from the support, $l = 18$; and from the equation, $d = 6$ in.; at 24 in. from the support, $l = 12$ and $d = 4.9$ in.; at 80 in. from the support, $l = 6$ and $d = 3.46$ in.; at 36 in. from the support, or at the end of the beam, $l = 0$ and $d = 0$.

The depths required to be given to the lever or beam at the point of support and at intervals of 6 inches along its

length, are found to be 8.49, 7.75, 6.93, 6, 4.90, and 3.46 inches, respectively.

The lever is shown in the figure; theoretically, it would taper to nothing at the end, as indicated by dotted lines, but practically sufficient metal must be added at that point to provide means of attaching the weight.

Note.—In the preceding examples the weight of the beam has been neglected. If, however, this weight is large in comparison with the weight or weights carried by the beam, it should be taken into account, considering it (when the cross-section of the beam is the same throughout) as a load uniformly distributed over the whole length of the beam.

COLUMNS.

To find the breaking strength of a column, use the following formula:

$$P = \frac{SA}{1 + q\,\dfrac{l^2}{G^2}}. \qquad (4)$$

S is taken from Table II, in the column for compression, G^2 from Table V, and q from the following table, according to the character of the ends.

TABLE VI.

Material.	Both Ends Flat or Fixed.	One End Round.	Both Ends Round.
Cast iron	$\dfrac{1}{5,000}$	$\dfrac{1.78}{5,000}$	$\dfrac{4}{5,000}$
Wrought iron	$\dfrac{1}{36,000}$	$\dfrac{1.78}{36,000}$	$\dfrac{4}{36,000}$
Steel	$\dfrac{1}{25,000}$	$\dfrac{1.78}{25,000}$	$\dfrac{4}{25,000}$
Wood	$\dfrac{1}{3,000}$	$\dfrac{1.78}{3,000}$	$\dfrac{4}{3,000}$

The breaking load of an elliptical wooden column 18 ft. long, having rounded ends, the diameters of the cross-section being 12 in. and 8 in., is

$$P = \frac{SA}{1 + q\,\dfrac{l^2}{G^2}} = \frac{8,000 \times (\tfrac{1}{4}\,\pi \times 12 \times 8)}{1 + \dfrac{4}{3,000} \times \dfrac{(18 \times 12)^2}{\dfrac{8^2}{16}}} = 36,442 \text{ lb.}$$

Using a factor of safety of 8, the column should support $\dfrac{36,442}{8} = 4,555$ lb with perfect safety.

SHAFTING.

The diameter of a shaft may be found by the following formulas. The first is used when great stiffness is required, and the shafts are very long; the second when strength only is required to be considered.

$d =$ diameter of shaft in inches;
$H =$ horsepower transmitted;
$N =$ number of revolutions per minute;
$c =$ constant in formula (5);
$k =$ constant in formula (6).

$$d = c\sqrt[4]{\frac{H}{N}}. \quad (5) \qquad d = k\sqrt[3]{\frac{H}{N}}. \quad (6)$$

$c = 5.29$ for cast iron; 4.92 for wrought iron; 4.7 for steel;
$k = 4.56$ for cast iron; 3.62 for wrought iron; 3.3 for steel.

NOTE.—To extract the fourth root, extract the square root twice.

PIPES AND CYLINDERS.

$p =$ pressure in pounds per square inch;
$d =$ diameter of pipe or cylinder in inches;
$t =$ thickness in inches;
$S =$ ultimate tensile strength taken from Table II;
$r =$ inside radius in inches;
$f =$ factor of safety, usually taken as 6 for wrought iron and 12 for cast iron.

For thin pipes; $\quad p\,d f = 2\,t S.$ \qquad (7)

For thick pipes or cylinders,

$$pf = \frac{S t}{r + t}. \qquad (8)$$

ROPES AND CHAINS.

$D =$ diameter of the rope in inches = diameter of iron from which the link in chain is made;
$W =$ safe load in tons of 2,000 lb.

For common hemp rope, $W = \frac{1}{4} D^2$.

For iron-wire rope, $W = \frac{2}{3} D^2$.

For steel-wire rope, $W = \frac{14}{4} D^2$.

For close-link wrought-iron chain, $W = 6 D^2$.

For stud-link wrought-iron chain, $W = 9 D^2$.

BOILERS.

BOILER DESIGN.

TO DEVELOP THE DOME OF A BOILER.

A side view of the dome, together with a section of the boiler, is shown in Fig. *A*. Draw Fig. *B*, the end view of the dome and of the boiler. Above the dome draw a circle *i n e'' m* of the same diameter as the dome. Divide the lower

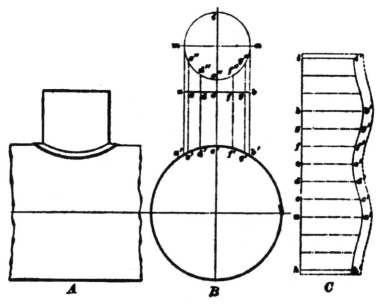

A   *B*   *C*

half of this circle, as *n e'' m*, into any number of equal parts, as *m c''*, *c' d''*, *d'' e''*, *e'' f''*, and *f'' g'*. The greater the number of these divisions, the more accurate will be the results. From the points of division *c''*, *d''*, *e''*, *f''*, and *g''*, draw lines parallel to the vertical center line of the boiler, as *c'' c'*, *d'' d'*, *f'' f'*, and *g'' g'*.

We are now ready to draw the templet of the dome, as shown in Fig. *C*. Draw a straight line of indefinite length, and on it lay off a distance *h i* equal to the circumference of

the dome. (The circumference of the dome is found by
multiplying the diameter $a\,b$ of the dome by 3.1416.) Divide
the distance $h\,i$ into twice the number of equal parts that the
semicircle above the dome in Fig. B has. In the figure it has
been divided into 6 equal parts; therefore, divide this line
into $2 \times 6 = 12$ equal parts, as $b\,g$, $g\,f$, $f\,e$, $e\,d$, etc., and
through these points of division draw lines at right angles to
the line $h\,i$, as shown; make the length of each of these lines
the same as the length of the line that corresponds to it in
Fig. B. Thus, $e\,e'$ is equal to $e\,e'$ in Fig. B, $d\,d'$ is equal to
$d\,d'$ in Fig. B, $a\,a'$ is equal to $a\,a'$ in Fig. B, etc. After having laid off the lengths of these lines, draw the curved line
$i'\,e'\,h'$. This being done, we have the templet of the dome
on the seam. The lap for riveting must be allowed, as shown •
by the dotted lines around the templet.

TO DEVELOP THE SLOPE SHEET $a\,b\,c\,d$ OF A BOILER, SHOWN AT A IN THE FIGURE BELOW.

Draw a straight line $a\,b$, as shown in Fig. B, and on it lay
off the distance $a\,d$, equal to $b\,c$, Fig. A. At a and d, erect

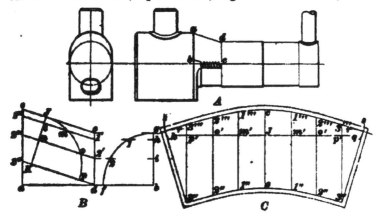

perpendiculars $a\,c$ and $d\,e$, respectively, making $a\,c$ equal to
$b\,a$, and $d\,e$ equal to $c\,d$, of Fig. A. With a point b on $a\,b$ as a
center, and a radius $d\,e$, describe the quadrant $f\,g$. Divide this
quadrant into any number of parts; the greater the number,

13

the more accurate will be the results. Here it is divided
into three, as g-1, 1-2, and 2-f. Through the points g, 1, and 2,
draw lines parallel to ab, intersecting the perpendicular de
in e, $1'$, and $2'$, and the perpendicular bg in h and i. Through
the points $1'$, $2'$, and d, draw lines parallel to ce. Through any
point, as J, on the line ce, draw JK perpendicular to ce, cutting the lines $1''$-$1'$, $2''$-$2'$, and $3''$-d in the points i, n, and K,
respectively. From the line JK lay off the distances im, no,
and Kp, equal to the distances $h1$, $i2$, and bf, respectively,
and pass the dotted curve $Jmop$ through the points. Now
draw Fig. C. Draw the straight line kq, and through the
point J draw ec perpendicular to it. Lay off on the line kq,
on each side of the line ce, points m' and m' at distances
from it equal to the length of Jm in Fig. B. Lay off, also,
points o' and o' at distances from m' and m' equal to mo in
Fig. B; also, points p' and p' at distances from o' and o' equal
to op of Fig. B. Through the points thus laid off, draw lines
parallel to ce. Lay off the distances Jc and Je from J, in
Fig. C, equal to Jc and Je, respectively, in Fig. B; the distances $m'1'''$ and $m'1''$ from m' equal to $i1''$ and $i1'$ in
Fig. B; $o'2'''$ and $o'2''$ from o' equal to $n2''$ and $n2'$; and
$p'3'''$ and $p'3''$ from p' equal to $K3''$ and Kd of Fig. B.
Through the points thus laid off draw the curved lines
$3'''c3'''$ and $3''e3''$. With the points $3''$ as centers and a
radius ad, Fig. B, describe the arcs r and r. With the points
$3'''$ as centers and a radius $3''a$, Fig. B, describe the arcs
s and s. From the points of intersection of these arcs, draw
lines to the points $3'''$ and $3''$. This being done, we have the
templet of the slope sheet on the seams. The laps for riveting must be allowed as shown by the dotted lines around the
templet.

TO DEVELOP THE SLOPE SHEET $lmno$ OF A BOILER, SHOWN AT A IN THE FIGURE ON THE FOLLOWING PAGE.

Draw the two views of the sheet as shown in Figs. B and C.
Suppose the seam to be at on, Fig. A, and the sheet to be
made in one piece. Divide the semicircles adg and $a'd'g'$,
Fig. C, into any number of equal parts; the greater the number

of these divisions, the more accurate will be the results.
Join the points b and b', c and c', d and d', e and e', and f and
f' by full lines, and join the points b and a', c and b', d and c',
e and d', f and e', and g and f' by dotted lines, as shown.
Then draw Figs. D and E. Draw at right angles to one
another the lines wa and wx, also the lines za' and zy.
Make the length of the line wx equal to r, Fig. B, and the

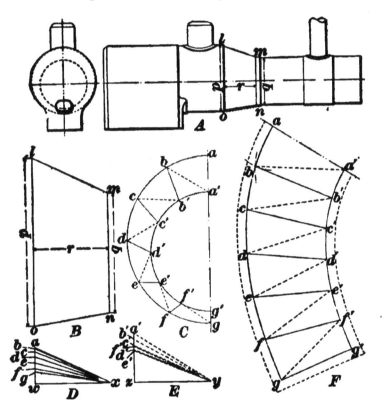

length of the line wa equal to aa', Fig. C. From w lay off on
the line wa, Fig. D, distances wb, wc, wd, we, wf, and wg,
respectively, equal to the lengths of the full lines bb', cc',
etc. of Fig. C, and draw the lines ax, bx, cx, dx, ex, fx, and
gx, as shown. Make the length of the line zy, Fig. E, the
same as that of wx, Fig. D. From z lay off on the line za',

Fig. *E*, distances *z a'*, *z b'*, *z c'*, *z d'*, *z e'*, and *z f'*, respectively, equal to the lengths of the dotted lines *b a'*, *c b'*, etc., in Fig. *C*, and draw the lines *a' y*, *b' y*, *c' y*, *f' y*, *d' y*, and *e' y*.

We are now ready to draw the templet of the slope sheet. Instead of drawing the whole templet, we will draw only one-half of it, as is shown in Fig. *F*, since the other half is exactly the same. Draw the line *a a'*, and make it equal in length to the distance *a x*, Fig. *D*. With *a'* as a center, and a radius *y a'*, Fig. *E*, describe an arc at *b*. With *a* as a center and a radius = arc *a b*, Fig. *C*, describe another arc intersecting the first arc in *b*. With *a'* as a center, and a radius = arc *a' b'*, Fig. *C*, describe an arc at *b'*. With *b* as a center, and a radius *x b*, Fig. *D*, describe an arc, intersecting the arc already drawn, at *b'*; draw the full line *b b'* and dotted line *b a'*. With *b'* as a center, and a radius *y b'*, Fig. *E*, describe an arc at *c*. With *b* as a center, and a radius = arc *c b*, Fig. *C*, describe an arc cutting the last arc at *c*. With *b'* as a center, and a radius = arc *c' b'*, Fig. *C*, describe an arc at *c'*. With *c* as a center, and a radius *x c*, Fig. *D*, describe an arc cutting the last arc at *c'*; draw the full line *c c'* and dotted line *c b'*.

Continue to construct the remaining portion of the half templet in a similar manner, taking the distances for the full lines from Fig. *D*, and those for the dotted lines from Fig. *E*. Through the points *a*, *b*, *c*, *d*, *e*, *f*, and *g*, and through the points *a'*, *b'*, *c'*, *d'*, *e'*, *f'*, and *g'*, draw the curved lines shown. Since this is the development of the slope sheet at the seam, the laps for riveting must be allowed; they are shown by the dotted lines around the templet in Fig. *F*.

CARE AND INSPECTION OF BOILERS.

POINTS TO BE OBSERVED.

Preliminary to a boiler inspection, the boiler, flues, mud-drum, ash-pit, and all connections should be thoroughly cleaned, to facilitate a careful examination. Blisters may occur in the best iron or steel, and their presence, and also that of thin places, is ascertained by going over all parts of the boiler with a hammer. When blisters are discovered, the plates should be repaired or replaced. Repairing a blister consists in cutting out the blistered space and riveting a "hard patch" over the hole on the inside of the boiler, if possible, to avoid forming a pocket for sediment. All seams, heads, and tube ends should be examined for leaks, cracks, corrosions, pitting, and grooving, detection of the latter possibly requiring the use of a magnifying glass. Uniform corrosion is a wasting away of the plates, and its depth can be determined only by drilling through the plate and measuring the thickness, afterwards plugging the hole. Pitting is due to a local chemical action, and is readily perceived. Grooving is usually due to buckling of the plates when under pressure, and frequently to the careless use of the sharp calking tool. Seam leaks are generally caused by overheating, and demand careful examination, as there may be cracks under the rivet heads. If such cracks are discovered, the seam should be cut out, and a patch riveted on. Loose rivets should be carefully looked for, and should be cut out and replaced, if found. Pockets, or bulging, and burns should be looked for in the firebox. The former are not necessarily dangerous, but if there are indications of their increasing, they should be heated and forced back into place or cut out and a patch put on. Burns are due to low water, the presence of scales, or to the continuous action of flames formed on account of air leaking through the brickwork. The burned spots should be cut out and patched as previously described. The conditions of all stays, braces, and their fastenings should be examined and defective ones replaced. The shell of the boiler should be thoroughly examined externally for evidences of corrosion, which is liable to set in on account of dampness, exposure to weather, leakage, etc., and may be serious. The boiler should be so set that joints and seams are accessible for inspection, and should have as little brickwork in contact with it as possible. The brickwork should be in good condition, and not have air holes in it, since they decrease the efficiency of the boiler and are liable to cause injury to the plates by burning, as above explained, and also by unevenly heating and distorting them. The mud-drum and its connections are liable to corrosion, pitting, and grooving, and should be examined as carefully as the boiler.

All valves about a boiler should be easy of access, and should be kept clean and working freely. Each boiler should have at least three gauge-cocks, properly located, and it is of the utmost importance that they be kept clean and in order, and the same may be said of the glass water gauge. The middle gauge-cock should be at the water level of the boiler, and the other two should be placed one above and one below it, at a distance of about 6 in.

The condition of the pumps or injectors should be looked into to make sure that they are in the best working order. The steam gauge should be tested to ascertain that it indicates correctly, and if it does not, it should be corrected. If the hydraulic test is to be used, the boiler should be tested to a pressure of 50% higher than that at which the safety valve will be set.

External Inspection When Boiler Is Under Steam.—The gauge-cocks, and also the gauge glass, should be tried, to make sure that they are not choked. The steam gauge should be taken down, if permissible, and tested, and corrected if necessary. The gauge pointer should move freely. Blowing out the gauge connection will show whether it is clear or not. The boiler connections should be examined for leaks. The safety valve should be lifted from its seat, to make sure that it does not stick from any cause, and it should be seen that the weight is in the right place. Observe from the steam gauge if the valve blows off at the pressure it is set for. See that all pumps and feed-apparatus are working properly, and that the blow-off and check-valves are in order. Blisters and bagging may sometimes be detected in the furnace. The condition of the brickwork is of considerable importance, since the existence of air holes is a source of trouble, as already explained.

Incrustation.—One of the chief sources of trouble to the boiler user is that of incrustation. All water is more or less impure; and as the water in the boiler is continuously evaporated, the impurities are left behind as powder or sediment. This collects on the plates, forming a scaly deposit, varying in nature from a spongy, friable texture to a hard, stony one. This deposit impedes the transmission of heat from the plates to the water and often causes overheating and injury to the plates. It is probable that $\frac{1}{16}$ in. of scale necessitates the consumption of 12% to 20% more fuel. The various impurities in the water may be either in suspension or solution. If the former, the water can be purified by filtration before going into the boiler. If the latter, the substances must first be precipitated and then filtered. Many impurities (sulphate and carbonate of lime, etc.) may be removed by heating the water before feeding it into the boiler.

The first thing to do, when dealing with a water supply, is to have an analysis of it made by a competent chemist. The fact that a water contains a certain amount of solid matter is no criterion as to its unfitness for boiler use. The presence of certain salts, as carbonate or chloride of sodium, even in large quantities (say 40 to 50 gr. per gal.), would not be serious if due attention were given to the blowing off. On the other hand, salts of lime in the above proportion would be very objectionable, requiring greatly increased attention in the matter of purification and blowing off or else cleaning out.

The various methods of dealing with impure water may be classed as follows:

1. *Filtration.*—Where the matter (sand, mud, etc.) is held in suspension, it can be removed, before the water enters the boiler, by the aid of settling tanks or by filtering, or by forcing the water up through layers of sand, broken brick, etc., or by using filtering cloths in a proper machine.

2. *Chemical Treatment.*—Clark's process, combined with a subsequent filtration (the joint process being known as the *Atkins* system), has been successfully applied on both small and large scales in the chalk districts of England. Lime water is mixed with the water to be purified, the amount used depending on the composition of the water, as determined by a careful analysis. The lime is thus precipitated, and the water is then filtered in a machine containing traveling cotton cloths. Not only is the carbonate of lime entirely removed, but it has been proved that any sulphate of lime that may be present is also prevented from incrusting. This ·is important, as the latter impurity forms, perhaps, the worst scale one has to contend with.

Various chemical compounds are in use for boilers. Carbonate of soda is perhaps the best general remedy. It forms the basis, in fact, of nearly all boiler compounds, whatever their name or appearance. This soda deals efficaciously both with the carbonate and the sulphate of lime. The precipitates thus thrown down do not form a hard crust; they can be washed out in the form of sludge or mud.

Carbonate of soda is also useful where condensers are employed, as it counteracts the effect of the grease, which is brought over with the exhaust steam. If used in too large quantities, it will cause priming. The best way to use it is to make a solution of it and connect with the feed, fixing a cock so as to regulate the amount fed in. Soda ash is cheaper, but more of it is required, and, besides, it is generally impure. Caustic soda removes lime scale quicker than ordinary soda does, but it is much stronger and liable to attack the plates. It should be used in smaller quantities than the ordinary kind.

Barks, molasses, vinegar, etc. develop acids that attack the plates. Animal and vegetable oils do the same, and also harden the deposits and make their removal more difficult. It is a good rule to keep all animal and vegetable matter out of boilers altogether.

Feed-Water Heaters.—Carbonates and sulphates of lime are precipitated by high temperatures. The heaters should be arranged so that the deposit forms chiefly on a series of plates that can be easily removed for cleaning. If the deposit gathers in pipes, however, it is simply transferring the evil from one vessel to another. A double advantage is gained by these heaters, for the feedwater is put into the boiler already heated, and so fuel is saved.

Mechanical Aids.—Deposits take place chiefly in sluggish places. Various devices to aid circulation have been brought out. With good attention and a not too impure water, they give satisfactory results.

Potatoes, linseed oil, molasses, etc. are sometimes put into the boiler with the idea of lessening scale formation, by forming a kind of coating round the particles of solid matter and so preventing their adhering together. This certainly takes place, but the substances are injurious, as already pointed out.

Whenever a boiler has been cleaned out, we may with advantage give the inside a thin coating of oil, or tallow and black lead; this arrests the incrustation to a great extent.

Sand, sawdust, etc. are often used, the idea being that their grains act as centers for the gathering together of the solid matter in the water, the resulting small masses not readily collecting together themselves and therefore being easily washed out. This may be so, but the cocks, valves, etc. are liable to suffer from the practice.

Kerosene is strongly recommended by some boiler users. There is no doubt that in many cases its use has given good results. It prevents incrustation, by coating the particles of matter with a thin covering of oil, the deposit thus formed being easily blown out. The oil also seems to act on the scale already formed, breaking it up and thus facilitating its removal. As already remarked, it is a good plan, when the boiler is empty, to give the inside a good coating of this oil, afterwards putting it in with the feed, the supply being regulated automatically. As to the quantity required, this will be found to vary in different cases, according to the nature of the water; an average of 1 qt. per day for every 100 horsepower will give good results in most cases.

In marine boilers, strips of zinc are often suspended; the deposit largely settles on them instead of on the boiler plates. Also, any scale that may be formed on the latter is less hard and compact and more easily broken up. Further, any acids formed by the oil and grease brought over from the condenser attack this zinc instead of the boiler plates.

Miscellaneous.—Acids are often introduced into boilers to dissolve the scale already formed, the solid matter then being washed out. This treatment should be adopted with great care, if at all, as the plates are likely to be affected.

Scale is often loosened and broken up by deliberately inducing sudden expansion or contraction in the boiler. In the former case, the expansion is brought about by blowing off the boiler, and then, when it is quite cooled down, turning on steam at as high a temperature as obtainable, thus causing the scale to expand more quickly than the plates and thus become loose.

In the second method, the boiler is blown off when the steam (and therefore the temperature) is at its highest and a stream of cold water then turned in. The fires are then drawn and the fire-hole doors, dampers, etc. opened, letting in a rush of cold air. All this cools the plates and, by the contraction thus brought about, loosens the scale. These two practices should be guarded against.

Foaming or priming is usually due either to forcing a boiler beyond its capacity for furnishing dry steam, or to the presence of foreign matter. It is dangerous if occurring to any great extent, since water may be carried along with steam into the engine, and a cylinder head knocked out. Foaming, when it cannot be checked by the use of the surface blow-out apparatus, may necessitate the emptying of the boiler, which must then be filled with fresh water; this rids the boiler of the impurities that have collected during the operation of the boiler.

HORSEPOWER OF BOILERS.

In actual practice, the result of a great many tests has shown that an evaporation of 30 lb. of water per hr. from a feedwater temperature of 100° F. into steam at 70 lb. gauge pressure is the equivalent of 1 horsepower, or that this steam, in a properly designed engine, will do the equivalent of $33,000 \times 60 = 1,980,000$ ft.-lb. of work per hr. In order, however, to have a more ready standard of comparison, the above evaporation has been reduced to another standard, and is found to be equal to the evaporation of 34.5 lb. of water from and at a temperature of 212° F. under atmospheric pressure, and it is on this latter quantity that the calculations of the horsepower of boilers are usually based.

In making an approximation of the horsepower of a given boiler, the square feet of water-heating surface of the boiler should first be determined, and in doing this the area of all the surfaces exposed to the fire and hot gases, which, on their opposite sides come in contact with the water in the boiler, should be taken into account.

EXAMPLE.—An externally-fired flue boiler, having a shell 38 in. in diameter, and containing two flue pipes 10 in. in

eter, is 22 ft. long without the smokebox. If the greatest
h of the water in the boiler is ¼ × 38 = 25.33 in., what is
total water-heating area of the boiler?

SOLUTION.—Six feet of the circumference of the boiler
lies below the water-line, as could be found by actual
surement, and the circumference of the two flues is
al to

$$\left(\frac{10 \times 3.1416}{12}\right) \times 2 = 5.24 \text{ ft.}$$

Therefore, the water-heating surface of the shell is 6 × 22
132 sq. ft., and that of the flues is 5.24 × 22 = 115.28 sq. ft.
water-heating surface of the heads of the shell (that is,
area below the water-line, minus the area of the flues,
ch could be obtained by direct measurement) is 4.5 × 2 =
q. ft. Therefore, the total water-heating surface of the
ler is the sum of all these, or 256.28 sq. ft.

Having determined the water-heating surface of a boiler,
approximate its horsepower:

Rule.—*Divide the total water-heating surface in square feet by
number of square feet of heating area, as given in the table
ow, required to produce an evaporation equivalent to 1
sepower in boilers of the given type.*

EXAMPLE.—The total water-heating surface of the above
ternally-fired flue boiler is 256.28 sq. ft. What is the horse-
wer of the boiler?

SOLUTION.—By referring to the table, we find that it takes
out 10 sq. ft. of heating surface to produce 1 horsepower;
erefore, the above boiler would be rated at about

$$\frac{256.28}{10} = 25.63 \text{ H. P.}$$

Type of Boiler.	Water-Heating Surface for 1 Horsepower. Square Feet.	Ratio of Water-Heating Area to Grate Area Required.
Cylindrical	9	From 12 to 15 : 1
Flue	10	From 20 to 25 : 1
Firebox tubular	12	From 25 to 35 : 1
Return tubular	15	From 25 to 35 : 1
Vertical	15	From 25 to 30 : 1
Water tube	11	From 35 to 40 : 1

The above rule must not be taken as furnishing anything but an approximate method, since the same boiler will give a different horsepower whenever the conditions under which it is operated are changed; or, in other words, the horsepower developed depends largely on the amount of coal burned per square foot of grate area per hour, the velocity and character of the furnace draft, and the quality of the coal used. In ordinary practice, however, we may expect an evaporation of from 8 to 11 lb. of water from and at 212° F. for each pound of good coal burned, where from 11 to 13 lb. of coal are consumed per sq. ft. of grate surface per hr., or about from 3 to 4 lb. per H. P. per hr.

CHIMNEYS.

The *chimney* serves the double purpose of creating a draft and carrying away obnoxious gases. The production of the draft depends on the fact that the furnace gases (the products of combustion) passing up the chimney have a high temperature, and are, consequently, lighter than an equal volume of outside air at the ordinary temperature; that is, the pressure within the chimney is slightly less than the pressure of the outside air. Consequently, the air will flow from the place of higher pressure to the place of lower pressure, that is, into the chimney through the furnace.

Suppose, for example, the average temperature of the gases in a chimney 150 ft. high is 500° F. A pound of the gases at 62° F. has a volume of 12.5 cu. ft.; its volume at 500° is, then,

$$\frac{12.5 \times (500 + 460)}{62 + 460} = 23 \text{ cu. ft.}$$ Therefore, a column of the

gases 1 ft. square and 150 ft. long would weigh $\frac{150}{23} = 6.52$ lb.

A similar column of air at 62° F. would weigh $\frac{150}{13.14} = 11.42$ lb.,

nearly. Hence, the pressure of the draft is $11.42 - 6.52 = 4.9$ lb. per sq. ft. $= .941$ in. of water. It is evident that the pressure of the draft depends on the temperature of the furnace gases and the height of the chimney. The higher the chimney, the lower may be the temperature of the gases to produce the same draft, and the greater will be the economy of the furnace. In general, chimneys are not built much less than 100 ft. in height.

The relation between the height of the chimney and the pressure of the draft in inches of water is given by the following formula:
$$p = H\left(\frac{7.6}{T_a} - \frac{7.9}{T_e}\right),$$
where p = draft in inches of water;

H = height of chimney in feet;

T_a = absolute temperature of outside air;

T_e = absolute temperature of chimney gases.

Absolute temperatures are found by adding 460° F. to the ordinary temperatures.

EXAMPLE.—What draft pressure will be produced by a chimney 120 ft. high, the temperature of the chimney gases being 600° F. and the external air 60° F.?

SOLUTION.—By the formula we find
$$p = H\left(\frac{7.6}{T_a} - \frac{7.9}{T_e}\right) = 120\left(\frac{7.6}{460 + 60} - \frac{7.9}{460 + 600}\right) = .86 \text{ in. of}$$
water.

The draft pressures ordinarily produced by chimneys vary from 0 to 2 in. of water. A water-gauge pressure of 1 in. is equivalent to .03617 lb. per sq. in. Wood requires least draft, and the small sizes of anthracite coal the greatest draft. To successfully burn anthracite, slack, or culm, a draft of 1¼ in. is necessary.

To find the height of chimney to give a specified draft pressure, the formula may be transformed:
$$H = \frac{p}{\dfrac{7.6}{T_a} - \dfrac{7.9}{T_e}}.$$

EXAMPLE.—Required the height of the chimney to produce a draft of 1¼ in. of water, the temperature of the gases and of the external air being, respectively, 550° and 62° F.

SOLUTION.—By the formula we find
$$H = \frac{p}{\dfrac{7.6}{T_a} - \dfrac{7.9}{T_e}} = \frac{1.125}{\dfrac{7.6}{522} - \dfrac{7.9}{1,010}} = 167 \text{ ft.}$$

The sizes of chimneys for boilers of various horsepowers are given in the following table:

SIZES OF CHIMNEYS AND HORSEPOWERS OF BOILERS.

Height of Chimney in Feet.											Actual Area in Sq. Ft.	Side of Sq. in In.	Diameter in In.
50	60	70	80	90	100	110	125	150	175	200			
Commercial Horsepower.													
23	25	27									1.77	16	18
35	38	41									2.41	19	21
49	54	58	62								3.14	22	24
65	72	78	83								3.98	24	27
84	92	100	107	113							4.91	27	30
	115	125	133	141							5.94	30	33
	141	152	163	173	182						7.07	32	36
		183	196	208	219						8.30	35	39
		216	231	245	258	271					9.62	38	42
			311	330	348	365	389				12.57	43	48
			363	427	449	472	503	551			15.90	48	54
			505	539	565	593	632	692	748		19.64	54	60
				658	694	728	776	849	918	981	23.76	59	66
				792	835	876	934	1,023	1,105	1,181	28.27	64	72
					995	1,038	1,107	1,212	1,310	1,400	33.18	70	78
					1,163	1,214	1,294	1,418	1,531	1,637	38.48	75	84
					1,344	1,415	1,496	1,639	1,770	1,893	44.18	80	90
					1,537	1,616	1,720	1,876	2,027	2,167	50.27	86	96

EXAMPLE.—A round chimney 100 ft. high is to be used for a battery of boilers of 550 H. P. What should be the internal diameter?

SOLUTION.—Looking under column 100 in "Height of Chimney in Feet" the nearest horsepower is 565, and the diameter corresponding is 60 in., which should be the internal diameter of the chimney.

Chimneys are usually built of brick, though in some cases iron stacks are preferred. The external diameter of the base should be $\frac{1}{10}$ of the height, in order to provide stability. The taper of a chimney is from $\frac{1}{8}$ to $\frac{1}{4}$ in. to the foot on each side. The thickness of brickwork is usually 1 brick (8 or 9 in.) for 25 ft. from the top, increasing $\frac{1}{2}$ brick for each 25 ft. from the top downward. If the inside diameter is greater than 5 ft., the top length should be 1$\frac{1}{2}$ bricks, and if under 3 ft., it may be

¼ brick in thickness for the first 10 ft. A round chimney is better than a square one, and a straight flue better than a tapering one. If the flue is tapering the area for calculation is measured at the top.

The flue through which the gases pass from the furnaces to the chimney should have an area equal to, or a little larger than, the area of the chimney. Abrupt turns in the flue or contractions of its area should be carefully avoided, as they greatly retard the flow of the gases. Where one chimney serves several boilers, the branch flue from each furnace to the main flue must be somewhat larger than its proportionate part of the area of the main flue.

SAFETY VALVES.

Balance the valve and lever over a sharp, knife-like edge, and measure the distance from the point of suspension to the fulcrum (center of pin on which the lever turns).

Let a = distance thus measured in inches;

b = distance from center of valve to fulcrum in inches;

x = distance of weight from fulcrum in inches;

W = weight in pounds hung on lever;

Q = weight of lever and valve in pounds;

A = area of safety valve in square inches;

p = pressure per square inch in the boiler.

Then, $x = \dfrac{A\,p\,b - Q\,a}{W}$; $W = \dfrac{A\,p\,b - Q\,a}{x}$; $p = \dfrac{W x + Q a}{A\,b}$.

EXHAUST HEATING.

Exhaust steam from non-condensing engines usually contains from 20% to 25% of water and oil, the latter being employed to lubricate the engine cylinders. Before exhaust steam is allowed to enter a heating system, the water and oil should be separated from it.

The effect of turning exhaust steam into a heating system is to form a back pressure on engine, which must be avoided as far as possible by using large steam-distributing pipes.

A direct connection to the steam boilers through a pressure-reducing valve must be employed, to automatically furnish steam to the heating system when the exhaust fails. A relief valve, also, should be placed upon the system, so that surplus exhaust steam may escape to the atmosphere.

To proportion an exhaust-heating system, it is necessary to know about how many square feet of radiating surface we should employ to properly condense the exhaust steam from the non-condensing engines. To do this we must first know the weight of steam that would be discharged from the engine.

Class of Non-Condensing Engine.	Water Used per Hour for Indicated Horsepower.
Compound automatic	25 lb.
Simple Corliss	30 lb.
Simple automatic	35 lb.
Simple throttling...................................	40 lb.

From this must be deducted about 10% for condensation in the cylinders, etc., in order to obtain the real available weight of steam for heating purposes.

APPROXIMATE RATIO BETWEEN CUBIC CONTENTS AND RADIATOR SURFACE FOR EXHAUST HEATING.

Class of Building.	Direct Radiation.	Indirect Radiation.	Blower System.
	sq.ft. cu.ft.	sq.ft. cu.ft.	sq.ft. cu.ft.
Dwellings	1 to 50	1 to 40	1 to 300
Offices	1 to 70	1 to 60	1 to 365
Stores and shops..............	1 to 100	1 to 80	1 to 500
Churches, etc...................	1 to 200	1 to 150	1 to 900

The figures in the foregoing tables simply form a reasonable average, and allowance must be made for exposure, etc.

Each square foot of direct radiating surface gives off to the air around it about $1\frac{1}{2}$ thermal units per hour per degree of difference between the temperature of the steam and that of the surrounding air. This is equivalent to about $\frac{1}{4}$ lb. of steam per hr., or, in other words, about 4 to $4\frac{1}{2}$ sq. ft. of surface to each pound of steam to be condensed.

MACHINE DESIGN.

BLUEPRINTS.

Blueprint paper for copying tracings of plans and other drawings may be prepared as follows: Dissolve 1 oz., avoirdupois, of ammonia citrate of iron in 6 oz. of water, and in a separate bottle dissolve the same quantity of potassium ferricyanide in 6 oz. of water. Keep these solutions separate, and in a dark place, or in opaque bottles.

To prepare the paper, mix equal quantities of the two solutions, and with a sponge spread it evenly over the surface. Let the paper remain in a horizontal position until the chemical has set on the surface, which will take but a few minutes; then hang the paper up to dry. In preparing the paper darken the room by pulling down the shades, as direct rays of light affect sensitized surfaces. The prepared paper should be kept in a closed drawer, well covered with heavy paper, so that no light can come in contact with the sensitized surface; otherwise it will lose much of its value.

To make a blueprint from a tracing, lay the tracing with ink side down against the glass of the printing frame, then take the prepared paper, and place the sensitized surface down on the tracing. On the top of the paper place the felt cushion, on top of which place the hinged back of the printing frame, after which expose to the sunlight. The exposure will vary in sunlight from about 3 to 10 minutes. After the exposure, wash the paper thoroughly in a trough of cold water for about 10 minutes, and hang it up to dry.

The print after washing should be of a deep-blue color, with clear white lines. If the color is a pale blue, this indicates that the print has not had sufficient exposure, and if the lines of the drawing are not perfectly clear and white, that the exposure has been too long.

Corrections may be made on the print with an ordinary writing or ruling pen and a solution of washing soda, caustic potash, strong ammonia, or any other alkali. When any of these are mixed with carmine ink, the marks on the print will be red, thus making the corrections clear.

14

MACHINE TOOLS.

SPEED OF EMERY WHEELS.

The speed most strongly recommended by their manufacturers is a peripheral velocity of 5,500 ft. per min. for all sizes. All things being considered, it is stated that no advantage is gained by exceeding this speed. If run much slower than this, the wear on the wheels is much greater in proportion to the work accomplished, and if run much faster, the wheel is likely to burst.

SPEED OF GRINDSTONES.

Grindstones used for grinding machinists' tools are usually run so as to have a peripheral speed of about 900 ft. per min., and those used for grinding carpenters' tools at about 600 ft. per min. With regard to safety, it may be stated in general that with any size of grindstone having a compact and strong grain, a peripheral velocity of 2,800 ft. per min. should not be exceeded.

SPEED OF POLISHING WHEELS.

Polishing wheels are run at about the following peripheral speeds:

Leather-covered wooden wheels..................7,000 ft. per min.
Walrus-hide wheels..8,000 ft. per min.
Rag wheels..7,000 ft. per min.

SPEED OF CUTS FOR MACHINE TOOLS.

Brass: Use high speeds, about the same as for wood.

Bronze: 6 to 18 ft. per min., according to alloy used.

Cast or wrought iron: 20 ft. per min. is a good average for all machines, except millers. 30 is about the maximum.

Machinery steel: 15 ft. on shapers, planers, and slotters. 20 to 45 on turret lathes, according to cut.

Tool steel: 8 to 10 ft.

Milling Cutters.—*Gun metal,* 80 ft. per min.; *cast iron,* 30; *wrought iron,* 35 to 40; *machinery steel,* 30. These are good speeds to adopt, with a view to economy, time required for regrinding, etc.

Twist Drills.—The best results are obtained when the rates of speed of twist drills are as given in the following table:

Diameter of Drills.	Revolutions of Drills per Minute.		
	Steel.	Iron.	Brass.
1/16	940	1,280	1,560
1/8	460	660	785
3/16	310	420	540
1/4	230	320	400
5/16	190	260	320
3/8	150	220	260
7/16	130	185	230
1/2	115	160	200
9/16	100	140	180
5/8	95	130	160
11/16	85	115	145
3/4	75	105	130
13/16	70	100	120
7/8	65	90	115
15/16	62	85	110
1	58	80	100
1 1/16	54	75	95
1 1/8	52	70	90
1 3/16	49	66	85
1 1/4	46	62	80
1 5/16	44	60	75
1 3/8	42	58	72
1 7/16	40	56	69
1 1/2	39	54	66
1 9/16	37	51	63
1 5/8	36	49	60
1 11/16	34	47	58
1 3/4	33	45	56
1 13/16	32	43	54
1 7/8	31	41	52
1 15/16	30	40	51
2	29	39	49

The following are recommended as the best rates of feed for twist drills:

Diameter of drill in inches......................	1/16	1/4	3/8	1/2	3/4	1	1 1/2
Number of revolutions per inch depth of hole......................	125	125	120 to 140		1 in. feed per min.		

CHANGE GEARS REQUIRED FOR CUTTING SCREW THREADS.

The pitch of a single-threaded screw is the distance between two adjacent threads, measured on a line parallel to the axis of the screw; or, in *any* screw, whether single- or multiple-threaded, it is the distance the nut is moved by 1 revolution of the screw. Usually, a screw is spoken of as having a certain number of threads to the inch, and this is equal to the number of revolutions the screw must make in order to move the nut a distance of 1 inch; so, whether the screw is single- or multiple-threaded, the pitch is always equal to 1 divided by the number of revolutions that the screw must make in order to move the nut 1 inch.

The Simple-Geared Lathe.—In Fig. 1 is shown the usual arrangement of the change gears of a simple-geared screw-cutting lathe. By a simple-geared lathe is meant a lathe in

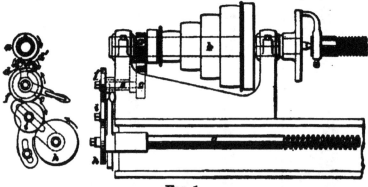

FIG. 1.

which the change gears are so arranged that the circumferential velocity of the change gear on the stud is the same as that of the change gear on the lead screw, which means that, when the change gear on the stud has rotated, say, 5 teeth, the change gear on the lead screw has also rotated 5 teeth, whatever the diameter of these gears, or of any intermediate gears between them, may be.

Referring to Fig. 1, the gear *a* is fastened to the spindle *b* and drives another gear *c* by means of either one of the reversing gears d, d'. The gear c is keyed to one end of the spindle e; this spindle is called the *stud*, and carries on its outer end a change gear f. The lead screw g carries a change gear h; and these two change gears f and h are connected by means of the *idler* gear i, so that gear f drives gear h, and with it, the lead screw g.

In making calculations for the change gears of a simple-geared screw-cutting lathe, the idler gear i is ignored, as it is only introduced to connect gears f and h. The gears d and d' are also ignored, since they are only used to change the direction of rotation of the gear c, their duty being to facilitate the cutting of either right-hand or left-hand threads; when d meshes with gear a, as shown in Fig. 1, a a right-hand thread is cut, and when d' meshes with gear a, a left-hand thread is cut.

The number of teeth in the gear a is not always the same as the number of teeth in the gear c; it is so in some lathes, but in others it is not; hence, in calculating the change gears for any lathe, the number of teeth in the gears a and c *must* be taken into account.

By the following formulas and rules, the number of teeth required in each change gear in order to cut a given number of threads to the inch, or the number of threads to the inch that given change gears will produce may be found.

Let a = number of teeth in the spindle gear a;

$\quad c$ = number of teeth in the gear c;

$\quad f$ = number of teeth in the change gear on stud;

$\quad h$ = number of teeth in the change gear on lead screw;

$\quad g$ = number of threads to the inch in the lead screw;

$\quad n$ = number of threads to the inch to be cut.

Then, $\quad n = \dfrac{g c h}{a f}$. $\quad$ (1) $\qquad h = \dfrac{n a f}{g c}$. $\quad$ (3)

$\qquad \dfrac{h}{f} = \dfrac{n a}{g c}$. $\quad$ (2) $\qquad f = \dfrac{g c h}{n a}$. $\quad$ (4)

Now, of the gears h, f, c, a, a and f are the *drivers*, and c and h being driven by a and f, are called the *driven* gears; remembering this, we deduce, from formula (1), the following rule for simple-geared screw-cutting lathes:

Rule.—*The number of threads to the inch to be cut is equal to the number of threads to the inch in the lead screw, multiplied by the product of the number of teeth in each driven gear, and divided by the product of the number of teeth in each driving gear.*

EXAMPLE.—If the lead screw *g* of a simple-geared lathe has 5 threads to the inch, and the gear *a* has 21 teeth, the gear *c* 42 teeth, the change gear *f* 60 teeth, and the change gear *h* 72 teeth, how many threads to the inch will be cut?

SOLUTION.—Using formula (1), we have

$$n = \frac{g\,c\,h}{a\,f} = \frac{5 \times 42 \times 72}{21 \times 60} = 12 \text{ teeth.}$$

From formula (2) we deduce the following rule for simple-geared screw-cutting lathes:

Rule.—*The number of teeth in the change gear on the lead screw, divided by the number of teeth in the change gear on the stud, is equal to the product of the number of threads to the inch to be cut and the number of teeth in the driving spindle gear, divided by the product of the number of threads to the inch in lead screw and the number of teeth in the fixed gear on the stud.*

EXAMPLE.—If the lead screw *g* of a simple-geared lathe has 8 threads to the inch, and the gear *a* has 16 teeth, and the gear *c* 32 teeth, how many teeth must there be in each of the gears *f* and *h* in order that the lathe may cut 10 threads to the inch?

SOLUTION.—Using formula (2),

$$\frac{h}{f} = \frac{n\,a}{g\,c} = \frac{10 \times 16}{8 \times 32} = \frac{5}{8},$$

and, if it were possible to have gears with 5 and 8 teeth, respectively, then a solution of the problem would be, *h* = 5, *f* = 8. It is evident that such gears are impracticable; but, as it does not change the value of a fraction to multiply both numerator and denominator by the same number, we may multiply 5 and 8, each by such a number that the resulting numbers of teeth in the gears are satisfactory. There is evidently, therefore, *more than one solution* to the problem—for if we multiply by 10 we, shall have *h* = 50, *f* = 80, which would give 12 threads to the inch; and if we multiply by 13, we shall have, as another solution, *h* = 65, *f* = 104, which would also give 12 threads to the inch, because $\frac{65}{104} = \frac{5}{8}$.

Having found that $\frac{h}{f} = \frac{1}{2}$, it is customary in practice to choose the change gears in the following manner: From the assortment of gears belonging to the lathe, choose one of convenient diameter, the number of whose teeth is divisible by either the numerator 5 or the denominator 8, and, after dividing by one of these numbers, multiply both numerator and denominator by the quotient.

EXAMPLE.—Given, $\frac{h}{f} = \frac{5}{8}$, to find the number of teeth in the two change gears h and f, respectively.

SOLUTION.—Choose a gear of convenient diameter, the number of whose teeth, say 60, is divisible by either 5 or 8, in this case by 5; divide 60 by 5, and the answer is 12. Then,

$$\frac{5 \times 12}{8 \times 12} = \frac{60}{96};$$

that is, h has 60 teeth, and f 96 teeth.

If one of the change gears is given, and it is desired to find the number of teeth in the other change gear in order to cut a given number of threads to the inch, use either formula (3) or formula (4) according as the number of teeth in gear h or in gear f is required. After the examples given, these formulas will not need explanation.

In a simple-geared screw-cutting lathe, it is often possible to cut a *fractional number of threads* to the inch, as is the case in the following example:

EXAMPLE.—If the lead screw g has 2 threads per inch, and the gear a has 20 teeth, and the gear c has 20 teeth, how many teeth must there be in each of the change gears f and h, in order to cut $5\frac{1}{4}$ threads to the inch?

SOLUTION.—Using formula (2),

$$\frac{h}{f} = \frac{n\,a}{g\,c} = \frac{5\frac{1}{4} \times 20}{2 \times 20} = \frac{5\frac{1}{4}}{2}.$$

Then, choosing a gear whose number of teeth, say 32, is divisible by 2, divide 32 by 2 and the quotient is 16. Then, $\frac{5\frac{1}{4} \times 16}{2 \times 16} = \frac{84}{32}$; that is, h has 84 teeth, and f 32 teeth. In many cases, however, it is impossible, out of the assortment of gears supplied with a simple-geared screw-cutting lathe, to find gears to cut a screw of the required number of threads to the inch. In such cases, it becomes necessary either to *make* suitable gears or to resort to a compound-geared lathe.

The Compound-Geared Lathe.—In Fig. 2 is shown the usual arrangement of the change gears of a compound-geared screw-cutting lathe. The difference between this and the simple-geared lathe lies in putting two change gears of different sizes on one spindle, in place of the idler between the gear on the stud and the gear on the lead screw. These two gears on one spindle are shown at i and j in Fig. 2, gear j meshing with gear h on the lead screw, and gear i meshing with gear f on the stud.

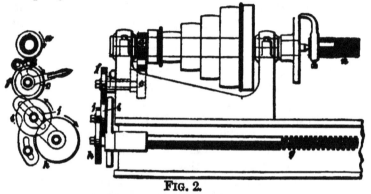

Fig. 2.

From the following formulas, the number of teeth in each change gear, or the number of threads per inch that can be cut with given change gears, can be found.

Let a = number of teeth in the spindle gear a;

c = number of teeth in the gear c;

f = number of teeth in the change gear f;

h = number of teeth in the change gear h;

i = number of teeth in the change gear i, which meshes with the change gear f;

j = number of teeth in the change gear j, which meshes with the change gear h;

g = number of threads to the inch in the lead screw;

n = number of threads to the inch to be cut.

Then,
$$n = \frac{g \times c\,h\,i}{a\,f\,j}. \qquad (5)$$

Now, remembering that gears a, f, and j are the drivers, and gears c, h, and i are the *driven* gears, and also that the idlers are ignored in all calculations, we can, from formula (5), deduce the following rule for compound-geared screw-cutting lathes:

Rule.—*The number of threads to the inch to be cut is equal to the number of threads to the inch in the lead screw, multiplied by the product of the number of the teeth in each of the driven gears, and divided by the product of the number of teeth in each of the driving gears.*

EXAMPLE.—If the lead screw g of a compound-geared lathe has 2 threads to the inch, and the gear a has 20 teeth, gear c 40 teeth, change gear f 48 teeth, change gear i 72 teeth, change gear j 36 teeth, and change gear h 96 teeth, how many threads to the inch will be cut?

SOLUTION.—Using formula (5), we have

$$n = \frac{g \times c\,h\,i}{a\,f\,j} = \frac{2 \times 40 \times 96 \times 72}{20 \times 48 \times 36} = 16 \text{ threads to the inch.}$$

If it is desired to find what combination of change gears will enable us to cut a given number of threads to the inch, the following formula may be used:

$$\frac{i}{j} = \frac{n\,a\,f}{g\,c\,h}. \qquad (6)$$

From this formula the following rule is deduced:

Rule.—*Of the change gears of a lathe, any driven gear divided by any driver gear is equal to the product of the numbers of teeth in each of the other driver gears and the number of threads to the inch to be cut, divided by the product of the numbers of teeth in each of the other driven gears and the number of threads to the inch in the lead screw.*

EXAMPLE.—In a compound-geared lathe, in which the lead screw has 5 threads to the inch, gear a 20 teeth, gear c 40 teeth, and the number of threads per inch to be cut is 8¼, what must be the number of teeth in each of the change gears h, i, j, f?

SOLUTION.—Using formula (6), we have

$$\frac{i}{j} = \frac{n\,a\,f}{g\,c\,h}.$$

From the assortment of gears belonging to the lathe, choose, for the driven gear h, one whose number of teeth, say 28, can be divided by the number of threads per inch to be cut, in this case $3\frac{1}{2}$; 28 is a multiple of $3\frac{1}{2}$, because it is obtained by multiplying $3\frac{1}{2}$ by 8. Substitute this value in place of h; then choose any gear of convenient size, say one having 40 teeth, and substitute 40 in place of f; we shall then have,

$$\frac{i}{j} = \frac{n\,a \times 40}{g\,c \times 28};$$

or, substituting the given values of n, a, g, and c,

$$\frac{i}{j} = \frac{3\frac{1}{2} \times 20 \times 40}{5 \times 40 \times 28} = \frac{1}{2}.$$

Choose, for j, a gear whose number of teeth, say 60, is divisible by 2; then, dividing the number of teeth in j by 2, we have $60 \div 2 = 30$. Now multiplying both terms of the fraction $\frac{1}{2}$ by 30,

$$\frac{i}{j} = \frac{1 \times 30}{2 \times 30} = \frac{30}{60};$$

that is, $i = 30$, and $j = 60$. Hence, one solution of the problem is, $h = 28$; $i = 30$; $j = 60$; $f = 40$.

HORSEPOWER OF ENGINES, BOILERS, AND PUMPS.

THEORETICAL HORSEPOWER.

The theoretical horsepower of any machine that uses a fluid (steam, gas, water, etc.) as a motive power, or that discharges a fluid (i. e., a pump or a fan), may be readily computed by the following formula, in which v is the volume of the fluid used or discharged in cubic feet per minute, and p is the average pressure in pounds per square inch:

$$\text{H. P.} = \frac{144\, v\, p}{33,000}.$$

If, in the above formula, allowance for friction, etc. is made, the final result will be the actual horsepower.

EXAMPLE.—A ventilating fan delivers 5,000 cu. ft. of air per min. at a pressure of .56 lb. above the atmospheric pressure; what is the theoretical horsepower required to drive the fan?

SOLUTION.—
$$\text{H. P.} = \frac{144\,v\,p}{33,000} = \frac{144 \times 5,000 \times .56}{33,000} = 12.218.$$

If all hurtful resistances are taken in this case as 20% of the total horsepower, the actual horsepower will be

$$12.218 \div (1-.20) = 12.218 \div .80 = 15.27 \text{ H. P.}$$

EXAMPLE.—The mean effective pressure computed from an indicator card taken from the air cylinder of an air compressor is 30.6 lb. per sq. in.; diameter of cylinder, 28 in.; stroke, 48 in.; number of strokes per minute, 108; what is the horsepower?

SOLUTION.—In this case
$$v = \frac{28^2 \times .7854 \times 48 \times 108}{1,728} \text{ cu. ft. per min.}$$

Hence,
$$\frac{144\,v\,p}{33,000} = \frac{144 \times 28^2 \times .7854 \times 48 \times 108 \times 30.6}{1,728 \times 33,000} = 246.66 \text{ H. P.}$$

HORSEPOWER OF AN ENGINE.

Let $P =$ mean effective pressure in pounds per square inch on the piston during one stroke;

$L =$ length of stroke in feet;

$A =$ area of piston in square inches;

$N =$ number of strokes per minute;

$D =$ diameter of piston in inches.

Then, to find the indicated horsepower,
$$\text{I. H. P.} = \frac{PLAN}{33,000} = \frac{238\,PLD^2N}{10,000,000}.$$

The actual horsepower may be taken as three-fourths of the indicated horsepower. The mean effective pressure may be found exactly by taking some indicator cards, finding the areas by means of a planimeter, and dividing the area by the length of the card. Multiply the result by the scale of the indicator spring, and the product will be the mean effective pressure, or M. E. P. If no planimeter is at hand, divide the card into 10 equal parts and measure each part in the middle, as shown by the dotted lines in the following figure.

Add all the dotted ordinates together, and divide by 10; this result, multiplied by the scale of the indicator spring, gives the M. E. P.

Thus, suppose a double-acting engine 26″ × 30″, making 80 rev. per min. (80 R. P. M.), gives an indicator card that, being divided up as shown in the figure and measured, gives, for the total length of the ordinates, 21.4 in. This divided by

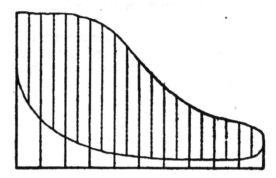

10 = 2.14 in. for the length of the mean ordinate. If a No. 40 spring is used in the indicator, every inch measured vertically on the diagram = 40 lb. per sq. in., and 2.14 × 40 = 85.6 lb. per sq. in. for the M. E. P. on the piston. Then the indicated horsepower, or I. H. P., equals

$$\frac{PLAN}{33,000} = \frac{85.6 \times \frac{49}{44} \times (.7854 \times 26^2) \times (2 \times 80)}{33,000} = 550.88.$$

The calculation is rendered much easier by using the second formula. Thus,

$$\text{I. H. P.} = \frac{238 \times 85.6 \times \frac{49}{44} \times 26^2 \times (2 \times 80)}{10,000,000} = 550.88.$$

If an indicator card cannot be obtained, a fair approximation to the M. E. P. may be obtained by adding 14.7 to the gauge pressure, and multiplying the number opposite the fraction indicating the point of cut-off in the following table by the boiler pressure. Subtract 17 from the product, and multiply by .9. The result is the M. E. P. for good simple non-condensing engines. If the engine is a simple condensing engine, subtract the pressure in the condenser instead of 17. The fraction indicating the point of cut-off is obtained by dividing the distance that the piston has traveled when the steam is cut off by the whole length of the stroke. Thus, if the stroke is 30 in., and the steam is cut off when the piston has traveled 20 in., the engine cuts off at $\frac{18}{20} = \frac{9}{10}$ stroke. For a $\frac{9}{10}$ cut-off, and 92-lb. gauge pressure in the boiler, the M. E. P. is $[(92 + 14.7) \times .943 - 17] \times .9 = 75.26$ lb. per sq. in.

Cut-off.	Constant.	Cut-off.	Constant.	Cut-off.	Constant.
$\frac{1}{4}$	.545	$\frac{3}{8}$	.772	$\frac{5}{8}$	.943
$\frac{7}{24}$	.590	.4	.794	.7	.954
$\frac{1}{3}$	.650	$\frac{1}{2}$	.864	$\frac{3}{4}$	.970
.3	.705	.6	.916	.8	.981
$\frac{1}{3}$	.737	$\frac{5}{8}$	.927	$\frac{7}{8}$	.993

THE SLIDE VALVE.

Figs. *A*, *B*, *C*, and *D* show sections of an ordinary D slide valve at different points of its travel. Fig. *A* shows the valve in its central position, with the center of the valve in line with the center line of the exhaust port. The names of the various parts are as follows: *p* and *p* are the *steam ports;* *e* is the *exhaust port;* *s, s* is the *valve seat;* the amount *o* by which the valve overlaps the outer edges of the steam ports is the *outside lap;* the amount *i* by which the valve overlaps the inside edges of the steam port is called the *inside lap;* the amount *l* (Fig. *C*) that the port is open when the piston is at the end of the stroke is called the *lead.* The valve travel is the total distance in one direction that the valve can be moved by the eccentric; it is the total distance between two extreme positions of the valve. The *displacement* of the valve is the distance that the valve has moved (in either direction) from its central position.

The line joining the center of the eccentric with the center of the crank-shaft is called the *eccentric radius.* When the eccentric radius makes a right angle with the center line of the crank, that is, when the eccentric radius is vertical (see *o e*, Fig. *E*), the valve is in its central position, provided the valve seat is horizontal, as is usually the case. When the crank is on a dead center, say *a*, Fig. *E*, the valve must be in the position shown in Fig. *C;* that is to say, the valve must have moved from its central position an amount equal to the outside lap plus the lead. In order that this may happen, the eccentric must be at *c*, Fig. *E.* The angle *eoc*, through which the eccentric must be moved from its vertical position when the crank is on a dead center, is called the *angle of advance.*

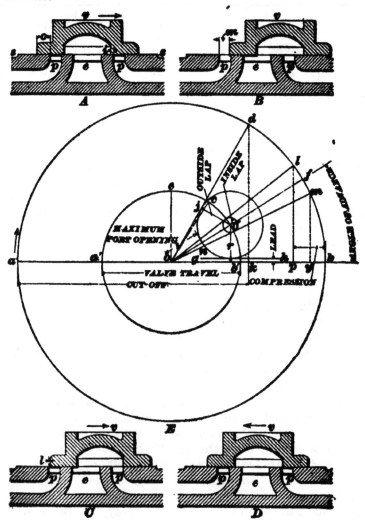

In Fig. *B*, the valve is shown in its extreme position at the right. The distance marked *m* is the *maximum port opening*. It matters not whether the outer edge of the valve travels beyond the inner edge of the port or falls short of it, as in the figure, the distance *m* between the edge of the valve and the edge of the port when the valve is in its extreme position is the maximum port opening. If, in Fig. *C*, the valve were shown moving to the left, a little farther movement would bring the left outer edge just even with the outer edge of the left steam port, and from here on to the end of the stroke no more steam could enter the left end of the cylinder; in other words, the valve *cuts off* at this point. A little farther movement of the valve to the left brings the valve to the position shown in Fig. *D*, with the right inner edge opposite the inner edge of the right steam port; it is at this point that compression begins.

When designing a valve for an engine, some of the above quantities are assumed and the remaining ones are required; these may be found by means of the diagram shown in Fig. *E*.

Let *a b*, Fig. *E*, drawn to any convenient scale, represent the stroke of the engine; then *a d b* will represent the crankpin circle. About *o*, the center of the crankpin circle, describe a circle *a' e b'*, whose diameter *a' b'* is equal to the actual travel of the valve. Draw the line *g h* parallel to *a b* and at a distance from it equal to the lead of the valve. Then, with a radius *o' f* equal to the outside lap of the valve, describe a circle, called the *outside lap circle*, tangent to the line *g h*, and having its center *o'* on the circle *a' e b'*. Draw the line *o o'*, and produce it to *f*; then *f o b = e o c = angle of advance*.

Now, draw any position of the crank center line, such as *a o*, and drop upon it, from the point *o'*, a perpendicular; the length of this perpendicular (marked *r* in Fig. *E*) is the displacement of the valve for that position of crank center line.

About the center *o'* with a radius equal to the inside lap of the valve, describe a circle; this is called the *inside lap circle*.

The radius *o d*, drawn from the point *o* tangent to the outside lap circle, is the position of the center line of crank at the point of cut-off. Drop a perpendicular from point *d*, meeting the line ab at k; then ak is the distance moved by piston before cut-off, and the fraction of the stroke at which the valve cuts off is represented by the fraction $\frac{ak}{ab}$.

Draw the radius ol tangent to the upper side of the inside lap circle, and it will be the position of the center line of the crank when *compression* commences; if a perpendicular is dropped from point l, meeting the line ab at p, the fraction of the stroke of piston at which compression begins will be represented by the fraction $\frac{ap}{ab}$.

In like manner, the radius om, drawn tangent to the lower side of the inside lap circle, is the position of the center line of the crank at the moment of *release;* and $\frac{ay}{ab}$ is the fractional part of the stroke at which the expanding steam is released.

The maximum steam-port opening is equal to on, n being the point of intersection of the outside lap circle with the angle of advance line of.

The essential features of the valve diagram having been given, the following examples will make clear its application in practice:

EXAMPLE 1.—Given, the point of cut-off, the point of release, the lead, and the maximum port opening, to find the valve travel, the outside and inside lap, the angle of advance, and the point of compression.

SOLUTION.—Draw to a convenient scale the crankpin circle adb, Fig. E, having its center at o, and its diameter ab equal to the stroke of the piston.

From the point a, lay off, on the line ab, the distances ak and ay, so that $\frac{ak}{ab}$ and $\frac{ay}{ab}$ are equal, respectively, to the fractions of the stroke at which cut-off and release are to occur. At k and y draw perpendiculars to the line ab, intersecting the crankpin circle at d and m, respectively; the radii od and om will represent the positions of the crank at cut-off and release, respectively. Now draw gh parallel to ab, and at a distance above it equal to the lead; then, about o as a center, and with a radius equal to the given maximum port opening, describe an arc. Find by trial a center o', from which a circle can be drawn tangent to this arc, and also to the radius $o\,d$, and to the line $g\,h$. The radius of this circle will be the required outside lap; and its center o' will be a point in the valve circle whose center is at o; this circle can now be drawn, since the radius $o\,o'$ is known.

The diameter $a'\,b'$ is equal to the required valve travel. Now, with o' as a center, draw a circle tangent to $o\,m$, and the radius of this circle will be the required inside lap. Draw $o\,f$ through o' and the angle $f\,o\,b$ is the required angle of advance. Draw the radius $o\,l$ tangent to the inside lap circle on its upper side, and $l\,p$ perpendicular to $a\,b$.

Then, $\dfrac{a\,p}{a\,b}$ represents the fraction of the stroke at which compression begins.

EXAMPLE 2.—Given, the valve travel, the angle of advance, the cut-off, and the point of compression, to find the lead and the outside and inside lap.

SOLUTION.—Draw the crankpin circle, as before, and the valve circle $a'\,e\,b'$; construct the angle $f\,o\,b$ equal to the angle of advance. By the same method as employed in the last example, locate the radii $o\,d$ and $o\,l$, representing the positions of the crank at the points of cut-off and compression, respectively.

About the point o', at which $o\,f$ intersects the valve circle, describe a circle tangent to $o\,d$, and the radius $o'\,j$ of this circle will be the required outside lap. Now draw the line $g\,h$ parallel to $a\,b$ and tangent to the outside lap circle; then, the perpendicular distance between $g\,h$ and $a\,b$ is the required lead. The radius of a circle drawn from o' tangent to $o\,l$ will be the inside lap.

EXAMPLE 3.—Given, the valve travel, outside lap, and the lead, to find the point of cut-off and angle of advance.

SOLUTION.—Draw the crankpin circle and the valve circle $a'\,e\,b'$ as before; draw a line parallel to $a\,b$, at a distance above it equal to the outside lap r plus the lead, intersecting the valve circle at the point o'. About o' as center, and with a radius equal to the given lap, describe a circle; draw $o\,d$

15

tangent to this circle, and drop a perpendicular from d, meeting line ab at a point k; then the required cut-off is represented by the fraction $\dfrac{ak}{ab}$. Draw the radius of through the point o' and the angle fob is the required angle of advance.

EXAMPLE 4.—Given, the outside lap, the lead, and the point of cut-off, to find the valve travel and the angle of advance.

SOLUTION.—Draw the crankpin circle as before, and by the same method as employed in Example 1 locate the radius od, the position of the crank at the point of cut-off. Draw gh parallel to ab, and at a distance above it equal to the lead. At a distance above the line ab equal to the lap plus the lead, draw another line parallel to ab; about a center o' on this line, and with a radius $o'j$ equal to the outside lap, describe a circle tangent to od and gh. Draw the radius of through o', then fob will be the required angle of advance. About o as a center, and with a radius oo', describe the valve circle $a'eb'$, and $a'b'$ will be the required valve travel.

LOCKNUTS.

A good method of locking a nut is shown in the figure.

The lower portion of the nut is turned down, and in the center of the circular portion a groove is cut. A collar is fastened by means of a pin to one of the pieces to be connected, and into this collar is fitted the circular part of the nut. The nut is then bound to the collar by a setscrew passing through the latter, the point of the setscrew engaging into the groove turned in the nut. The following proportions have proved very satisfactory, in which d, the diameter of the bolt, is taken as the unit. All dimensions are in inches:

$$a = 1\tfrac{1}{4} d - \tfrac{1}{16}''; \qquad f = \tfrac{1}{2}d + \tfrac{1}{4}'';$$
$$b = 1\tfrac{1}{4} d + \tfrac{1}{4}''; \qquad g = \tfrac{1}{2}d + \tfrac{1}{16}'';$$
$$c = \tfrac{1}{2}d + \tfrac{1}{4}''; \qquad h = \tfrac{1}{2}d + \tfrac{1}{4}''.$$
$$e = \tfrac{1}{2}d;$$

PROPORTION OF KEYS.

In common designing, the sizes of keys are determined by empirical formulas, which give an excess of strength. For an ordinary sunk key, these proportions may be adopted:

t = thickness of key in inches;

b = breadth of key in inches;

d = diameter of shaft in inches;

$b = \frac{1}{4} d$;

$t = \frac{2}{3} b = \frac{1}{6} d$.

LINE SHAFTING.

The speed of a shaft is fixed largely by the speed of the driving belt or the diameters of the pulleys upon it. In general, machine-shop shafts run about 120 to 150 rev. per min.; shafts driving wood-working machinery, about 200 to 250 rev. per min.; in cotton mills, the practice is to make the shaft diameter smaller and run at a higher speed. Line shafts should generally not be less than $1\frac{3}{4}$ in. in diameter.

The distance between the bearings should not be great enough to permit a deflection of more than $\frac{1}{100}$ in. per foot of length; hence, the bearings must be closer when the shaft is heavily loaded with pulleys.

The maximum distances between bearings of different sizes of continuous shafts used for transmitting power are:

DISTANCES BETWEEN BEARINGS.

Diameter of Shaft. Inches.	Distance Between Bearings in Feet.	
	Wrought-Iron Shaft.	Steel Shaft.
2	11	11.50
3	13	13.75
4	15	15.75
5	17	18.25
6	19	20.00
7	21	22.25
8	23	24.00
9	25	26.00

Pulleys that give out a large amount of power should be placed as near a hanger as possible.

SHAFT COUPLINGS.

A *box*, or *muff*, *coupling* is shown in the figure. It consists

of a cast-iron cylinder that fits over the ends of the shaft. The two ends are prevented from moving relatively to each other by the sunk key. The keyway is cut half into the box and half into the shaft ends. Quite commonly the ends of the shafts are enlarged to allow the keyway to be cut without weakening the shaft.

The key may be proportioned by the formula already given. For the other dimensions, take

$$l = 2\tfrac{1}{2}d + 2''$$
$$t = .4d + .5''$$

EXAMPLE.—Find the dimensions of a muff coupling for a shaft $2\tfrac{1}{4}$ in. in diameter.

SOLUTION.—For the key we use the formula previously given,

$$b = \tfrac{1}{4}d = \tfrac{1}{4} \times 2\tfrac{1}{4} = \tfrac{9}{16}''$$
$$t = \tfrac{1}{8}d = \tfrac{1}{8} \times 2\tfrac{1}{4} = \tfrac{7}{16}''$$

For the muff,

$$l = 2\tfrac{1}{2}d + 2'' = 2\tfrac{1}{2} \times 2\tfrac{1}{4} + 2'' = 8\tfrac{1}{8}''$$
$$t = .4d + .5'' = .4 \times 2\tfrac{1}{4} + .5'' = 1\tfrac{2}{5}''$$

A *flange coupling* is shown in the following figure. Cast-

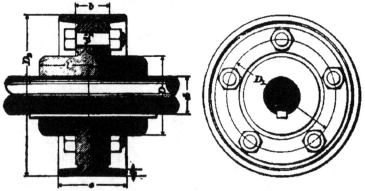

iron flanges are keyed to the ends of the shafts. To insure a perfect joint the flange is usually faced in the lathe after being keyed to the shaft. The two flanges are then brought face to face and bolted together.

Sometimes the ends of the shafts are enlarged to allow for the keyway. To prevent the possibility of the shafts getting out of line, the end of one may enter the flange of the other.

The following proportions may be used for this form of flange coupling:

d = diameter of shaft; n = number of bolts.

$$D = 1\tfrac{3}{4}d + 1''$$
$$D_1 = 2\tfrac{1}{2}d + 2''$$
$$l = 1\tfrac{1}{2}d + 1''$$
$$n = 3 + \frac{d}{2}$$

(Take the nearest whole number for n.)

$$d_1 = \frac{d}{n} + \tfrac{1}{4}''$$
$$D_2 = 1.4\,D_1$$
$$b = \tfrac{1}{4}d + \tfrac{1}{2}''$$
$$e = 2b$$
$$t = \tfrac{1}{4}d$$

The proportions for the key have already been given.

In the accompanying figure is shown a flexible coupling, or *universal joint*. These joints, when constructed of wrought iron, may have the following proportions in terms of the diameter d of the shaft:

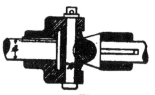

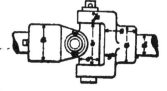

$a = 1.8\,d$	$g =$	$.6\,d$
$b = 2\,d$	$h =$	$.5\,d$
$c = d$	$k =$	$.6\,d$
$e = 1.6\,d$		

PEDESTALS.

The names *pedestal, pillow-block, bearing,* and *journal-box* are used indiscriminately. They are all a form of bearing, and indicate a support for a rotating piece.

A form of journal-box frequently used for small shafts is shown in Fig. 1. It consists of two parts: (1) the box that supports the journal, and (2) the cap that is screwed down to the box. In this journal-box the seats are of babbitt, or, as it is commonly expressed, the box is *babbitted*. The cap is held in place by what are called *capscrews*. This is invariably done in small pedestals.

The proportioning of a pedestal is largely a matter of

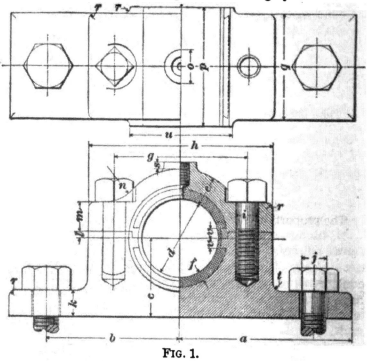

Fig. 1.

experience. Few or none of the parts are calculated for strength.

All the proportions of the pedestals that follow are based on the diameter of the journal *d* as the unit; the length of the seats is the same as that of the journal.

For the journal-box shown in Fig. 1, the following proportions may be used for sizes of journals from ½ in. to 2 in. diameter, inclusive. The diameter of shaft *d* is the unit.

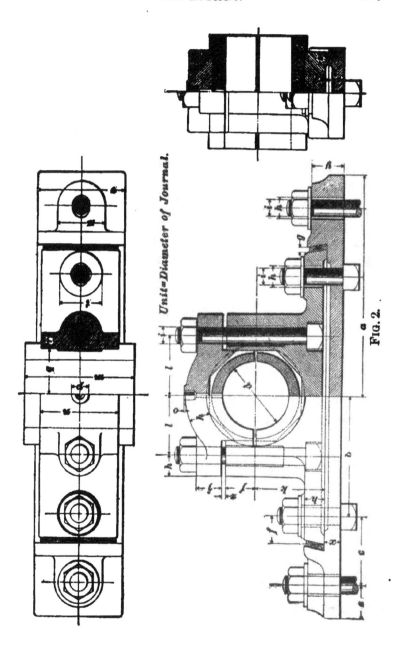

Unit=Diameter of Journal.

FIG. 2.

$a = 2.25\,d;$

$b = 1.75\,d;$

$c = d;$

$e = .375\,d;$

$f = .08\,d + .0625'';$

$g = 1.75\,d;$

$h = 2.45\,d;$

$i = .3\,d;$

$j = .33\,d;$

$k = .25\,d + .125'';$

$l = .08\,d;$

$m = .25\,d + .1875'';$

$n = .5\,d;$

$o = .625''$ (constant);

$p = 1.5\,d;$

$q = 1.333\,d;$

$r = .08\,d;$

$s = .125''$ (constant);

$t = .16\,d;$

$u = 1.333\,d;$

$v = .125\,d.$

In Fig. 2 is shown a common form of pedestal that is used for somewhat larger journals than the one shown in Fig. 1.

It consists of (1) a foundation plate that is bolted to the foundation on which the pedestal rests; the plate is essential when the pedestal rests on brickwork or masonry, but may be dispensed with when the pedestal rests on the frame of the machine; (2) the block that carries the seats and supports the journal; (3) the cap that is screwed down over the seats. The bolt holes in both foundation plate and block are oblong, so that the pedestal may be readily adjusted.

The following proportions may be used for this kind of pedestal, having journals from 2 in. to 6 in., inclusive. An oil cup having a ¼ in. pipe-tap shank may be used on pedestals for journals having diameters from 3 in. to 4 in., and ⅜ in. pipe-tap shank for larger sizes up to 6 in. diameter.

NOTE.—The shanks of oil cups and grease cups bought in the market are made with a ⅛", ¼", ⅜", or ½" pipe thread. The amount of oil or grease the cup holds when filled is usually expressed in ounces.

The diameter of journal d is the unit.

$a = 3.25\,d;$

$b = 1.75\,d;$

$c = d;$

$e = .5\,d;$

$f = .4375\,d;$

$g = .09\,d;$

$h = .3125\,d;$

$i = .25\,d;$

$j = .375\,d;$

$k = 1.0625\,d;$

$l = .875\,d;$

$m = 1.75\,d;$

$n = 1.25\,d;$

$o = .125''$ (constant);

$p = .875''$ (constant);

$q = .625\,d;$

$r = .25\,d;$

$s = .1875\,d;$

$t = .65\,d;$

$u = .75\,d;$

$v = 1.375\,d;$

$x = .25\,d;$

$y = .5\,d;$

$z = .0625\,d.$

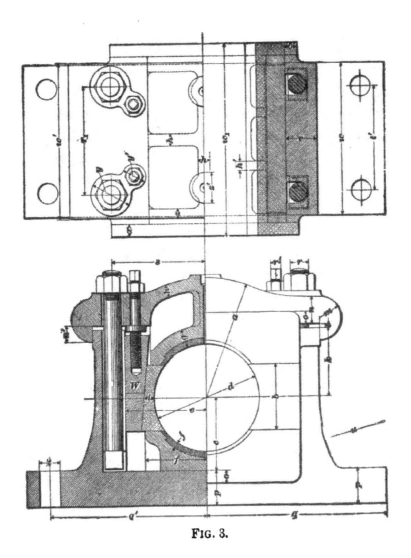

FIG. 3.

Fig. 3 shows a pedestal suitable for the crank-shaft of a horizontal engine with journals from 8 in. to 20 in. in diameter. The block may be complete in itself, as shown in the figure, but more often it forms part of the engine bed.

The seats are in three parts, and may be adjusted horizontally by means of the wedges W. The lower seat may be raised by placing packing pieces under it. To obtain its dimensions, use the following proportions, which are based on the unit d = the diameter of the crank-shaft journal.

$$a = d + 1''; \qquad q' = 1.5\,d;$$
$$b = .5\,d + 1''; \qquad r = .15\,d;$$
$$c = .66\,d; \qquad r' = .1\,d;$$
$$e = .825\,d - .25''; \qquad r_1 = d;$$
$$f = .6\,d; \qquad s = .9\,d;$$
$$g = .1\,d + .5625''; \qquad t = 15\,d + .375'';$$
$$h = .1\,d + .25''; \qquad t' = .9\,d;$$
$$h' = .08\,d; \qquad u = 1.5\,d;$$
$$i = .11\,d; \qquad v = .25\,d + .375'';$$
$$j = .625'' \text{ (constant)}; \qquad w = 1.45\,d;$$
$$k = .5\,d + 1.25''; \qquad w' = 1.47\,d;$$
$$l = .375'' \text{ (constant)}; \qquad w_1 = 1.75\,d;$$
$$m = .175\,d + .31.25''; \qquad x = .1\,d;$$
$$n = .25\,d + 25''; \qquad y = .3\,d + .75'';$$
$$n' = .1\,d + .375''; \qquad y' = .2\,d + .5'';$$
$$o = 1'' \text{ (constant)}; \qquad z = .09\,d;$$
$$p = .25\,d + .625''; \qquad z' = 2.5'' \text{ (constant)}.$$
$$q = 1.75\,d;$$

Taper of adjusting wedge, 1 : 10.

Further details of the bottom seat and the cap are shown in Fig. 4, in which the unit is the same as in Fig. 3, and the proportions are as follows:

$$a = 1'' \text{ (constant)}; \qquad c = .08\,d;$$
$$b = 1.65\,d - .5''; \qquad d = .1\,d.$$

The foundation casting, or the bed casting, is shown in Fig. 5, and has dimensions to suit the pedestal that is shown in Fig. 3. The proportions of the casting are given in connection with Fig. 5, on page 201. The diameter d of the crank-shaft journal is taken as the unit.

$a = 2.45\,d + 7.25''$;
$b = 2.3\,d + 5.25''$;
$c = .5\,d + 3.5''$;
$e = 3.5\,d + 2''$;
$f = .25\,d + .5''$;
$g = .25\,d + 1.75''$;
$h = .25\,d + 2.25''$;
$i = .05\,d + .5''$;
$j = .05\,d + 1.125''$;
$k = .05\,d + .75''$;
$l = .25\,d + .75''$;
$m = .4\,d$;
$m' = .6\,d$;
$n = 1.55\,d + 2.5''$;
$o = .25\,d + 2''$;
$o' = .25\,d + .5''$;
$o'' = .5\,d + 4.5''$;
$p = .08\,d$;
$q = 1.5\,d$;
$r = .15\,d + .375''$;
$s = .15\,d + .375''$;
$t = .9\,d$;
$u = .15\,d + .875''$;
$v = .2\,d$;
$w = 1.5\,d$;
$x = 1.65\,d$.

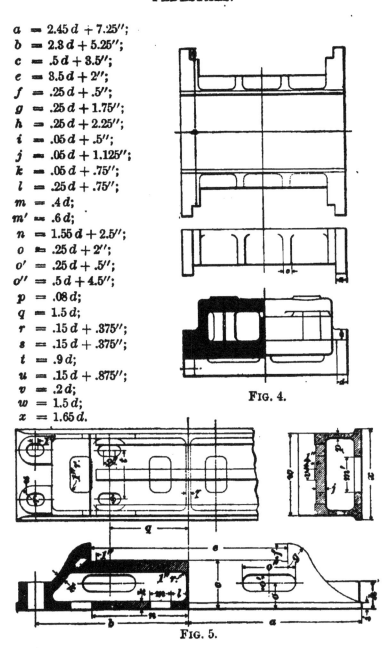

FIG. 4.

FIG. 5.

HANGERS.

A hanger is used when a shaft bearing is to be suspended from the ceiling. The figure on page 203 shows a form of hanger made by a leading manufacturing company.

The frame of the hanger is divided and the parts are connected by bolts. With such a form, the shaft may be more easily removed than when the hanger frame is a solid piece.

The units for determining the leading dimensions of a shaft hanger are the diameter d of the shaft and the drop D of the hanger.

The following proportions are suitable for shafts ranging from $1\frac{1}{2}$ in. to $4\frac{1}{2}$ in. in diameter:

$A = 6 d + .45 D$;

$A_1 = 2 d + .03 D$;

$B = 4 d + .35 D$;

$C = 2 d + .3 D$;

$E = 2 d + .25 D$;

$F = .5 d + .01 D$;

$F_1 = 1.5 d + .05 D$;

$G = 1.25 d$;

$H = 2 d$;

$I = .4 d$;

$J = .125 d + .01 D$;

$K = .5 d + .5''$

$L = .25 d + .5''$;

$M = .75 d + .6875''$;

$N = .25 d + .375''$;

$O = 1.25 d$;

$O_1 = .094 d + .002 D$;

$P = .375 d + .008 D$;

$Q = .375 d + .008 D$;

R and R_1 (see note);

$S = .25 d + .005 D$;

$S_1 = .125 d + .003 D$;

$T = .125 d + .01 D$;

$T_1 = $ (see note);

$U = 2 d$;

$V = .5 d$;

$W = .75 d$;

$X = .375 d$;

$Y = .25 d + .125''$;

$Z = .625 d$;

$a = .15 d + .375''$;

$a_1 = 2.4 d + .3125''$;

$b = .08 d$;

$c = .125 d + .0625''$;

$e = .2 d$;

$e_1 = .4 d$;

$e_2 = .2 d$;

$f = .375 d + 1''$;

$f_1 = .09 d + .25''$;

$g = .75 d$;

$g_1 = 1.3125 d + .125''$;

$h = 1.25 d + .1875''$;

$i = .1 d$;

$j = .25 d + .25''$;

$j_1 = .125 d + .0625''$;

$k = 2.2 d$;

$l = 4 d$;

$m = 1.4 d + .375''$;

$n = d$;

$o = .25 d$;

$o_1 = .0625 d$;

$p = d$;

$p_1 = .0625 d$;

$q = .4 d$;

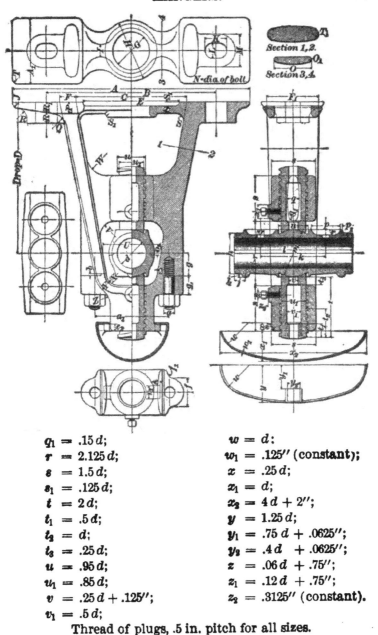

Section 1, 2.

Section 3, 4.

N-dia of bolt

$q_1 = .15\,d;$

$r = 2.125\,d;$

$s = 1.5\,d;$

$s_1 = .125\,d;$

$t = 2\,d;$

$t_1 = .5\,d;$

$t_2 = d;$

$t_3 = .25\,d;$

$u = .95\,d;$

$u_1 = .85\,d;$

$v = .25\,d + .125'';$

$v_1 = .5\,d;$

$w = d:$

$w_1 = .125''$ (constant);

$x = .25\,d;$

$x_1 = d;$

$x_2 = 4\,d + 2'';$

$y = 1.25\,d;$

$y_1 = .75\,d + .0625'';$

$y_2 = .4\,d + .0625'';$

$z = .06\,d + .75'';$

$z_1 = .12\,d + .75'';$

$z_2 = .3125''$ (constant).

Thread of plugs, .5 in. pitch for all sizes.

NOTE.—To find R_1, draw the arc J; also, draw the arc Q tangent to P; then, draw a straight line tangent to these arcs, and R_1 will be the distance along the center line determined by B included between this tangent and the upper face of the hanger. Having found R_1, make R equal to it.

The radius T_1 is made equal to three-eighths of the thickness at the middle.

The steps of the ball-and-socket bearings are of cast iron, and are bored to fit the journal without using either lining or brasses. The ball and the recesses in the ends of the plugs, into which the ball is fitted, should be faced. The screw threads on the plugs may be cast on the plugs or turned, the latter being preferable. It is customary to use 2 threads per inch for all sizes of plugs.

BELT PULLEYS.

The accompanying table gives the dimensions of a set of cast-iron belt pulleys ranging from 6 in. to 72 in. in diameter, as

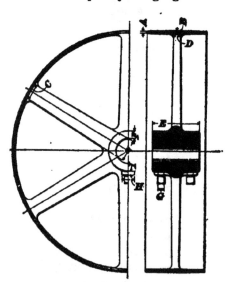

made by a well-known manufacturing company. These pulleys are so designed that the number of patterns may be kept within reasonable limits, and at the same time have the dimensions correspond as nearly as possible with well-established rules.

The letters over the columns of dimensions given in the table correspond to the letters in the figure.

In all cases the number of arms is 6, and the arms increase in size toward the hub, the taper being ¼ in. per ft.

In order to prevent heavy stresses in shafts and bearings, pulleys that are to run at high speeds must be carefully balanced. Perfect balance involves two conditions: (a) the center of gravity of the pulley must lie in the center line of the shaft, (b) the straight line joining the centers of gravity of any pair of opposite halves of the pulley must be perpendicular to the center line of the shaft.

The usual method of balancing a pulley is to rivet a weight to the light side and test the balance by putting the pulley on a mandrel that is placed on two carefully leveled ways on which it can roll with very little friction. If the center of gravity of the pulley lies in the center of the shaft, the pulley will stay in position when stopped with any point of its circumference over the mandrel; if, however, one side of the pulley is heavier, the mandrel will roll until the heavy side is at the lowest possible point.

While the above method does not determine whether or not the second condition of perfect balance is fulfilled, it is generally sufficient for pulleys running at ordinary limits of speed and reasonably well made.

In some cases, however, a failure to meet the requirements of the second condition of perfect balance may result in unsatisfactory running and severe stresses in the shaft and its bearings. Consider a pulley in which the center of gravity of one half is at the right of a line perpendicular to the center line of the shaft while the center of gravity of the opposite half is on the left of the perpendicular. This condition will not affect the balance of the pulley when tested by the mandrel rolling on the ways; when, however, the pulley revolves around the center line of the shaft, the centrifugal forces of the two halves act in opposite directions and along different lines. These forces thus form a couple that tends to bend the shaft. Since the centrifugal force is proportional to the square of the number of revolutions, it is apparent that, at high speeds, the bending effect may be considerable, even though the lack of symmetry is not very great.

It is usually considered unsafe to run a cast-iron pulley, gear-wheel, or flywheel at a higher rim speed than 100 ft. per sec. Since the centrifugal force increases in direct proportion to the cross-section of the rim, it is evident that it is useless to try to provide against it by putting more material in the rim.

Proportions of Pulleys.

Diam.	Face.	Rim.		Arm.		Hub.		Boss.		
		A	*B*	*C*	*D*	*E*	*F*	*G*	*H*	*I*
6″	4	1/8	3/16	3/4	7/16	3	3/8	1/2	1	1/4
	6	......	......	3/4	7/16	3 1/2	1/2	1/2	1	1/4
	8	......	......	3/4	7/16	3 1/2	1/2	1/2	1	1/4
	10	......	......	3/4	7/16	4	1/2	1/2	1	1/4
	12	......	......	3/4	7/16	4	1/2	1/2	1	1/4
8	4	1/8	3/16	11/16	7/16	3	3/8	1/2	1	1/4
	6	......	......	......	......	3 1/2	1/2	......	......	......
	8	5/32	1/4	1 1/16	3/16	4 1/2	......	......	......	......
	10	......	......	......	......	5 1/2	......	......	......	......
	12	......	......	......	......	......	......	......	......	......
10	4	1/8	3/16	15/16	3/16	3	1/2	1/2	1	1/4
	6	5/32	1/4	1 1/16	......	3 1/2	......	......	......	......
	8	......	......	......	......	4 1/2	......	......	......	......
	10	......	......	1 5/16	5/8	5 1/2	5/8	5/8	1 1/4	3/8
	12	......	......	......	......	......	......	......	......	......
12	4	5/32	1/4	1	7/16	3 1/4	1/2	1/2	1	1/4
	6	......	......	1 1/4	1/2	4	......	......	......	......
	8	......	......	......	......	5	5/8	5/8	1 1/4	3/8
	10	3/16+	3/16	1 1/2	3/4	5 1/2	......	......	......	......
	12	......	......	......	......	6 1/2	......	......	......	......
14	4	5/32+	1/4	1 1/8	1/2	3 1/2	1/2	1/2	1	1/4
	6	......	......	......	......	4 1/2	5/8	5/8	1 1/4	3/8
	8	3/16	5/16	1 5/16	9/16	5	......	......	......	......
	10	......	......	......	......	6	......	......	......	......
	12	......	......	1 11/16	13/16	6 1/2	......	......	......	......
16	4	5/32+	1/4	1 3/8	9/16	3 1/2	1/2	1/2	1	1/4
	6	......	......	......	......	4 1/2	5/8	3/4	1 1/4	5/8
	8	3/16+	3/16	1 7/16	5/8	5	......	......	......	......
	10	......	......	......	......	6	......	......	......	......
	12	7/32	11/32	......	......	6 1/2	3/4	......	......	......
	16	......	......	1 7/8	15/16	8 1/4	7/8	3/4	1 3/4	......
18	4	3/16	3/16	1 5/16	9/16	4	5/8	5/8	1 1/4	3/8
	6	......	......	......	......	4 1/2	......	......	......	......
	8	7/32	11/32	1 1/2	11/16	5 1/2	3/4	......	......	......
	10	......	......	......	......	6	......	......	......	......
	12	......	......	......	......	7 1/4	7/8	3/4	1 3/8	......
	16	1/4	3/8	2 1/4	1 1/4	8	......	......	......	......
	20	......	......	......	......	9	......	......	......	......
20	4	3/16+	5/16	1 3/8	5/8	4	5/8	5/8	1 1/4	3/8
	6	......	......	......	......	4 1/2	......	......	......	......
	8	......	......	......	......	5	3/4	......	......	......
	10	7/32	11/32	1 5/8	3/4	6	......	......	......	......
	12	......	......	......	......	7	......	......	......	......
	16	9/32	7/16	2 1/4	1 1/8	8	7/8	3/4	1 3/4	......
	20	......	......	......	......	10	1	......	......	......

TABLE—(*Continued*).

Diam.	Face.	Rim.		Arm.		Hub.		Boss.		
		A	*B*	*C*	*D*	*E*	*F*	*G*	*H*	*I*
22″	4	3/16	5/16	1 1/2	5/8	4	5/8	5/8	1 1/4	3/8
	6	......	......	......	......	4 1/2	3/4	......	......	......
	8	......	......	......	......	5	......	......	......	......
	10	7/32+	11/32	1 3/4	13/16	6 1/2	......	......	......	......
	12	......	......	......	......	......	7/8	3/4	1 3/8	......
	16	......	......	......	......	8 3/4	1	......	......	......
	20	9/32+	7/16	2 1/2	1 1/4	11	1 1/8	7/8	1 1/2	......
24	4	7/32	11/32	1 9/16	11/16	4	5/8	5/8	1 1/4	3/8
	6	......	......	......	......	4 3/4	3/4	......	......	......
	8	......	......	......	......	5 1/2	......	......	......	......
	10	1/4	3/8	1 7/8	......	7	7/8	3/4	1 3/8	......
	12	......	......	......	......	......	......	......	......	......
	16	......	......	......	......	9 1/2	1	......	......	......
	20	5/16	15/32	2 3/4	1 3/8	......	1 1/8	7/8	1 1/2	......
	24	......	......	......	......	11	......	......	......	......
26	4	7/32	11/32	1 11/32	3/4	4 1/4	3/4	5/8	1 1/4	3/8
	6	......	......	......	......	5	......	......	......	......
	8	......	......	......	......	6	7/8	3/4	1 3/8	......
	10	1/4	3/8	2	7/8	7	......	......	......	......
	12	......	......	......	......	7 1/2	......	......	......	......
	16	......	......	......	......	10	1 1/8	7/8	1 1/2	......
	20	5/16+	15/32	2 15/16	1 7/16	10 1/2	......	......	......	......
	24	......	......	......	......	11	......	......	......	......
28	4	7/32+	15/32	1 3/4	3/4	4 1/2	3/4	5/8	1 1/4	3/8
	6	......	......	......	......	5 1/2	7/8	3/4	1 3/8	......
	8	......	......	......	......	7	......	......	......	......
	10	......	......	......	......	7 1/2	......	......	......	......
	12	1/4+	3/8	2 1/8	13/16	8	1	......	......	......
	16	......	......	......	......	10	......	......	......	......
	20	5/16	15/32	......	......	11	1 1/8	7/8	1 1/2	......
	24	11/32	1/2	3 1/2	1 1/2	......	......	......	......	......
30	4	7/32+	11/32	1 7/8	13/16	4 1/2	3/4	5/8	1 1/4	3/8
	6	......	......	......	......	5 1/2	7/8	......	......	......
	8	......	......	......	......	6 1/4	......	3/4	1 3/8	......
	10	......	......	......	......	6 1/2	......	......	......	......
	12	9/32	7/16	2 1/4	1	8	1	......	......	......
	16	......	......	......	......	8 1/2	......	......	......	......
	20	......	......	......	......	11 1/2	1 1/4	7/8	1 1/2	......
	24	3/8	9/16	3 5/16	1 5/8	13	......	......	......	......
32	4	1/4+	3/8	2 1/8	13/16	4 1/2	7/8	3/4	1 3/8	3/8
	6	......	......	......	......	5 1/2	......	......	......	......
	8	......	......	......	......	6 1/2	1	......	......	......
	10	......	......	......	......	7 1/2	......	......	......	......

TABLE—(Continued).

Diam.	Face.	Rim. A	Rim. B	Arm. C	Arm. D	Hub. E	Hub. F	Boss. G	Boss. H	Boss. I
	12	$\frac{7}{16}$	$\frac{15}{32}$	$2\frac{7}{16}$	$1\frac{1}{16}$	8	$1\frac{1}{8}$	$\frac{7}{8}$	$1\frac{1}{2}$	......
	16					$9\frac{1}{2}$				
	20					11	$1\frac{1}{4}$			
	24					13				
34″	4	$\frac{1}{4}+$	$\frac{3}{8}$	$2\frac{1}{8}$	$1\frac{1}{16}$	$4\frac{1}{2}$	$\frac{7}{8}$	$\frac{3}{4}$	$1\frac{3}{8}$	$\frac{3}{8}$
	6					$5\frac{1}{2}$				
	8					$6\frac{1}{2}$	1			
	10					$7\frac{1}{4}$				
	12	$\frac{7}{16}$	$\frac{15}{32}$	$2\frac{7}{16}$	$1\frac{1}{16}$	$7\frac{3}{4}$	$1\frac{1}{8}$	$\frac{7}{8}$	$1\frac{1}{2}$	
	16					$9\frac{1}{2}$				
	20					12	$1\frac{1}{4}$			
	24					13				
36	4	$\frac{1}{4}+$	$\frac{3}{8}$	$2\frac{3}{16}$	$1\frac{1}{16}$	$4\frac{1}{2}$	$\frac{7}{8}$	$\frac{3}{4}$	$1\frac{3}{8}$	$\frac{3}{8}$
	6					$5\frac{1}{2}$				
	8					$6\frac{3}{4}$				
	10					$7\frac{1}{4}$				
	12	$\frac{7}{16}$	$\frac{15}{32}$			$7\frac{3}{4}$	$1\frac{1}{8}$	$\frac{7}{8}$	$1\frac{1}{2}$	
	16			$2\frac{3}{16}$	$1\frac{1}{8}$	$10\frac{1}{4}$	$1\frac{1}{4}$			
	20					12				
	24					$13\frac{1}{2}$	$1\frac{3}{8}$	1	$1\frac{3}{4}$	$\frac{1}{2}$
40	8	$\frac{7}{16}$	$\frac{15}{32}$	$2\frac{1}{16}$	1	$6\frac{3}{4}$	1	$\frac{3}{4}$	$1\frac{5}{8}$	$\frac{5}{8}$
	12					$7\frac{3}{4}$	$1\frac{1}{8}$	$\frac{7}{8}$	$1\frac{1}{2}$	
	16	$\frac{15}{32}$	$\frac{1}{2}$	$2\frac{3}{4}$	$1\frac{1}{4}$	10	$1\frac{1}{4}$			
	20					$11\frac{1}{2}$	$1\frac{3}{8}$	1	$1\frac{3}{4}$	$\frac{1}{2}$
	24					$15\frac{1}{2}$	$1\frac{1}{2}$			
44	8	$\frac{9}{32}$	$\frac{7}{16}$	$2\frac{1}{2}$	$1\frac{1}{4}$	$6\frac{3}{4}$	$1\frac{1}{2}$	$\frac{7}{8}$	$1\frac{1}{2}$	$\frac{3}{8}$
	12					8				
	16	$\frac{15}{32}$	$\frac{1}{2}$	3	$1\frac{1}{16}$	10	$1\frac{1}{4}$			
	20					12	$1\frac{3}{8}$	1	$1\frac{3}{4}$	$\frac{1}{2}$
	24			$3\frac{1}{2}$	$1\frac{3}{4}$	15				
48	8	$\frac{9}{32}+$	$\frac{7}{16}$	$2\frac{3}{4}$		$7\frac{1}{2}$	$1\frac{1}{2}$	$\frac{7}{8}$	$1\frac{1}{2}$	$\frac{3}{8}$
	12					$8\frac{3}{4}$	$1\frac{1}{2}$			
	16	$\frac{3}{8}$	$\frac{7}{16}$	$3\frac{1}{4}$	$1\frac{7}{16}$	10	$1\frac{3}{8}$	1	$1\frac{3}{4}$	$\frac{1}{2}$
	20					12				
	24					15	$1\frac{1}{2}$			
54	12	$\frac{7}{16}+$	$\frac{15}{32}$	3	$1\frac{1}{16}$	$9\frac{3}{4}$	$1\frac{3}{8}$	1	$1\frac{3}{4}$	$\frac{1}{2}$
	16					$11\frac{1}{4}$	$1\frac{1}{2}$			
	20	$\frac{15}{32}$	$\frac{15}{32}$	$3\frac{5}{8}$	$1\frac{5}{8}$					
	24					15	$1\frac{3}{4}$	$1\frac{1}{4}$	2	
60	12	$\frac{15}{32}$	$\frac{1}{2}$	$3\frac{1}{16}$	$1\frac{1}{16}$	10	$1\frac{3}{8}$	1	$1\frac{3}{4}$	$\frac{1}{2}$
	16					$11\frac{1}{4}$	$1\frac{1}{2}$			
	20	$\frac{7}{16}$	$\frac{5}{8}$	$3\frac{1}{8}$	$1\frac{3}{4}$	$12\frac{1}{2}$	$1\frac{3}{4}$	$1\frac{1}{4}$	2	
	24					15				

TABLE—(*Continued*).

Diam.	Face.	Rim.		Arm.		Hub.		Boss.		
		A	*B*	*C*	*D*	*E*	*F*	*G*	*H*	*I*
66″	12	$\frac{11}{32}$	$\frac{1}{2}$	$3\frac{9}{16}$	$1\frac{9}{16}$	10	$1\frac{1}{2}$	1	$1\frac{3}{4}$	$\frac{1}{2}$
	16	......	......	......	......	$11\frac{1}{2}$	$1\frac{5}{8}$	$1\frac{1}{4}$	2	......
	20	$\frac{1}{2}$	$\frac{3}{4}$	$4\frac{1}{4}$	$1\frac{15}{16}$	$13\frac{1}{2}$	$1\frac{7}{8}$	......	......	......
	24	......	......	......	......	15	......	......	......	......
72	12	$\frac{3}{8}$	$\frac{9}{16}$	$3\frac{7}{8}$	$1\frac{11}{16}$	$10\frac{1}{2}$	$1\frac{5}{8}$	$1\frac{1}{4}$	2	$\frac{1}{2}$
	16	......	......	......	......	$12\frac{1}{2}$	$1\frac{3}{4}$	......	......	......
	20	$\frac{9}{16}$	$\frac{13}{16}$	$4\frac{5}{8}$	$2\frac{1}{16}$	$13\frac{1}{2}$	$1\frac{7}{8}$	......	......	......
	24	......	......	......	......	15	2	......	......	......

ROPE BELTING.

There is a growing tendency toward the substitution of hemp and cotton ropes for belting and line shafting as a means of transmitting power in large factories and shops. The advantages claimed for the rope-driving system are:

1. Economy; for a rope system is cheaper to install than either leather belting or shafting.

2. In the rope system there is less loss of power by slipping.

3. Flexibility; that is, the ease with which the power is transmitted to any distance and in any direction.

In this country, a single rope is carried round the pulley as many times as is necessary to produce the required power, and the necessary tension is obtained by passing the rope round a tension pulley weighted to give the desired tension.

The ropes used in rope transmission are either of hemp, manila, or cotton. Manila ropes are mostly used in this country. They are of three strands, hawser laid, and may be from $\frac{1}{2}$ in. to 2 in. in diameter.

The weight of ordinary manila or cotton rope is about $.3\,D^2$ lb. per ft. of length, where D represents the diameter of the rope in inches. Letting w = the weight per foot of length, $w = .3\,D^2$.

The breaking strength of the rope varies from 7,000 to 12,000 lb. per sq. in. of cross-section. The average value may be taken as $7{,}000\,D^2$, when D is the diameter of rope.

For a continuous transmission, it has been determined by experiment that the best results are obtained when the tension in the driving side of the rope is about $\frac{1}{5}$ of the breaking strength. That is,

$$T_1 = \text{tension in tight side} = \frac{7,000\,D^2}{35} = 200\,D^2.$$

The ropes run in V-shaped grooves, and the coefficient of friction is, of course, greater than on a smooth surface. The coefficient for grooves with sides at an angle of 45° may be taken at from .25 to .33.

The horsepower that can be transmitted by a single rope running under favorable conditions is given by the formula

$$H = \frac{v\,D^2}{825}\left(200 - \frac{v^2}{107.2}\right),$$

in which H = horsepower transmitted;

      D = diameter of rope in inches;

      v = velocity of rope in feet per second.

The maximum power is obtained at a speed of about 84 ft. per sec. For higher velocities, the centrifugal force becomes so great that the power is decreased, and when the speed reaches 145 ft. per sec. the centrifugal force just balances the tension, so that no power at all is transmitted. Consequently, a rope should not run faster than about 5,000 ft. per min., and it is preferable on the score of durability to limit the velocity to 3,500 ft. per min.

EXAMPLE.—A rope flywheel is 26 ft. in diameter, and makes 55 rev. per min. The wheel is grooved for 35 turns of $1\frac{1}{2}''$ rope. What horsepower may be transmitted?

SOLUTION.—Velocity in feet per second =

$$v = \frac{26 \times \pi \times 55}{60} = \frac{4,492}{60} = 74.9 \text{ ft.}$$

Applying the formula,

$$H = \frac{v\,D^2}{825}\left(200 - \frac{V^2}{107.2}\right),$$

the horsepower transmitted by one rope or turn is

$$\frac{74.9 \times (1\frac{1}{2})^2}{825}\left(200 - \frac{(74.9)^2}{107.2}\right) = 30.16.$$

Then, $30.16 \times 35 = 1,055.6$ = horsepower transmitted by the 35 ropes.

EXAMPLE.—How many times should a 1″ rope be wrapped around a grooved wheel in order to transmit 200 horsepower, the speed being 3,500 ft. per min.?

SOLUTION.— 3,500 ft. per min. $= \dfrac{3,500}{60} = 58\frac{1}{3}$ ft. per sec.

Applying the formula, the horsepower transmitted with one turn is,

$$H = \frac{58\frac{1}{3} \times 1^2}{825}\left[200 - \frac{(58\frac{1}{3})^2}{107.2}\right] = 11.9.$$

Hence, $200 \div 11.9 = 16.8$, say 17 turns.

Rope pulleys differ from belt pulleys only in their rims. The inclination of the sides of the grooves may vary from 30° to 60°. The more acute the angle, the greater the coefficient and, consequently, the wear on the rope.

A section of a grooved rim in which the sides of the grooves are formed with circular arcs is shown in the figure.

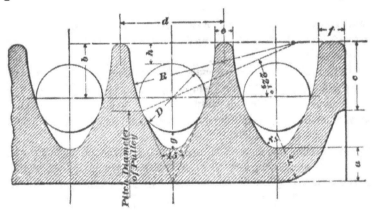

The proportions for this rim are as follows, using the diameter D of the rope as a unit:

$a = \frac{1}{2}D;$　　　　　　$e = \frac{1}{2}D + \frac{1}{16}'';$

$b = \frac{3}{4}D + \frac{1}{16}'';$　　　$f = \frac{1}{2}D + \frac{1}{16}'';$

$c = D;$　　　　　　　$g = \frac{1}{2}D;$

$d = 1.6\,D;$　　　　　$h = \frac{1}{2}D + \frac{1}{16}''.$

The radii r_1 and r_2 are to be found by trial; they should be of such lengths as to make the curves drawn by them tangent to the required lines.

The long rádius R is determined by drawing a line through the center of the rope at an angle of $22\frac{1}{2}°$ with the horizontal, and producing it until it intersects a line drawn through the tops of the dividing ribs; then, with this point of intersection as a center, draw the curve forming the side of the groove tangent to the circumference of the rope.

The advantage claimed for this groove is that the rope will turn more freely in it, thus presenting new sets of fibers to the sides of the grooves and increasing the life of the rope.

The diameter of a rope pulley should be at least 30 times the diameter of the rope. Good results are obtained when the diameters of pulleys and idlers on the driving side are 40 times, and those on the driven side 30 times, the rope diameter. Idlers used simply to support a long span may have diameters as small as 18 rope diameters, without injuring the rope.

When possible, the lower side of the rope should be the driving side, for in that case the rope embraces a greater portion of the circumference of the pulley, and increases the arc of contact.

When the continuous system of rope transmission is used, the tension pulley should act on as large an amount of rope as possible. It is good practice to use a tension pulley and carriage for every 1,200 ft. of rope, and have at least 10% of the rope subjected directly to the tension.

Aside from the grooved rim, rope pulleys are constructed the same as other pulleys. They may be cast solid, in halves, or in sections. The pulley grooves must be turned to exactly the same diameter; otherwise, the rope will be severely strained.

TRANSMISSION OF POWER BY WIRE ROPE.

Wire rope for transmitting power is made up of 6 strands twisted about a hemp core, each strand being composed of either 7 or 19 wires, according to the size of the sheaves, the 19-wire rope being employed in cases where it is impracticable to use the larger sheaves required by the 7-wire rope. Where the conditions, however, do not preclude the use of the proper size of sheaves, the 7-wire rope is to be recommended in preference to the other, except sometimes on very short spans, where 19-wire rope is to be preferred, composed of the same size of wires as the smaller 7-wire rope, such as would ordinarily be used to transmit the power, and run under a tension corresponding to the smaller rope, or considerably below the maximum safe tension of the rope used. This is done in order to avoid stretching, which would otherwise occur, and the consequent use of mechanical appliances for preserving the necessary tension.

In flying transmission, where the rope makes a single half lap at each end, the sheaves are usually made of cast iron, with rims having grooves lined with segments of rubber and leather, dipped in tar, and laid in alternately, upon which the rope tracks. The diameters of the minimum sheaves, corresponding to a maximum efficiency, are as follows, according to a prominent manufacturer:

Diam. of sheave for 7-wire steel rope, 77 times diam. of rope.
Diam. of sheave for 19-wire steel rope, 46 times diam. of rope.
Diam. of sheave for 7-wire iron rope, 160 times diam. of rope.
Diam. of sheave for 19-wire iron rope, 96 times diam. of rope.

In long-distance transmissions, where the rope makes 2 or more half laps at each end about a pair of drums or several sheaves, the rims may be lined with wood or the rope may be run in plain turned grooves.

The horsepower capable of being transmitted is determined by the general formula:

$$N = [c\,D^2 - .000006\,(w + g_1 + g_2)]v,$$

in which

D = diameter of rope in inches;
v = velocity of rope in feet per second;
w = weight of rope in pounds;
g_1 = weight of terminal sheaves and shafts;
g_2 = weight of intermediate sheaves and shafts;
c = constant depending on the material of which rope is made, the character of the filling or surface material in the sheaves or drums upon which the rope tracks, and the number of half laps at each end.

The values of c for from 1 up to 6 half laps for steel rope are given in the following table:

c for Steel Rope on	Number of Half Laps at Each End.					
	1	2	8	4	5	6
Iron.	5.61	8.81	10.62	11.65	12.16	12.56
Wood.	6.70	9.93	11.51	12.26	12.66	12.88
Rubber and Leather.	9.29	11.95	12.70	12.91	12.97	13.00

The values of c for iron ropes are one-half the above. It is apparent from this table that, when more than 3 half laps are made, the character of filling or surface in contact is immaterial so far as slipping is concerned.

Where the distance is comparatively short, as in most flying transmissions, the effect of the weight of the rope and sheaves is so slight that it may be neglected, and we have the general rule, that *the actual horsepower capable of being transmitted by a wire rope approximately equals c times the square of the diameter of the rope in inches, multiplied by the speed of the rope in feet per second.*

The tension of the rope is measured by the amount of sag or deflection at the center of the span, and the deflection corresponding to the maximum safe working tension is determined by the following formulas, in which s represents the span in feet:

Deflection.	Steel Rope.	Iron Rope.
Still rope at center, in ft............	$h = .00004 s^2$	$h = .00008 s^2$
Driving portion, running, in ft...	$h_1 = .000025 s^2$	$h_1 = .00005 s^2$
Slack portion, running, in ft...	$h_2 = .0000875 s^2$	$h_2 = .000175 s^2$

In very long transmissions it often happens that the conditions will not allow of the required amount of tension to drive properly with but a single half lap on the pulley. In such cases it is customary to give the rope a sufficient number of half turns around successive grooves in the driving pulley and a series of guide pulleys that serve to lead the rope from one groove on the driving pulley to the next.

With this arrangement a guide pulley at one end of the

line is usually made *to* serve the purpose of a *tension pulley* by being mounted in a movable frame that can be drawn by means of a screw or a weight so as to give the rope the desired tension.

PIPE FLANGES.

The figure shows the method of flanging and bolting the ends of two cast-iron pipes. The dimensions of the flanges for the various sizes of pipes are given in the following table:

STANDARD PIPE FLANGES. n = number of bolts.

a	b	c	d	n	e	f	g
2.0	.409	5/8	2.000	4	5/8	4.75	6.00
2.5	.429	5/8	2.250	4	11/16	5.50	7.00
3.0	.448	5/8	2.500	4	3/4	6.00	7.50
8.5	.466	5/8	2.500	4	3/4	7.00	8.50
4.0	.486	3/4	2.750	4	3/4	7.50	9.00
4.5	.498	3/4	3.000	8	3/4	7.75	9.25
5	.525	3/4	3.000	8	3/4	8.50	10.00
6	.563	3/4	3.000	8	1	9.50	11.00
7	.600	3/4	3.250	8	1 1/8	10.75	12.50
8	.639	3/4	3.500	8	1 1/8	11.75	13.50
9	.678	3/4	3.500	12	1 1/8	13.25	15.00
10	.713	7/8	3.625	12	1 1/8	14.25	16.00
12	.790	7/8	3.750	12	1 1/4	17.00	19.00
14	.864	1	4.250	12	1 3/8	18.75	21.00
15	.904	1	4.250	16	1 3/8	20.00	22.25
16	.946	1	4.250	16	1 1/8	21.25	23.50
18	1.020	1 1/8	4.750	16	1 1/8	22.75	25.00
20	1.090	1 1/8	4.750	20	1 1/4	25.00	27.50
22	1.180	1 1/4	5.500	20	1 1/4	27.25	29.50
24	1.250	1 1/4	5.500	20	1 7/8	29.50	32.00
26	1.300	1 1/4	5.750	24	2	31.75	34.25
28	1.380	1 1/2	6.000	28	2 1/8	34.00	36.50
30	1.480	1 3/8	6.250	28	2 1/8	36.00	38.75
36	1.710	1 3/8	6.500	32	2 3/8	42.75	45.75
42	1.870	1 1/2	7.250	36	2 5/8	49.50	52.75
48	2.170	1 1/2	7.750	44	2 3/4	56.00	59.50

LINING FOR SEATS.

Seats for large bearings are óften lined with Babbitt metal, or anti-friction metal. It has been found by experience that a bearing will run cooler when so lined, probably because the Babbitt metal, being softer, accommodates itself to the journal more readily than the more rigid gun metal.

Some of the common methods of lining the seats are shown in the figure. At (a) the Babbitt metal is shown cast

into shallow helical grooves; at (b), into a series of round holes; and at (c), into shallow rectangular grooves. Consequently, the journal rests partly on the brass and partly on the Babbitt metal.

In cheap work, very frequently the seats are made entirely of Babbitt metal. A mandrel the exact size of the journal is placed inside the bearing, and the melted Babbitt metal is poured around it. In better work a smaller mandrel is used, and the metal is hammered in, the bearing being then bored out to the exact size of the journal.

CYLINDERS AND STEAM CHESTS.

Fig. 1 shows a cylinder designed for a simple slide-valve engine. The front head A is cast solid with the cylinder. The method of fastening to the frame B is clearly shown.

The principal dimensions of this cylinder may be determined from the following proportions:

D = diameter of cylinder;

L = length of stroke + thickness of piston + twice the piston clearance;

C = length of stroke + distance from outer edge to outer edge of piston rings $- (.01 D + .125'')$;

a = 5.5 i;

$b = 4.2i$;

$c = i$;

$d = i$;

e' = net area of a single cylinder-head bolt whose nominal
diameter is $e = \dfrac{A\,P}{4{,}000\,n}$,

where A = area of cylinder head in square inches;

P = steam pressure;

n = number of bolts.

The pitch of the bolts may be from 4.5 to 5.5 in., but should never be more than $5f$.

$f = 1.5i$;

$g = .04\,D + .125''$. Take the nearest nominal size pipe tap.

h = twice the outside diameter of drain pipe.

$i = .0003\,PD + .375''$, where P is the steam pressure. If the steam pressure is less than 100 lb., make $P = 100$.

$j = .85i$;

$k = 4i$;

$l = .75i$;

$m = 1.01\,D + .125''$;

$n = m + 6e$, never less. Here, e is the nominal diameter of the bolt.

o = the nominal diameter of steam-chest bolts. The net area of a single steam-chest bolt $= \dfrac{A'P}{4{,}000\,n'}$,

where A' = area of steam chest;

n' = number of bolts in steam chest.

$p = 2.75\,o$;

$q = 1.5\,r$;

$r = 1.25\,i$;

$s = i$. This is required only when the length of the port is greater than 12 in.

$t = 1.25\,i$. When D is greater than 24 in., use 4 bolts in the standard and make $t = 1.1\,i$.

$u = 1.5\,i$;

$v = .25''$ (constant).

The dimensions of the *steam ports*, *exhaust ports*, and other steam passages depend on the velocity of the flow of steam. The ports and passages must be large enough to allow the steam to follow up the advancing piston without loss of

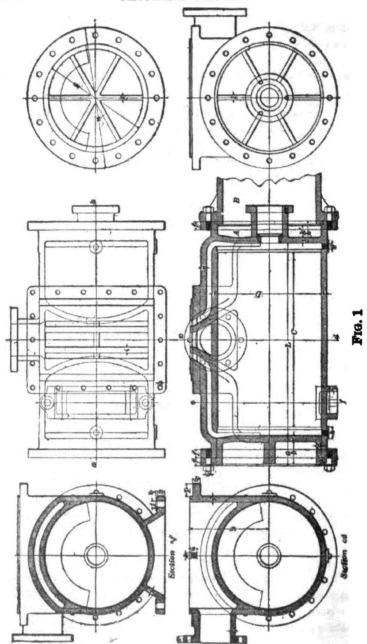

FIG. 1

pressure. The maximum allowable velocity of the steam in the passages, when they are short, is about 160 ft. per sec. But, with the ordinary ratio between the length of connecting-rod and length of crank, the average velocity is about five-eighths of the maximum. Hence, the allowable average velocities are 100 to 125 ft. per sec. for long and short passages, respectively.

Let l = length of port in inches;
$\quad b$ = breadth of port in inches;
$\quad A$ = area of cylinder;
$\quad S$ = average piston speed in feet per second;
$\quad v$ = average velocity of steam in feet per second.

Then, area of port $\times$ velocity of steam = area of piston $\times$ velocity of piston, or $l\,b\,v = A\,S$; whence,

$$l b = \frac{A\,S}{v}.$$

For long indirect passages, take v = 100; and for short direct passages, take v = 125.

The constant 100 may be used for v, when designing plain slide-valve engines of the ordinary type, which cut off late in the stroke, and 125 may be used for high-speed engines with early cut-off, and for the Corliss type.

The area of the exhaust port or ports may be from 1¼ to 2¼ times the area of a steam port.

The area of the cross-section of the steam pipe is approximately equal to the area of the steam port; likewise, the area of the exhaust pipe should be equal to that of the exhaust port.

The length l of the port may be .6 D to .9 D for slide-valve engines, and about .9 D to D for the Corliss type.

The height w, Fig. 1, of the valve seat must be such that the area of the most contracted part of the exhaust port is not less than 75% of the area of the steam port.

THE STEAM CHEST.

Fig. 2 shows a steam chest for the cylinder illustrated in Fig. 1. The principal dimensions are to be determined by the following proportions, which are based on the thickness t of the cylinder walls, and on the travel and dimensions of the valve:

a = length of valve + travel of valve + twice the clearance between the valve and the steam chest at ends of valve travel;

b = breadth of valve + twice the clearance between one valve and steam chest;

c = .75 t;

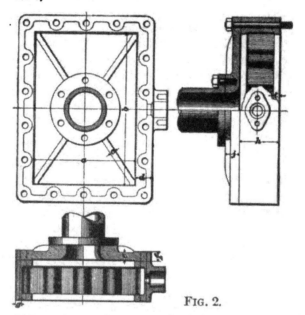

FIG. 2.

d = 2.75 o, where o is the nominal diameter of the steam-chest bolts, as in Fig. 1;

e = .04 $\sqrt{A'}$ + .125″ for all areas above 100 sq. in. A' = area of steam chest, outside measurement, in square inches;

f = 1.3 e;

g = .85 t;

h = height of valve + necessary clearance;

t = .85 t;

j = 2.5 t.

NOTE.—When the area of the steam-chest cover is less than 100 sq. in., its thickness e may be made equal to t. If the area of the steam-chest cover exceeds 600 sq. in., the height of the ribs should be 3.5 t, and their number should be increased.

Fig. 3 shows a design for a steam-chest cover when the steam-pipe flange is on one side of the steam chest. Determine the thickness e by the same formula and rules as for the cover in Fig. 2. The other dimensions are found as follows:

$$c = .75e; \qquad\qquad j = 2.6c;$$
$$f = 1.3e; \qquad\qquad r = 6e.$$

p should never exceed the distance in inches given by the formula $p = \sqrt{\dfrac{40\,e_1^2}{p_g}}$, where e_1 is the numerator of the frac-

tion expressing the thickness of the cover in sixteenths of an inch, and p_g is the gauge boiler pressure in pounds per square inch.

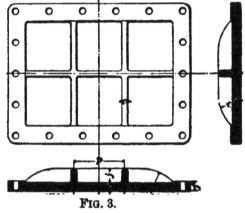

EXAMPLE.—Find the maximum pitch of the ribs for a cover $1\frac{5}{8}$ in. thick, subjected to a steam pressure of 160 lb. per sq. in.

FIG. 3.

SOLUTION.—Substituting in the formula for p, we have

$$p = \sqrt{\frac{40 \times e_1^2}{p_g}} = \sqrt{\frac{40 \times 15^2}{160}} = 7.5 \text{ in.}$$

Fig. 4 shows a Corliss engine cylinder that may be designed according to the following proportions:

$$D = \text{diameter of cylinder.}$$

$a = 1.21\,D + 2e + 1.22''$;

$b = 2\,D + 1.125''$;

$c = .048\,D$;

$c' = .079\,D$;

$d = .17\,D$;

$e = .0003\ P\ D + .375''$, if boiler pressure is above 100 lb.; otherwise, $e = .03\,D + .375''$:

$f = .82\,e$:

$g = .9\,e$;

$h = b + 2(c + g)$;

$h' = h$;

$i = 1.8\,e$;

$j = e$;

$k = 1.2\,e$;

$l = 1.7\,x + 2'' - 1.2\,e$, where $x = $ diameter of piston rod;

$l' = .32\,D$, about;

$m = .25 D$;

$n = .82 D$;

$o = 1.25 e$;

$p = 1.3 e$;

$q = .25 D$;

$q' = .82 D$;

$r = 1.2 e$;

$s = 1.5 e$;

$i = $ (see note);

$u = e$; take diameter nearest standard size bolt;

$v = 1.2 e$; take diameter nearest standard size bolt;

$w = 1.7 x + 2.25''$, where $x = $ diameter of piston rod;

$y = D$;

$z = 1.5 e$.

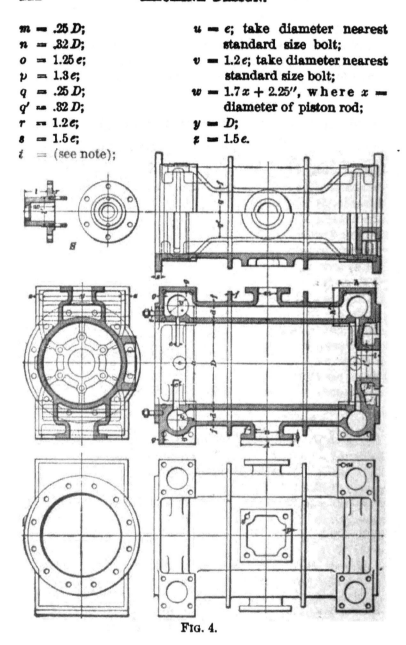

FIG. 4.

A is to be made according to proportions given on page 215.

Bolts to be made according to the same table.

NOTE.—The bolts for cylinder heads are to be calculated from the formula given for cylinder-head bolts in connection with Fig. 1.

In this cylinder the stuffingbox *S* is a separate piece that is to be bolted to the cylinder head.

CRANK-SHAFTS.

For high-speed, automatic short-stroke engines, the following formula corresponds with good practice:

$$d = .44\,D + \tfrac{1}{2}'',$$

where *d* is the diameter of shaft and *D* is the diameter of cylinders.

For the Corliss type, in which the stroke is equal to or greater than twice the diameter,

$$d = .34\,D + 2\tfrac{1}{2}'',$$

when *D* is equal to or greater than 16 in. When *D* is less than 16 in., $d = \tfrac{1}{2}D.$

PISTONS.

A form of piston that is much used is shown in the following figure. It consists simply of a hollow circular disk of cast iron.

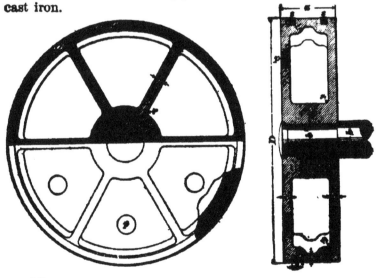

17

The packing rings s, s are made of cast iron, and are split and sprung into place. Their elasticity causes them to press against the cylinder walls and thus prevent the leakage of steam.

The following proportions will give dimensions suitable for this piston:

D = diameter of cylinder in inches;

$a = .2D + 1.5''$; $e = .75c$;

b = diameter of piston rod; $f = .5c$;

$b' = 2b$; p = core plug;

$c = .18\sqrt{2D} - .1875''$; number of ribs $= .08(D + 34)$.

CONNECTING-RODS.

The figure shows a *strap-end* connecting-rod. The straps c_1 and c_2 are fastened to the ends of the rods by means of the gibs a_1 and a_2 and the cotters b_1 and b_2. The cotters are held in place by the setscrews s_1 and s_2. Small steel blocks shown between the ends of the setscrews and the cotters are used to prevent injury of the cotter by the setscrews.

The rod, cotters, gibs, and straps may be made of either wrought iron or steel. The crankpin brasses are shown babbitted and wristpin brasses without babbitt. The brasses are adjusted by means of the cotters, which draw the straps farther on to the rod when they are driven in.

The dimensions for the rod are given by the following proportions:

For wristpin end:

D = diameter of cylinder;

$d = .2D$ = diameter of wristpin;

$n = .155D + .0625''$;

$x = \frac{\pi}{4}n^2$ = a factor for use in finding proportions below;

$a = .75d + .125''$;

$a' = .75d + .125''$;

$b = \sqrt{2.5x}$;

$c = .25b$;

$e = .125d$;

$f = .26D + .5''$ for cylinders to 26'' in diameter, and $f = .28D$ for cylinders above 26'' in diameter;

$g = 1.3n$;

$h = \dfrac{.5x}{g - c}$;

$i = \dfrac{.32x}{h}$;

$$k = \frac{x}{1.8\,d};$$

$$l = .375\,b;$$

$$o = .25\,b;$$

$m = 1.35\,d$ for wristpins up to 3.5'' in diameter, and $m = 1.48\,n$ for pins above 3.5'' in diameter;

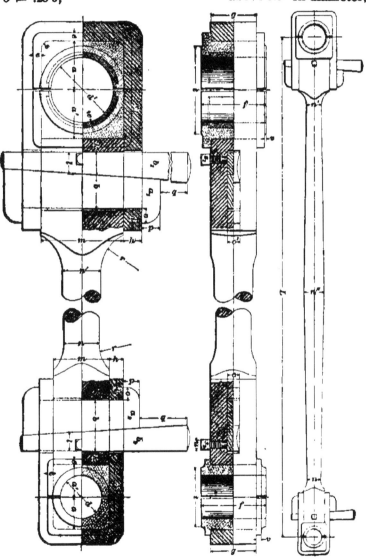

$p = .33\,b;$

$q = 1.125\,d$ for wristpins up to 3.5'' in diameter, and

$q = 4''$, constant, for pins above 3.5'' in diameter;

The taper of the cotter is $\frac{3}{4}$ in. per foot.

Proportions for the crankpin end:

$D =$ diameter of cylinder in inches;

$d' = .28\,D =$ diameter of crankpin;

$n' = 1.1\,n;\ (n = .155\,D + .0625'');$

$x' = \frac{\pi}{4}\,n'^{2} =$ a factor used below;

$a = .75\,d';$

$a' = .75\,d';$

$b = \sqrt{2.5\,x'};$

$c' = .25\,b;$

$e = .125\,d';$

$f = .26\,D$ for cylinder diameters up to 26'', and $f = .28\,D$ for cylinders above 26'' in diameter;

$g = 1.3\,n =$ same as wristpin end;

$h = \dfrac{.5\,x'}{g - c'};$

$r = n;$

$s = .125\,d;$

$t = 1.35\,d;$

$u = .02\,D + .25'';$

$v = .125\,d.$

$i = \dfrac{.32\,x'}{h};$

$k = \dfrac{x}{1.8\,d}$ same as wristpin end;

$l = .375\,b;$

$m = 1.3\,d';$

$o = .25\,b;$

$p = .33\,b;$

$q =$ same values as for wristpin end;

$r = 1.1\,n;$

$s = .125\,d;$

$t = 1.35\,d';$

$v = .125$ (constant);

$w = .02\,D + .0625'';$

$n'' = n\left(\sqrt{\dfrac{L}{S}} - .22''\right),$

where $L =$ length of rod, and $S =$ stroke, both in inches.

The taper of the cotter is $\frac{3}{4}$ in. per foot.

ECCENTRIC AND STRAP.

The figure shows an eccentric sheave and strap, both of cast iron. The eccentric sheave is cast solid, and must be slipped over end of shaft. The eccentric rod is held in a boss on the strap by a cotter. For eccentrics used with valve stems $\frac{3}{4}$ in. in diameter or less, holes for bolts j are not to be cored. $A =$ boss for oil cup; $B =$ cross-section of rib r.

The proportions are as follows:

$D =$ diameter of valve stem;

$d =$ diameter of shaft;

$a = d + 2q + 2h$;

$b = 2D + .125''$;

$b' = 2.25D + .125''$;

$c = 1.5D$;

$e = .75D$;

$e' = .75D$;

$f = .7D$;

$g = 1.25d$;

$h = D + .125''$;

$i = .25D + .0625''$;

$j =$ area of bolt at root of thread $= .38D^2$; use the nearest standard size bolt;

$j' = j + .1875$;

$k = 4D$;

$l = j$;

$m = \dfrac{d + 2q + 2h + 2f}{2}$.

$m' = m$;

$n = D + .125''$;

$n' = D + .125''$;

$o = .75j$;

$p = D$;

$q =$ eccentricity;

$r = D$;

$s = 1.25D$;

$t = 2.25D + 1.25''$;

$u = D$;

$v = 2.25D$;

$v' = 1.125D$;

$w = 2.5D$;

$x = 2.25j$.

STUFFINGBOXES.

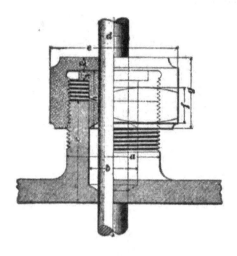

The stuffingbox of the form shown in the figure is generally used for small work, such as the spindles of valves, etc. The outside of the stuffingbox is threaded to receive a hexagonal nut that fits over the gland. As the nut is screwed down, the gland is pressed downwards and compresses the packing.

The proportions used are:

d = diameter of rod;

$a = 2.5\,d + .5''$;

$b = 1.5\,d + .125''$;

$c = 3\,d + .25''$;

$e = 3.5\,d + 625''$;

$f = d + .125''$;

$g = 2\,d + .25''$;

$h = 1.5\,d + .25''$;

$i = .25\,d + .0625''$;

$k = .5\,d$.

This design may be used for rods up to $1\frac{1}{4}$ in. in diameter.

Make the number of threads per inch the same as for a bolt whose diameter is equal to the diameter of the rod.

GEARING.

The *circular pitch* of a gear-wheel is the distance in inches measured on the pitch circle from the center of one tooth to the center of the next tooth.

If the distance of the teeth of a gear thus measured were $2\frac{1}{4}$ in., we would say that the circular pitch was $2\frac{1}{4}$ in.

Let P = circular pitch;

D = diameter of pitch circle, in inches;

C = circumference of pitch circle, in inches;

N = number of teeth;

π = 3.1416.

Then, $P = \dfrac{C}{N}$ or $\dfrac{\pi D}{N}$. $N = \dfrac{C}{P}$ or $\dfrac{\pi D}{P}$.

$C = PN$ or πD. $D = \dfrac{PN}{\pi}$ or $\dfrac{C}{\pi}$.

Addendum $= .3\,P$. Root $= .4\,P$.

The thickness of the teeth for a cut gear is equal to $.5\,P$, and for a cast gear $.48\,P$.

The *diametral pitch* of a gear-wheel is the name given to the quotient that is obtained by dividing the number of teeth in the wheel by the diameter of the pitch circle in inches; or, the diametral pitch may be defined as the number of teeth on the circumference of the gear-wheel for 1 in. diameter of pitch circle.

A gear with a pitch diameter of 5 in., and having 40 teeth is 8 pitch; one with the same pitch diameter and having 70 teeth is 14 pitch.

In the gear of 8 pitch there are 8 teeth on the circumference for each inch of the diameter of the pitch circle; and in one of 14 pitch there are 14 teeth on the circumference for each inch of the diameter of the pitch circle.

Let P = diametral pitch;

D = diameter of pitch circle, in inches;

N = number of teeth;

d = outside diameter;

l = length of tooth;

t = thickness of tooth;

$P = \dfrac{N}{D}$. $D = \dfrac{N}{P}$. $N = PD$. $d = \dfrac{N+2}{P}$. $l = \dfrac{2.157}{P}$. $t = \dfrac{1.57}{P}$.

The circular pitch corresponding to any diametral pitch may be found by dividing 3.1416 by the diametral pitch; and the diametral pitch corresponding to any circular pitch may be found by dividing 3.1416 by the circular pitch.

(*a*) If the diametral pitch of a gear is 6, what is the corresponding circular pitch?

(*b*) If the circular pitch is 1.5708 in., what is the corresponding diametral pitch?

(*a*) $\dfrac{3.1416}{6} = .5236$ in. (*b*) $\dfrac{3.1416}{1.5708} = 2.$

DIAMETRAL PITCHES WITH THEIR CORRESPONDING CIRCULAR PITCHES.

Diametral Pitch, or Teeth, per Inch in Diameter.	Corresponding Circular Pitch.	Diametral Pitch, or Teeth, per Inch in Diameter.	Corresponding Circular Pitch.
1	3.1416	8	.3927
2	1.5708	9	.3491
3	1.0472	10	.3142
4	.7854	12	.2618
5	.6283	14	.2244
6	.5236	16	.1963
7	.4488	20	.1571

ELECTRICITY.

PRACTICAL UNITS.

The *volt* is the practical unit of electromotive force or electrical pressure. It is that electromotive force which will maintain a current of *1 ampere* in a circuit whose resistance is *1 ohm*.

The *electromotive force* of a Daniell's cell is 1.072 volts.

The *ampere* is the practical unit denoting the strength of an electric current, or the rate of flow of electricity. It is that strength of current or rate of flow which would be maintained in a circuit whose resistance is *1 ohm* by an electromotive force of *1 volt*.

One ampere decomposes .00009342 gram of water (H_2O) per second; or deposits .001118 gram of silver per second.

The *ohm* is the practical unit of resistance. It is that resistance which will limit the flow of an electric current under an electromotive force of *1 volt* to *1 ampere*.

The *legal ohm* is the resistance of a column of mercury 106 centimeters long and 1 square millimeter sectional area at 0° C.

One mile of pure copper wire $\frac{1}{16}$ in. in diameter has a resistance of 13.59 ohms at a temperature of 59.9° F.

To make the significance of these units clearer, take the analogous case of water flowing through a pipe under a pressure of a column of water. The force that causes the water to flow is due to the pressure or head; the flow or current of water is measured in *gallons per minute;* and the resistance that opposes or resists the flow of water is caused by the friction of the water against the inside of the pipe.

In electrotechnics, the electromotive force or electrical potential expressed in volts corresponds to the pressure or head of water; and the resistance in ohms to the friction in the pipe.

The unit that expresses the *rate of transmission of electricity per second* is called the *ampere*, while the flow of water is expressed in gallons per minute.

In either case the strength of current or rate of flow depends on the ratio between the pressure and the resistance; for, as the pressure increases, the current increases proportionately; and as the resistance increases, the current diminishes.

This relation, as applied to electricity, was discovered by Dr. G. S. Ohm, and has since been called *Ohm's law*.

Ohm's Law.—*The strength of the current in any circuit is directly proportional to the electromotive force in that circuit and inversely proportional to the resistance of that circuit, i. e., is equal to the quotient arising from dividing the electromotive force by the resistance.*

Let E = electromotive force in volts;

        R = resistance in ohms;

        C = strength of current in amperes.

Then $C = \dfrac{E}{R}. \quad R = \dfrac{E}{C}. \quad E = CR.$

EXAMPLE.—The electromotive force of a circuit is 110 volts, and its resistance is 55 ohms; what is the strength of current?

SOLUTION.— E = 110 volts. R = 55 ohms. $C = \dfrac{E}{R} = \dfrac{110}{55}$

= 2 amperes.

The unit by which electrical power is expressed is called the *watt*. It is that *rate of doing work* when a current of 1 ampere is passing through a conductor under an electromotive force of 1 volt, and is equal to $\frac{1}{746}$ of a horsepower.

Let E = electromotive force in volts;

       C = strength of current in amperes;

       R = resistance in ohms;

       W = power in watts;

   H. P. = horsepower.

$$W = E \times C = C^2 \times R = \frac{E^2}{R}.$$

$$\text{H. P.} = \frac{E \times C}{746} = \frac{C^2 \times R}{746} = \frac{E^2}{R \times 746} = \frac{W}{746}.$$

One *kilowatt* is equal to 1,000 watts: sometimes abbreviated to K. W.

Watt hour is a unit of work. It is used to indicate the expenditure of an electrical power of 1 watt for 1 hour.

EXAMPLE.—The resistance of a lighting circuit is 5 ohms and the electromotive force is 110 volts. (*a*) What is the amount of electrical power in watts required for this current? (*b*) What is the equivalent horsepower?

SOLUTION.— $E = 110$. $R = 5$.

$$\frac{E^2}{R} = \frac{110^2}{5} = 2,420 \text{ watts.}$$

$$\frac{E^2}{R \times 746} = \frac{110^2}{5 \times 746} = 3.244 \text{ H. P.}$$

Conductivity is the name given to the reciprocal of the resistance of any conductor. There is no unit by which to express conductivity.

NOTE.—The reciprocal of any number is unity divided by that number. Thus, the reciprocal of 2 is $\frac{1}{2}$ or .5.

CURRENTS.

RULES FOR DIRECTION OF CURRENT, ETC.

To determine the direction of a current in a conductor by the aid of a compass:

Rule.—*If the current flows from the south pole over the needle to the north, the north end of the needle will point towards the west, as in Fig. 1. If the compass is placed over the conductor so that the current will flow from the south under the needle to the north, the north end of the needle will point towards the east, as in Fig. 2.*

To determine the polarity of an electromagnet:

Rule.—*In looking at the face of a pole* (Fig. 3), *if the current*

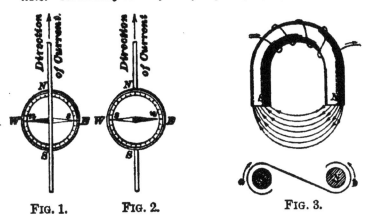

FIG. 1.   FIG. 2.   FIG. 3.

flows in the direction a, of the hands of a watch, it will be a south pole, and if in the opposite direction b, it will be a north pole.

To determine the direction of an induced current in a conductor that is moving in a magnetic field:

Rule.—*Place thumb, forefinger, and middle finger of right hand, each at a right angle to the other two, as shown in Fig. 4; if the forefinger shows direction of lines of force and the thumb the direction of motion of conductor, then the middle finger*

FIG. 4.

will show the direction of the induced current.

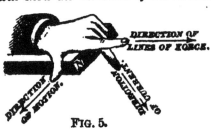

FIG. 5.

NOTE.—The above rule will give the polarity of a dynamo.

To determine the direction of motion of a conductor carrying a current when placed in a magnetic field:

Rule.—*Place thumb, forefinger, and middle finger of the left hand, each at a right angle to the other two, as shown in Fig. 5; if the forefinger shows the direction of the lines of force and the middle finger shows the direction of the current, then the thumb will show the direction of motion of the conductor.*

NOTE.—The above rule will give the polarity of a motor.

DERIVED OR SHUNT CIRCUITS.

A circuit divided into two or more branches, each branch transmitting part of the current, is said to be a *derived circuit;* the individual branches are in multiple-arc, or parallel with each other.

To find the joint resistance of a derived circuit:

Rule.—*As the conductivity of any conductor is equal to the reciprocal of its resistance, then the joint conductivity of two or more circuits in parallel is equal to the sum of the reciprocals of their separate resistances. The joint resistance of two or more circuits in parallel is equal to the reciprocal of their joint conductivity.*

In a derived circuit of three branches, let r_1, r_2, and r_3 be the resistances of the three branches, respectively. Their joint conductivity, or the sum of the reciprocals of their resistances, is

$$\frac{1}{r_1} + \frac{1}{r_2} + \frac{1}{r_3}, \text{ or } \frac{r_2 r_3 + r_1 r_3 + r_1 r_2}{r_1 r_2 r_3}$$

Their joint resistance is, therefore,

$$\frac{1}{\dfrac{r_2 r_3 + r_1 r_3 + r_1 r_2}{r_1 r_2 r_3}}, \text{ or } \frac{r_1 r_2 r_3}{r_2 r_3 + r_1 r_3 + r_1 r_2}.$$

The joint resistance of a derived circuit with but two branches in parallel may be thus expressed:

$$\frac{\text{product of their resistances}}{\text{sum of their resistances}}$$

EXAMPLE.—The resistances of two branches of a derived circuit are 20 and 30 ohms, respectively. Find their joint resistance.

SOLUTION.—

$$\frac{\text{product of their resistances}}{\text{sum of their resistances}} = \frac{600}{50} = 12 \text{ ohms.}$$

To find the strength of current in the separate branches of a derived circuit:

Rule.—*A current is divided among the branches of a derived circuit in proportion to their conductivities—i. e., to the reciprocal of their resistances.*

EXAMPLE.—If the resistances of the two branches A and B of a derived circuit are 20 and 30 ohms, respectively, and the total current in the main circuit is 60 amperes, what is the current in each? The conductivity of A is $\frac{1}{20}$ and of B $\frac{1}{30}$.

SOLUTION.—If C_1 represents the current in A, and C_2 represents the current in B,

then, $$C_1 : C_2 = \tfrac{1}{20} : \tfrac{1}{30}.$$

Hence, $$\frac{C_1}{C_2} = \frac{\frac{1}{20}}{\frac{1}{30}}, \text{ or } \frac{C_1}{C_2} = \frac{30}{20} = \frac{3}{2}.$$

Now, $$C_1 + C_2 = 60, \text{ or } C_2 = 60 - C_1.$$

Substituting, $$\frac{C_1}{60 - C_1} = \frac{3}{2};$$

$$C_1 = 36, \text{ and } C_2 = 24.$$

WIRING.

INTERIOR WIRING.

A *mil* is a unit of length used in measuring the diameters of wires, and is equal to .001 in.

A *circular mil* is a unit of area used in measuring the cross-sections of wires, and is equal to $\dfrac{.7854}{10^6}$ sq. in.

The sectional area of a wire expressed in circular mils is equal to the square of its diameter in mils.

Let   c. m. = circular mils;

   C = total current in amperes;

   c = current in amperes to each lamp;

   n = number of lamps in multiple;

   v = volts lost in line;

   r = resistance per foot of wire;

   d = distance from dynamo to lamps.

The resistance of 1 ft. of commercial copper wire, 1 mil in diameter, at a temperature of 75° F., is 10.8 ohms.

A 16 c. p. (candlepower) 110-volt lamp takes about .5 ampere; a 16 c. p. 55-volt lamp takes about 1 ampere.

All calculations for size of wire must be checked by comparing with a table of safe carrying capacity (see table on pages 238 and 239), and the current value there given must not be exceeded.

To find the size of wire for 110-volt circuit with 16 c. p. lamps:

$$r = \frac{v}{n\,d}.$$

For large cables,   c. m. $= \dfrac{10.8\,n\,d}{v}$.

EXAMPLE.—Find the size of wire necessary for a circuit supplying current to 50 110-volt 16 c. p. lamps, 300 ft. from the dynamo, allowing a loss of 5% in line.

SOLUTION.—Volts at dynamo $= \dfrac{110}{.95} = 115.8$.

Volts lost in line $= 115.8 - 110 = 5.8 = v$.

Then, $r = \dfrac{v}{n\,d} = \dfrac{5.8}{50 \times 300} = .000386$ ohm per ft.,

$= .386$ ohm per 1,000 ft.

The nearest size of wire, as given in the table on page 238, is No. 6 B. & S., and its current capacity is 35 amperes; therefore it is safe.

To find the size of wire for a 55-volt circuit with 16 c. p. lamps:

$$r = \frac{v}{2\,n\,d}.$$

For large cables,   c. m. $= \dfrac{21.6\,n\,d}{v}$.

EXAMPLE.—What size of wire should be used for supplying current to 75 16 c. p. lamps on a 55-volt circuit, the distance from dynamo being 230 ft., and line loss, 4 volts?

SOLUTION.—

$$r = \frac{v}{2\,n\,d} = \frac{4}{2 \times 75 \times 230} = .000116 \text{ ohm per ft.,}$$

$= .116$ ohm per 1,000 ft.

By referring to the table, (page 238) the nearest wire is found to be No. 1 B. & S., and its carrying capacity is greater than the current (75 amperes) that it is to conduct.

To find the size of wire for any circuit on a 2-wire system:

In general, $$r = \frac{v}{C \times 2\,d};$$

or, $$\text{c. m.} = \frac{10.8 \times 2\,d \times C}{v}.$$

EXAMPLE.—What wire should be used to carry 450 amperes a distance of 600 ft., the allowable drop being 6%, and the E. M. F. at the end of the circuit 115 volts?

SOLUTION.—Volts at dynamo $= \dfrac{115}{.94} = 122.3.$

Volts lost in line $= 7.3.$

Then, c. m. $= \dfrac{10.8 \times 2 \times 600 \times 450}{7.3} = 798,900.$

Comparing this number with the table on page 239, giving current capacity of cables, it will be seen that it is within the prescribed limits.

These formulas may be used for feeders, mains, branch mains, service mains, and inside wiring on continuous-current circuits, and for secondary wiring on alternating systems.

To find the size of wire for a 110-volt circuit, 3-wire system, 16 c. p. lamps:

$$r = \frac{4\,v}{n\,d} \text{ for each wire.}$$

For large cables,

$$\text{c. m.} = \frac{2.7\,n\,d}{v} \text{ for each wire.}$$

In checking for carrying capacity, remember that the wire carries only one-half the current that would be used on a 2-wire system, as the voltage between the outside conductors is double the voltage at the terminal of 1 lamp.

EXAMPLE.—What should be the size of the conductors for a 3-wire system, when 132 110-volt, 16 c. p. lamps are installed at a distance of 210 ft. from the source of supply, the loss being 4 volts?

SOLUTION.—

$$r = \frac{4 \times 4}{132 \times 210} = .000577 \text{ ohm per ft.},$$

$$= .577 \text{ ohm per 1,000 ft.}$$

This would call for a wire between Nos. 7 and 8. The

current will be $\dfrac{132 \times .5}{2}$ = 33 amperes; but this is too much for the wire to carry, and No. 6 B. & S. wire should be used, notwithstanding the somewhat less drop in volts that will result.

For continuous-current circuits, 5% loss is usually allowed, with full current from the dynamo to the lamps. For long distances a larger line loss may be allowed, if the dynamo is wound for that loss.

DIMENSIONS, WEIGHT, AND RESISTANCE OF COPPER WIRE.

B. & S. Gauge.	Diameter in Mils (d). 1 mil = .001 in.	Area. Circular Mils (d²).	Weight and Length. Lb. per 1,000 Ft.	Weight and Length. Ft. per Lb.	Resistance at 75° F. Ohms per 1,000 ft.	Current. Amperes. Exposed.	Current. Amperes. Concealed.	B. & S. Gauge.
0000	460.000	211,600.0	639.33	1.56	.049	300	175	0000
000	409.640	167,805.0	507.01	1.97	.062	245	145	000
00	364.800	133,079.0	402.09	2.49	.078	215	120	00
0	324.950	105,592.0	319.04	3.13	.098	190	100	0
1	289.300	83,694.0	252.88	3.95	.124	160	95	1
2	257.630	66,373.0	200.54	4.99	.156	135	70	2
3	229.420	52,634.0	159.03	6.29	.197	115	60	3
4	204.310	41,742.0	126.12	7.93	.248	100	50	4
5	181.940	33,102.0	100.01	10.00	.313	90	45	5
6	162.020	26,250.0	79.32	12.61	.395	80	35	6
7	144.280	20,817.0	62.90	15.90	.498	67	30	7
8	128.490	16,509.0	49.88	20.05	.628	60	25	8
9	114.430	13,094.0	39.56	25.28	.792			9
10	101.890	10,381.0	31.37	31.88	.999	40	20	10
11	90.742	8,234.1	24.88	40.20	1.260			11
12	80.808	6,529.9	19.73	50.69	1.589	30	15	12
13	71.961	5,178.4	15.65	63.91	2.003			13
14	64.084	4,106.8	12.41	80.59	2.526	22	10	14
15	57.068	3,256.7	9.83	101.65	3.186			15
16	50.820	2,582.9	7.80	128.17	4.017	15	5	16
17	45.257	2,048.2	6.19	161.59	5.066			17
18	40.303	1,624.3	4.91	203.76	6.388	10		18
19	35.890	1,288.1	3.89	257.42	8.055			19
20	31.961	1,021.5	3.08	324.12	10.158	5		20

CARRYING CAPACITY OF CABLES.

Area. Circular Mils.	Current. Amperes.		Area. Circular Mils.	Current. Amperes.	
	Exposed.	Concealed.		Exposed.	Concealed.
200,000	299	200	1,200,000	1,147	715
300,000	405	272	1,300,000	1,217	756
400,000	503	336	1,400,000	1,287	796
500,000	595	393	1,500,000	1,356	835
600,000	682	445	1,600,000	1,423	873
700,000	765	494	1,700,000	1,489	910
800,000	846	541	1,800,000	1,554	946
900,000	924	586	1,900,000	1,618	981
1,000,000	1,000	630	2,000,000	1,681	1,015
1,100,000	1,075	673			

To find the size of wire on primary circuits for alternating system:

$$c.\ m. = \frac{10.8 \times 2\,d \times C^1}{v}; \quad r = \frac{v}{C^1 \times 2\,d}.$$

C^1 = the total current in amperes on primary circuit, and may be determined by dividing the total current on the secondary circuit by the product of the ratio and efficiency of conversion.

The ratio is generally 20 to 1 on a 1,000-volt apparatus when using 52-volt lamps, and 10 to 1 when using 100- to 110-volt lamps.

The efficiency of conversion can be taken as 95% in ordinary transformers.

EXAMPLE.—If the loss is 5%, find the size of wire necessary on a 1,000-volt primary circuit when the distance between the dynamo and transformer is 2,000 ft., and the dynamo is supplying current for 500 16 c. p. 52-volt lamps.

SOLUTION.—

$$\text{Volts at dynamo} = \frac{1,000}{.95} = 1,052, \text{ nearly.}$$

$$\text{Volts lost in line} = 52.$$

18

Assume the lamp efficiency to be 3.6 watts per c. p. Then, since the product of amperes and volts gives watts,

$$\text{Current to each lamp} = \frac{3.6 \times 16}{52} = 1.11 \text{ amperes.}$$

Current on secondary $= 1.11 \times 500 = 555$ amperes.

$$\text{Total current on primary is } \frac{555}{.95 \times 20} = 29.21 \text{ amperes.}$$

Therefore,

$$\text{c. m.} = \frac{10.8 \times 2d \times C^1}{v} = \frac{10.8 \times 4,000 \times 29.21}{52} = 24,267.$$

And $r = \dfrac{v}{C^1 \times 2d} = \dfrac{52}{29.21 \times 4,000} = .000445$ ohm per ft., or .445 ohm per 1,000 ft. This gives No. 6 B. & S. See page 238.

For alternating systems under ordinary conditions, 5% loss at full load from dynamo to transformer on primary circuit is a maximum, although some dynamos are specially wound for 10% loss. A loss of from 1% to 2% may be allowed on secondary circuits from transformer to lamps.

INCANDESCENT LAMPS.

Let c = current in amperes to each lamp;

 E = electromotive force in volts;

 R $= \dfrac{E}{c}$ = resistance of lamp when hot;

 c. p. = candlepower of lamp;

W. per c. p. = watts per c. p. (often called *lamp efficiency*).

$$\text{W. per c. p.} = \frac{c \times E}{\text{c. p.}}.$$

The number of candles per electrical H. P. $= \dfrac{746}{\text{W. per c. p.}}$.

$$c = \frac{\text{W. per c. p.} \times \text{c. p.}}{E}.$$

As the commercial efficiency of good dynamos is about 90%, the calculations of candles per electrical H. P. must be multiplied by .90 to give the number of candles per mechanical H. P.

LAMP EFFICIENCIES.

3.1 watts per c. p., or 12 lamps, 16 c. p., to 1 mechanical H. P.

3.6 watts per c. p., or 10 lamps, 16 c. p., to 1 mechanical H. P.

4.0 watts per c. p., or 8 lamps, 16 c. p., to 1 mechanical H. P.

NOTE.—Lamps of an efficiency of 3.1 watts per c. p. should not be used where the voltage averages, for any length of time, more than 2⅚ high; lamps of 3.6 watts per c. p. should not be used where the voltage averages more than 4⅚ high; and lamps of an efficiency of 4 watts per c. p. should be used where the regulation of the plant receives little or no attention. If these cautions are not followed, the life of the lamp will be greatly diminished.

Size of Wire for Arc-Light Circuits.—For ordinary distances, or small currents, use No. 8 B. & S. wire. For longer distances, or large currents, use No. 6 B. & S. wire.

BELL WIRING.

The simple bell circuit is shown in Fig. 1, where p is the push button, b the bell, and c, c the cells of the battery con-

nected up in series. When two or more bells are to be rung from one push button, they may be joined up in parallel across the battery wires as in Fig. 2 at a

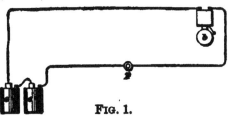

FIG. 1.

and b, or they may be arranged in series as in Fig. 3. The battery B is indicated in each diagram by short parallel lines,

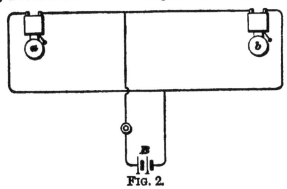

FIG. 2.

this being the conventional method. In the parallel arrangement of the bells, they are independent of each other, and the failure of one to ring would not affect the others; but in the series grouping all but one bell must be changed to a single-stroke action, so that each impulse of current will produce only one movement of the hammer. The current is then

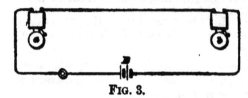

FIG. 3.

interrupted by the vibrator in the remaining bell, the result being that each bell will ring with full power. The only change necessary to produce this effect is to cut out the circuit-breaker on all but one bell by connecting the ends of the magnet wires directly to the bell terminals.

When it is desired to ring a bell from one of two places some distance apart, the wires may be run as shown in Fig. 4. The pushes p, p' are located at the required points, and the battery and bell are put in series with each other across the wires joining the pushes.

A single wire may be used to ring signal bells at each end of a line, the connections

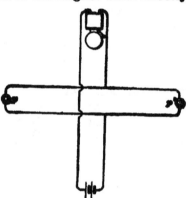

FIG. 4.

being given in Fig. 5. Two batteries are required, B and B', and a key and bell at each station. The keys k, k' are of the double-contact type, making connections normally between

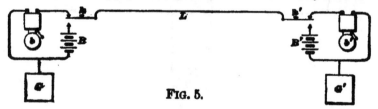

FIG. 5.

bell b or b' and line wire L. When one key, as k, is depressed, a current from B flows along the wire through the upper contact of k' to bell b' and back through ground plates G', G.

When a bell is intended for use with burglar-alarm apparatus, a constant-ringing attachment may be introduced, which closes the bell circuit through an extra wire as soon as the trip at door or window is disturbed. In the diagram, Fig. 6, the main circuit, when the push *p* is depressed, is through the automatic drop *d* by way of the terminals *a*, *b* to the bell and battery. This

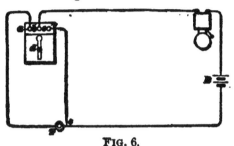

FIG. 6.

current releases a pivoted arm which, on falling, completes the circuit between *b* and *c*, establishing a new path for the current by way of *e*, independent of the push *p*.

For operating electric bells, any good type of open-circuit battery may be used. The Leclanché cell is largely used for this purpose, also several types of dry cells.

ANNUNCIATOR SYSTEM.

The wiring diagram for a simple annunciator system is shown in Fig. 1. The pushes *1, 2, 3*, etc. are located in the various rooms, one side being connected to the battery wire

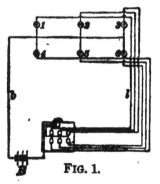

FIG. 1.

b, and the other to the leading wire *l* in communication with the annunciator drop corresponding to that room. A battery of 2 or 3 Leclanché cells is placed at *B* in any convenient location. The size of wire used throughout may be No. 18 annunciator wire.

A return-call system is illustrated in Fig. 2, in which there is one battery wire *b*, one return wire *r*, and one leading wire *l*, *l₂*, etc. for each room. The upper portion of the annunciator board is provided with the usual drops, and below these are the return-call pushes. These are double-contact buttons, held normally against the upper contact by a spring. When in this position, the closing of the circuit by the push button in any room, such as No. 4, rings the office bell and releases No. 4 drop, the path of the current in this case being from

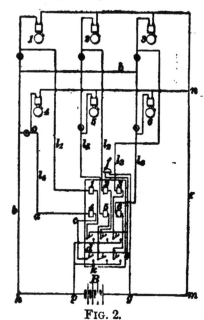

FIG. 2.

push 4 to a–c–d–e–f–g–B–h–b back to the push button. On the return signal being made by pressing the button at the lower part of the annunciator board, the office-bell circuit is broken at d, and a new circuit formed through k as follows: From the battery B to g–m–r–n–o–a–c–k–p to battery, the room bell being in this circuit. A general fire-alarm may be added to this system, consisting of an automatic clockwork apparatus for closing all the room-bell circuits at once, or as many at a time as a battery can ring. When this system is installed, the battery wire should be either No. 14 or No. 16. Four or five Leclanché cells are usually required in this case.

It will be seen that the connections are so arranged that the room bell will ring when the push in that room is pressed. If this be not desired, a double-contact push may be substituted, so that the room-bell circuit is broken at the same time that the circuit is made through the annunciator. This double push should be so connected that the circuit is normally complete through the bell, the leading wire being connected to the tongue and the battery wire being connected to the second contact point, which is normally out of circuit.

Incandescent Wires.—Conducting wires, carried over or attached to buildings, must be (a) at least 7 ft. above the highest point of flat roofs, and (b) 1 ft. above the ridge of pitch roofs; (c) when in proximity to other conductors likely to divert any portion of the current, they must be protected by guard irons or wires, or a proper additional insulation, as the case may require.

For entering buildings, (a) wires with an extra-heavy waterproof insulation must be used; (b) they must be protected by drip loops; (c) also protected from abrasion by awning frames; (d) be at least 6 in. apart; (e) the holes through which they pass in the outer walls of such buildings must be bushed with a non-inflammable, waterproof, insulating tube, and (f) should slant upward toward the inside.

(a) Wires must never be left exposed to mechanical injury, or to disturbance of any kind. (b) Wires must not be fastened by metallic staples. (c) When wires pass through walls, floors, partitions, timbers, etc., glass tubing, or so-called "floor insulators," or other moisture-proof, non-inflammable insulating tubing must be used. (d) At all outlets to and from cut-outs, switches, fixtures, etc., wires must be separated from gas pipes or parts of the building by porcelain, glass, or other non-inflammable insulating tubing, (e) and should be left in such a way as not to be disturbed by the plasterers. (f) Wires of whatever insulation must not in any case be taped, or otherwise be fastened, to gas piping. (g) If no gas pipes are installed at the outlets, an approved substantial support must be provided for the fixtures.

In crossing any metal pipes, or any other conductor, (a) wires must be separated from the same by an air space of at least ¼ in., where possible, and (b) so arranged that they cannot come in contact with each other by accident. (c) They should go *over* water pipes, where possible, so that moisture will not settle on the wires.

In unfinished lofts, between floors and ceilings, in partitions, and other concealed places, wires must (a) be kept free of contact with the building; (b) be supported on glass, porcelain, or other non-combustible insulators; (c) have at least 1 in. clear air space surrounding them; (d) be at least 10 in. apart, when possible; and (e) should be run singly on separate timbers or studding. (f) When thus run in perfectly dry places, not liable to be exposed to moisture, a wire having simply a non-combustible insulation may be used.

Soft rubber tubing is not desirable as an insulator.

Care must be taken that the wires are not placed above each other in such a manner that water could make a cross-connection.

On all loops of incandescent circuits, safety catches must be used on both sides of the loop, and switches on such loops should be double-poled.

Wires must not be fished (a) for any great distance, and (b) only in cases where the inspector can satisfy himself that the above rules have been complied with. (c) Twin wires must never be employed in this class of concealed work.

Dynamo Machines.—Dynamo machines must be located in dry places, not exposed to flyings or easily combustible material, and insulated upon wooden foundations. The machines must be provided with devices that shall be capable of controlling any changes in the quantity of the current; and if the governors are not automatic, a competent person must be in attendance near the machine whenever it is in operation.

Each machine must be used with complete wire circuits; and connections of wires with pipes, or the use of circuits in any other method, are absolutely prohibited.

The whole system must be kept insulated, and tested *every day* with a magneto for ground connections in ample time before lighting, to remedy faults of insulation, if they are discovered; and proper testing apparatus must in each case be provided. This applies to both central station and isolated plants.

Testing circuits for grounds with a battery and bell is not considered a reliable test.

Preference is given to switches constructed with a lapping connection, so that no electric arc can be formed at the switch when it is changed; otherwise the stands of switches, where powerful currents are used, must be made of some incombustible substances that will withstand the heat of the arc when the switch is changed.

Motors.—Wires for motors should be run exactly as for lamps on similar circuits.

On low-tension circuits, where motors are run in multiple, safety catches must be used on each side of the circuit.

On high-tension circuits the same restrictions apply as for arc lamps, and suitable cut-outs must be provided.

Motors must be treated as dynamos as regards insulation, flyings, dampness, etc.

NOTE.—If the regulations of the Underwriters' Association are not followed in wiring buildings, the wiring is liable to be condemned by the Insurance Inspectors and the policy canceled.

WIRE TABLES.

WEIGHT OF UNDERWRITERS' LINE WIRE, INSULATED.

No. B. & S.	Pounds per 1,000 Feet.	Feet per Pound.
0000	800	1.25
000	666	1.50
00	500	2.00
0	363	2.75
1	313	3.20
2	250	4.00
3	200	5.00
4	144	6.9
5	125	8.0
6	105	9.5
7	87	11.5
8	69	14.5
10	50	20.0
12	31	32.0
14	22	45.0
16	14	70.0
18	11	90.0

EQUIVALENT SECTIONAL AREA OF WIRES, B. & S. GAUGE.

Gauge No.	No. of Wires. Gauge No.	No. of Wires. Gauge No.	No. of Wires. Gauge No.	No. of Wires. Gauge No.	No. of Wires. Gauge No.	No. of Wires. Gauge No.	Gauge No. Gauge No.
0000	2- 0	4- 3	8- 6	16- 9	32-12	64-15	
000	2- 1	4- 4	8- 7	16-10	32-13	64-16	
00	2- 2	4- 5	8- 8	16-11	32-14	64-17	1 and 3
0	2- 3	4- 6	8- 9	16-12	32-15	64-18	2 and 3
1	2- 4	4- 7	8-10	16-13	32-16		3 and 5
2	2- 5	4- 8	8-11	16-14	32-17		4 and 6
3	2- 6	4- 9	8-12	16-15	32-18		5 and 7
4	2- 7	4-10	8-13	16-16			6 and 8
5	2- 8	4-11	8-14	16-17			7 and 9
6	2- 9	4-12	8-15	16-18			8 and 10
7	2-10	4-13	8-16				9 and 11
8	2-11	4-14	8-17				10 and 12
9	2-12	4-15	8-18				11 and 13
10	2-13	4-16					12 and 14
11	2-14	4-17					13 and 15
12	2-15	4-18					14 and 16
13	2-16						15 and 17
14	2-17						16 and 18
15	2-18						

The above table indicates the number of smaller wires required to give a sectional area equal to one larger size wire, the figures between the horizontal lines corresponding to each other. For example: It requires two wires, No. 0, or 4 wires, No. 3, etc., to give a sectional area equal to 1 wire, No. 0000. Again: it requires two wires, No. 13, or 4 wires, No. 16; or 2 wires, 1 No. 12 plus 1 No. 14, to give a sectional area equal No. 10.

COMPARATIVE SIZES OF WIRES, B. & S. AND BIRMINGHAM GAUGES.

Diameter. Inches.	B. & S.	Birmingham.
.460	0000	
.454		0000
.425		000
.4096	000	
.380		00
.3648	00	
.340		0
.3249	0	
.3000		1
.2893	1	
.284		2
.259		3
.2576	2	
.238		4
.2294	3	
.22		5
.2043	4	
.203		6
.1819	5	
.18		7
.165		8
.162	6	
.148		9
.1443	7	
.134		10
.1285	8	
.12		11
.1144	9	
.109		12
.1019	10	
.095		13
.0907	11	
.083		14

COMPARATIVE SIZES OF WIRES, B. & S. AND BIRMINGHAM GAUGES—(*Continued*).

Diameter, Inches.	B. & S.	Birmingham.
.0808	12	
.0720	13	15
.0650		16
.0641	14	
.0580		17
.0571	15	
.0508	16	
.0490		18
.0453	17	
.0420		19
.0403	18	
.0359	19	

NOTE.—B. & S. gauge is generally used in America.

COMPARISON OF PROPERTIES OF ALUMINUM AND COPPER.

	Aluminum.	Copper.
Conductivity (for equal sizes)	.54 to .63	1.
Weight (for equal sizes)	.33	1.
Weight (for equal length and resistance)	.48	1.
Price (per pound) Aluminum, 29c.; Copper, 16c. (bare wire)	1.81	1.
Price (equal length and resistance, bare line wire)	.868	1.
Temperature coefficient per degree F.	.002138	.002155
Resistance of mil-foot (20° C.)	18.73	10.5
Specific gravity	2.5 to 2.68	8.89 to 8.93
Breaking strength (equal sizes)	1.	1.

RESISTANCE OF PURE COPPER WIRE.

No. & B. & S.	Resistance at 75° F.			
	R. Ohms per 1,000 Feet.	Ohms per Mile.	Feet per Ohm.	Ohms per Pound.
4-0	.04904	.25891	20,392.90	.00007653
3-0	.06184	.32649	16,172.10	.00012169
00	.07797	.41168	12,825.40	.00019438
0	.09827	.51885	10,176.40	.00030734
1	.12398	.65460	8,066.00	.00048920
2	.15633	.82543	6,396.70	.00077784
3	.19714	1.04090	5,072.50	.00123700
4	.24858	1.31248	4,022.90	.00196660
5	.31346	1.65507	3,190.20	.00312730
6	.39528	2.08706	2,529.90	.00497280
7	.49845	2.63184	2,006.20	.00790780
8	.62849	3.31843	1,591.10	.01257190
9	.79242	4.18400	1,262.00	.01998530
10	.99948	5.27726	1,000.50	.03170460
11	1.26020	6.65357	793.56	.05054130
12	1.58900	8.39001	629.32	.08036410
13	2.00370	10.57980	499.06	.12778800
14	2.52660	13.34050	395.79	.20318000
15	3.18600	16.82230	313.87	.32307900
16	4.01760	21.21300	248.90	.51373700
17	5.06600	26.74850	197.39	.81683900
18	6.38800	33.72850	156.54	1.29876400
19	8.05550	42.53290	124.14	2.06531200
20	10.15840	53.63620	98.44	3.28437400

CONNECTIONS FOR DYNAMO-ELECTRIC MACHINES.

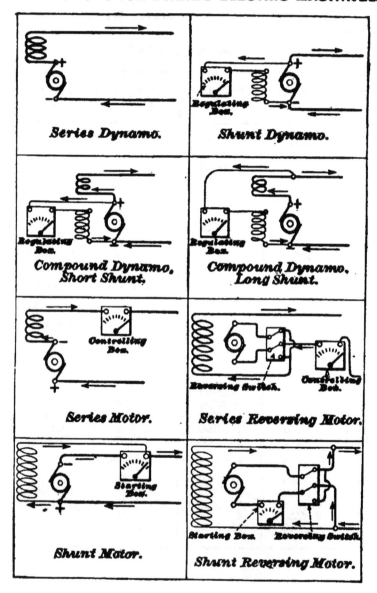

In the diagrams showing the connections of dynamo-electric machines, the heavy coils represent the series winding on the field magnets through which the entire current of the machine passes; the lighter coils represent the shunt winding on the field magnets through which only part of the main current passes.

Lamps connected in series.

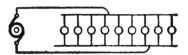

Lamps connected in multiple-arc or parallel.

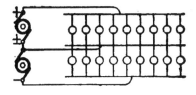

Edison three-wire system.

DYNAMOS AND MOTORS.

MOTOR CIRCUITS.

To find the size of wire on stationary motor circuits:

Let c. m. = circular mils;

e = E. M. F. of motor in volts;

v = loss of volts in line;

d = distance from generator to motor in feet;

k = efficiency of motor;

10.8 ohms is the resistance of 1 ft. of commercial copper wire 1 mil in diameter.

$$c. m. = \frac{H. P. \text{ of motor} \times 746 \times 2\,d \times 10.8}{e\,v\,k}.$$

APPROXIMATE MOTOR EFFICIENCY.

$\frac{1}{4}$ to $1\frac{1}{4}$ H. P. inclusive = 75% efficiency.

3 to 5 H. P. inclusive = 80% efficiency.

$7\frac{1}{2}$ to 10 H. P. inclusive = 85% efficiency.

25 H. P. and upwards = 90% efficiency.

Under ordinary circumstances, 10% loss from generator to motor is a maximum on stationary motor circuits.

EXAMPLE.—What is the size of wire necessary for a circuit on which a 10 H. P. 500-volt motor is running, when the distance between the motor and generator is 2,000 ft. and the loss is 5%?

SOLUTION.—Volts at generator, $\frac{500}{.95}$ = 526, nearly.

Volts lost in line, 526 − 500 = 26.

In the table on page 258, the approximate efficiency of a 10 H. P. motor is given as 85%.

$$c. m. = \frac{10 \times 746 \times 4,000 \times 10.8}{500 \times 26 \times .85} = 29,165.$$

In the table on page 238, the nearest size of wire corresponding to this area is No. 6 B. & S. gauge.

The approximate weight and resistance per mile of round bare wire when d is the diameter in mils, are, for copper wire, $\frac{d^2}{62.5}$ lb. and $\frac{56,970}{d^2}$ ohms; for iron wire, $\frac{d^2}{72}$ lb. and $\frac{380,060}{d^2}$ ohms.

Copper wire is approximately 1⅛ times the weight of an iron wire of the same diameter.

In determining the size of wire to be used for inside work, after finding the c. m., always refer to the table on page 238, and see that the wire obtained by the formula is sufficiently large to carry the current; if not, use larger wire, regardless of per-cent. loss. *For pole-line construction, never use wire smaller than No. 8 B. & S. gauge.*

DYNAMO DESIGN.

The fundamental principle of dynamo design is expressed by the formula

$$E = \frac{N C n}{10^8 \times 60},$$

in which

E = electromotive force in volts given by the dynamo;

N = number of lines of force used to magnetize the armature;

C = number of conductors in a bipolar machine, measured all round the outside of the armature (whether in one or more layers), or in a multipolar machine, as measured from a point opposite one north pole to a corresponding point opposite the next succeeding north pole;

n = number of revolutions per minute of the armature.

For example, a 2-pole dynamo has 2,000,000 lines of force passing from the north pole through the armature to the south pole; there are 200 conductors on the surface of the armature, and the speed is 1,500 rev. per min. The electromotive force generated will then be

$$E = \frac{2,000,000 \times 200 \times 1,500}{100,000,000 \times 60} = 100 \text{ volts.}$$

If a 4-pole dynamo were used, having a 4-circuit armature and 4 sets of brushes, with 1,000,000 lines of force passing through any one pole piece, then the total number would be 2,000,000, because the same lines of force pass into a south pole that emerge from a north pole. With the same armature as above, the number of conductors to be counted is only 100, as taken from one north pole to the next, and the electromotive force is

$$E = \frac{2,000,000 \times 100 \times 1,500}{100,000,000 \times 60} = 50 \text{ volts.}$$

For determining the number of lines of force required in a specific case, the above formula may be reversed, and we have

$$N = \frac{E \times 10^8 \times 60}{C n}.$$

These lines of force have a circuit to traverse composed of three different paths. One of these is through the field magnet and yoke M, Fig. 1; next, through a double air gap G; and, lastly, through the armature core A. A given density of lines of force may not be exceeded, this limit being for ordinary cast iron about 50,000 lines per square inch; for wrought-iron forgings or cast steel, about 90,000; and for soft sheet iron, 110,000.

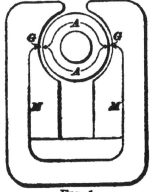

FIG. 1.

The ratio of magnetization to magnetizing force is called

19

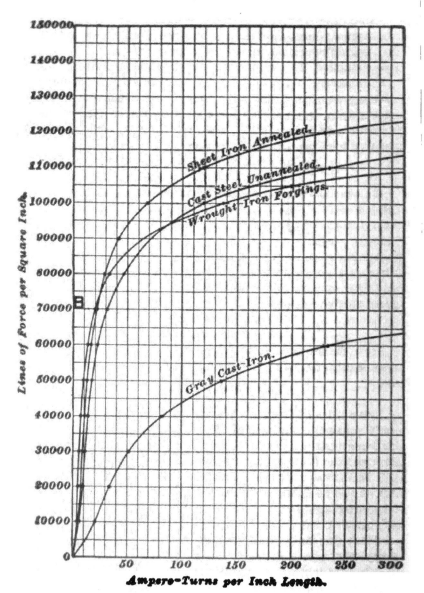

FIG. 2.

the *permeability*. The permeability of air is very low, the intensity of magnetization being a direct measure of the magnetizing force required; therefore, the air gap is usually made short.

In order to drive the lines of force through the magnetic circuit, magnetizing coils are wound on the cores at *M, M*. A certain number of ampere-turns will be required, depending on the density of the lines of force and the permeability of the different portions of the circuit. The number of turns may be found by taking a convenient current value, and dividing the ampere-turns by this. Reference to a wire table will then determine whether the resistance of the wire will be such that the terminal E. M. F. of the machine will supply the proper current. A margin should be allowed for regulating, and for the increase in resistance due to rise in temperature, which is about .4% for every degree centigrade, or .222% for every degree Fahrenheit above 75° F.

In the saturation curves of Fig. 2 are represented graphically the different values of the induction (**B**) in lines of force per square inch, corresponding to the magnetizing force expressed in ampere-turns per inch of length of circuit. Thus, to send 70,000 lines of force through a cast-steel core 1 sq. in. in cross-sectional area, would require about 30 ampere-turns for every inch in length of core. The 30 ampere-turns might be obtained by using a coil of 30 turns carrying 1 ampere, or 300 turns of $\frac{1}{10}$ ampere, etc. The number of lines of force N for any particular case being known, and also the allowable density **B**, which will vary somewhat with different samples of iron, the cross-sectional area $A = \dfrac{N}{B}$.

The ampere-turns to be added to the magnetizing coils to overcome the resistance of the air gap is

$$A.\,T. = \frac{H \times l}{3.192},$$

where **H** = number of lines of force per square inch;
and l = length of air gap (the two sides added together) in inches, usually a fraction.

It is necessary, in calculating the ampere-turns for the field circuit, to allow for leakage of lines of force through the surrounding air, as the total number generated does not pass through the armature core. This leakage may amount to 30% or 40% of the whole, but is much less in well-designed machines.

For example, a bipolar dynamo has magnet cores having a mean length, with pole pieces, of 10 in. each; the yoke of the magnet is 13 in.; air gap, $\frac{3}{8}$ in. each side; armature core, 10 in. The magnetic density in the core is 85,000; air gap, 46,000; yoke, 65,000; armature core, 90,000 lines of force per square inch. If the fields are wrought-iron forgings, and the armature is built up of soft sheet iron, then the ampere-turns necessary will be:

	Length.	B	A.-T. per In.	Ampere-Turns.
Magnet cores	20 in.	85,000	44	880
Yoke	13 in.	65,000	16	208
Armature	10 in.	90,000	40	400
Air gap	⅜ in.	46,000		5,425
Total ampere-turns				6,913

In determining the size of wire to be used in the armature winding, a certain density of current may be assumed as the limit. This is usually expressed in circular mils or thousandths of an inch per ampere. For most purposes of design, a density of 600 circular mils per ampere may be allowed. In estimating the current passing through the armature, it must be remembered that the current of the outside circuit divides on reaching the armature, and passes through it along two paths in parallel with each other.

FAULTS OF DYNAMOS.

Reversal of Field.—Run the machine up to speed, and hold a small compass near each pole piece in succession. Their polarity should alternate all the way round.

Failure to Build Up.—This is probably due to reversal of shunt connections. Rock the brushes around until any one set occupies a position formerly occupied by the next set. If this should remedy the trouble, and such position is inconvenient, move them back and reverse connections of shunt windings. If the failure of machine is due to want of residual magnetism, send a current from some external source through the fields. If it is due to a broken circuit, each coil may be tested separately with a battery and galvanometer or low-reading Weston voltmeter. Failure to generate may be due to the brushes being out of the neutral plane, which may be tested by moving them into different positions.

Heating.—This may be caused by a short-circuited armature coil. Allow the machine to cool, then run for a few minutes with no load, and stop. The defective coil will be found to be much hotter than the rest. It should be marked, and the armature taken out, when the coil may be rewound or otherwise repaired. If the heating is even, the load may be excessive and should be reduced. The effect may be due to eddy currents in the armature core, but this is a question of design in the first instance.

Sparking at Commutator.—If this be due to overload, the sparking cannot be cured except by reducing the load. The trouble may be due to improper position of brushes. Move the rocker-arm to one side or the other to determine this. If copper brushes (tangential) are used, they may be unevenly spaced round the commutator; each set of brushes should have the same relative position with regard to the respective pole tips. Sparking may be caused by an uneven commutator, in which case it should be smoothed with sandpaper (never emery) or turned down in the lathe. A broken connection at the commutator leads will produce flashing at each revolution, and one of the bars will show a burn extending nearly across it. The loose wire should be secured, or if broken, the commutator bars may be connected together with a piece of wire or a drop of solder as a temporary repair. As soon as possible a new coil should be put in. Sparking may also occur, in a multipolar machine, from the wearing away of the bearings, which produces eccentricity of the armature with respect to field, and consequent unequal magnetic induction at different points. A slight sparking at the brushes of the machine is not detrimental.

OUTPUT AND EFFICIENCY OF MOTORS.

A dynamo, when supplied with current from an external source, becomes a motor, turning the electrical energy into mechanical energy. The ratio between these two quantities, that is, between the input and output, determines the efficiency of the motor. The input may be found by measuring the current C with an ammeter, and the voltage E with a voltmeter, their product giving the power supplied in watts, $W = CE$. This quantity, divided by 746, gives the electrical horsepower, or E. H. P. $= \dfrac{W}{746}$.

The output is measured by means of a Prony brake (see figure). The motor pulley P is clamped between two blocks

of wood B, B, their pressure being regulated by the thumbscrews N, N, on the long bolts which hold them together. The lower block is extended to form an arm A of convenient length, and furnished with a sharp lagscrew C at the end. The lagscrew presses on the platform of a set of scales S, whereby its pressure may be determined. A counterbalance at W neutralizes the weight of the arm. When the pulley is revolved in the direction shown, the pressure on the scale will indicate the torque, or twisting power, developed, which is expressed as the product of the pressure on the scale into the distance between the center of pulley and the point of the screw. If the length of arm $R = 2$ ft., and the pressure is 50 lb., the torque $T = 100$ ft.-lb. The horsepower may be determined by the following formula:

$$\text{H. P.} = \frac{2 \times 3.1416\, T S}{33,000},$$

in which S is the speed of motor in revolutions per minute.

APPLICATIONS OF ELECTRIC MOTORS.

The same varieties of field and armature connections are used for motors as for dynamos, namely, series, shunt, and compound, and each type has distinguishing characteristics. The series motor is especially suitable for use in cases where a very high starting torque is required in order to obtain rapid acceleration under load, as, for instance, in street-railway work. Torque may be defined as the reaction of the current in the armature or moving part against the magnetic lines of force in the field magnets or stationary part. Strength of field is obtained by the current circulating through the magnet coils; consequently, the torque in a series motor will be a maximum when the current passing through is a maximum, as the same amount flows through armature and field. The opposition to the flow of current is the resistance of the circuit and the counter E. M. F. of the armature. When the current is applied, its value is determinable by Ohm's law for the first moment, supposing self-induction to be eliminated. The resistance of a series motor is usually so low that an additional resistance must be used at starting in order to prevent an excessive flow of current; but, as soon as the armature begins to revolve, the counter E. M. F. opposes and cuts down the current, and, consequently, the torque. The speed will continue to increase and the torque to decrease until the mechanical resistance to rotation balances the torque. If the motor is running light, the speed will rise continually, the counter E. M. F. will also increase and cut down the current, and the consequent reduction of field strength will require a still higher speed in order to develop the necessary counter E. M. F., the final result being, probably, the bursting of the armature. The speed of a series motor under a constant load may be regulated by the somewhat wasteful method of introducing a resistance in series to reduce the speed, and by cutting out or shunting part of the field coils, to increase it. When two motors are used, they may be put in series at starting and connected in parallel for higher speeds. The series motor is well adapted for electric cranes, because it will automatically regulate its speed to the weight to be raised, exerting a very powerful torque at low speed for a heavy load.

The shunt motor will give a nearly constant speed for any variation in load, as long as the potential of current supply (the applied E. M. F.) is constant. This condition produces a constant field, as the shunt winding is directly across the main leads, and the speed of the motor will then be such that the difference between the E. M. F. of supply and the counter E. M. F., divided by the resistance of the armature, will be equal to the current passing through the armature. A change in the current will then produce but a relatively · small change in the required counter E. M. F. of the motor, and the speed will only vary to that extent. As the load is put on, the motor tends to slow down; but this, by decreasing the counter E. M. F., allows more current to flow, thereby producing more torque to overcome the added mechanical resistance. Change of speed may be produced by varying the strength of the magnetic field, the weaker the field the higher the speed. If the load is constant, the torque will be decreased, but, if the load be correspondingly increased, the torque will remain nearly constant. Considerable weakening of the field is inadvisable, as it will cause destructive spark- ing at the commutator. The theoretically perfect method of speed regulation for a shunt motor is to provide a constant and independent field, and effect change of speed by varying the applied E. M. F. at the armature terminals without insertion of extra resistance. In this case the torque will always be proportional to the load, and the efficiency will be constant and independent of speed and torque. In the operation of such a system, certain complications are introduced, inasmuch as it is necessary to install in connection with each motor a special dynamo with variable field, and this condition may therefore constitute a serious objection when the first cost of the plant is required to be low.

A differential compound winding may be used when a more nearly constant speed is wanted. The series turns on the field magnets are so connected as to oppose the shunt turns, and when an increase of load tends to cut down the speed, the additional current through the series turns weakens the field slightly, so that the same speed as before is required to generate the lower counter E. M. F.

Shunt motors are especially useful for machine tools, which require a constant speed irrespective of load, and may also be used on printing presses and similar machines where the load is more nearly uniform. When a variation in speed with load is immaterial, a cumulative compound winding may be employed, in which the series turns act with the shunt, thereby increasing the torque at starting, and affording some of the characteristics of both the shunt and series windings.

BATTERIES.

The simple primary battery consists of two elements, the *anode*, which is usually zinc, and the *cathode*, which may be carbon, both immersed in an exciting liquid called the *electrolyte*. The chemical action incident to the generation of current dissolves the zinc and liberates free hydrogen at the cathode, which adheres to the surface and reduces the E. M. F. of the battery. To overcome this effect, called *polarization*, a depolarizer is used which will take up the hydrogen as it is formed.

Depolarizers may be solid or liquid. When solid, the material is usually packed round the cathode, as in the case of the Leclanché cell; when the depolarizer is liquid, it may be prevented from mixing with the electrolyte by a porous partition, or, if their specific gravities differ considerably, they will remain separated one over the other in the jar. The following table gives the elements and depolarizers for different cells, with the E. M. F. in volts:

Name.	Anode.	Electrolyte.	Cathode.	Depolarizer.	E.M.F.	Remarks.
Volta	Zinc	Sulphuric acid (dilute)	Copper	None	.9	Polarizes rapidly.
Law	Zinc	Sulphuric acid (dilute)	Carbon	None	1.35	
	Zinc	Sodium chloride (common salt)	Carbon	None	1.08	
Grenet	Zinc	Sulphuric acid 4, Potassium bi-chromate 3, Water 18	Carbon	Electrolyte	1.9 to 2	For large currents.
Pabst	Wrought iron	Ferric chloride	Carbon		.78	Non-polarizing electrolyte.
Bunsen	Zinc	Sulphuric acid (dilute)	Carbon	Nitric acid	1.89	Cathode and depolarizer in porous cup.
Fuller	Amalgamated zinc in mercury	Sulphuric acid (very dilute) or water	Carbon	Electropoion fluid diluted one-half	2.14	Anode in porous cup.
Partz	Zinc	Sodium chloride or magnesium sulphate	Carbon	Bichromate solution (sulphochromic salt)	1.9 to 2	Gravity cell. Resistance with sodium chloride, .5 ohm; with magnesium sulphate, 1 ohm.

Name.	Anode.	Electrolyte.	Cathode.	Depolarizer.	E.M.F.	Remarks.
D'Arsonval......	Zinc.......	Caustic soda.........	Carbon ...	Ferric chloride	2.7	Porous cup used; pores become filled with ferric hydrate, an insoluble conductor.
Daniell ...	Zinc.......	Zinc sulphate........	Copper....	Copper sulphate with copper-sulphate crystals.....	1.07	Cathode and depolarizer in porous cup.
Gravity Daniell.	Zinc.......	Zinc sulphate. Sp. Gr. 1.10	Copper....	Copper sulphate with copper-sulphate crystals.....	1.07	For closed-circuit work only; resistance 3 ohms.
Leclanché	Zinc.......	Ammonium chloride (saturated)...	Carbon ...	Peroxide of manganese	1.48	Carbon and depolarizer in porous cup; resistance 4 ohms.
Lalande and Chaperon	Zinc.......	Caustic potash.......	Iron or Copper.	Cupric oxide.........	.7	Surface of electrolyte covered with layer of oil.
Edison-Lalande	Zinc.......	Caustic potash.......		Molded plates of cupric oxide and magnesic chloride held in copper frames....	.7	Surface of electrolyte covered with layer of oil; resistance .07 ohm.
Chloride- of mercury cell	Zinc.......	Sal ammoniac (ammonium chloride)	Carbon ...	Paste of mercurous chloride............	1.45	Cathode and depolarizer in porous cups. For small currents.

Name.	Anode.	Electrolyte.	Cathode.	Depolarizer.	E.M.F.	Remarks.
Chloride-of-silver cell.	Zinc.	Ammonium chloride (dilute)	Silver wire or plate with chloride of silver.		1.03	For medical work and testing.
Chloride-of-silver cell.	Zinc.	Zinc chloride.	Silver wire or plate with chloride of silver.		1.02	
Chloride-of-silver cell.	Zinc.	Sodium chloride.	Silver wire or plate with chloride of silver.		.97	
....	Zinc.	Sulphuric acid (dilute)	Carbon.	Mercuric sulphate, mercurous sulphate, or turpeth mineral	1.3 to 1.5	Depolarizer in form of paste. Poisonous.
Latimer-Clark.	Zinc.	Paste of mercurous sulphate formed with zinc sulphate	Mercury.	Electrolyte.	1.442	Standard cell, for very minute currents.
Gouy.	Zinc.	Zinc sulphate 10% solution	Mercury.	Oxide of mercury.	1.39	Standard cell.
Weston.	Cadmium amalgam	Cadmium sulphate	Mercury, with sulphate of mercury.		1.019	Standard cell. No temperature coefficient.
Baille and Fery.	Amalgamated zinc.	Zinc chloride.	Lead, with crystals of lead chloride.		.5	Standard cell.

STORAGE BATTERIES.

Storage batteries or accumulators are composed of plates of prepared lead, placed side by side in glass cells or wooden boxes lined with rubber or lead, alternate plates being connected together, thus forming two sets, which constitute the positive and negative elements. The plates are entirely submerged in dilute sulphuric acid, specific gravity 1.17. The charging E. M. F. is about 2.5 volts per cell, so that, if 10 cells are connected in series, the required E. M. F. will be 25 volts. The discharging E. M. F. is usually taken as 1.9 volts, so that an installation to supply current at 115 volts should consist of $\frac{115}{1.9} = 61$ cells, with a few added to replace any that are out of order or to serve as regulators to vary the E. M. F. As soon as the battery is set up and the electrolyte added, the charging should commence, the first charge being continued a long while at a comparatively slow rate. Observe that the direction of current through the cell in charging is from the positive or brown plate to the negative or gray one. Discharging should be at a low rate, as rapid discharge leads to deterioration of the positive plates.

The rating of the capacity of accumulators is usually made on the basis of a discharge current that will cause the E. M. F. to fall to 1.8 volts in 10 hours, but it is well to stop discharging when the E. M. F. falls to 1.9 volts.

Storage-Battery Regulation.—In electric-lighting plants, an equalization of load on the dynamos is sometimes obtained by installing accumulators or storage batteries. Automatic or hand regulation may be employed, the usual method being to cut out one or more cells when the load is light and change the remainder, these cells being connected in again when the load rises. The following method obviates the many disadvantages of this system.

A shunt dynamo d, Fig. 1, supplies current to the lighting mains m, n, this current passing through the fields c of a low-voltage dynamo or *booster* b, driven by a shunt motor and connected across the mains in series with the battery B. The E. M. F. of the dynamo d is a little greater than that of the battery, so that it will charge the battery when there is no

external load. When all the lights are turned on, the booster
field will be fully energized, and the E. M. F. of the booster

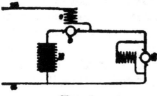

will be added to that of the battery, thereby causing the battery
to discharge and assist the dynamo. At a medium load, the
battery will be neutral, neither
taking current nor discharging,
while the dynamo is running
at full load. Any increase that

FIG. 1.

may be made in the load will then be taken up by the
battery.

In electric-railway plants the dynamos are usually over-
compounded, thus giving a higher E. M. F. at the brushes
at full load than at light load. In a case of this kind, a
differential winding is employed, as shown in Fig. 2, which

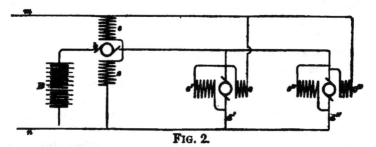

FIG. 2.

causes the booster to work both ways. On light loads
a differential winding will assist the dynamos d' and d'' to
charge the battery, raising the E. M. F. to the required
value; but on heavy loads the series winding c will over-
power the shunt s, and the battery will discharge into the
outer circuit. The shunt field must be regulated so that the
total charging and discharging that is done within a given
time will balance each other, as the battery will otherwise
tend either to overcharge or to undercharge. If the shunt
field is strengthened, it will cause the batteries to charge,
while if the field is weakened, it will cause the batteries to
discharge at a lower value of the external load than
before.

ELECTRODEPOSITION.

For electrodeposition of metals, low-resistance primary batteries giving from 2 to 10 volts may be used when the work is on a small scale. For larger work, accumulators may be employed, or the current may be taken directly from a low-voltage dynamo. The electroplating bath consists of a solution that has little or no chemical action on the objects to be plated, and that are suspended in it and electrically connected to the negative pole of the battery. The anode is a plate of the metal that it is desired to transfer; it is also submerged in the solution and connected through a resistance, if necessary, to the positive pole of the battery. For deposition of copper, the bath is made by taking 4 parts saturated solution of sulphate of copper mixed with 1 part of water containing one-tenth its volume of sulphuric acid. The current used must not exceed 18 amperes per square foot of surface of cathode. For nickel, use the double sulphate of nickel and ammonia, specific gravity 1.03; the current density must be low, and the solution should be neutral or slightly alkaline, as an acid bath will cause the nickel to peel off. For silver, the bath is a solution of cyanide of silver dissolved in cyanide of potassium. For gold, use cyanide of gold dissolved in cyanide of potassium. This solution is kept at 150° F. while in use.

ELECTRIC GAS LIGHTING.

The arrangement of the apparatus required for electric gas lighting is shown in the figure. A battery of about 6

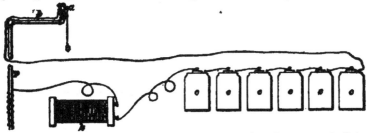

Leclanché cells *c, c,* etc., joined up in series, is connected to one terminal of a spark coil *k,* the other terminal of which is soldered to a gas pipe *p.* The wire from the free end of the battery is carried up through the house, and branches are run to the burners as at *b*, wherever needed. The insulation of this wire must be very thorough, special precautions being taken when it is carried through or along the fixtures. The burners are provided with a chain *a* attached to a movable contact spring, which is drawn past the burner, producing a spark of sufficient intensity to ignite the gas if it is previously turned on.

In multiple gas lighting, a fine wire is run from one burner to another of a group, as on a chandelier, leaving a small air gap at each one, and a current of very high tension is used, generated by a small frictional machine, causing a spark at each burner. The last contact in a series of burners is connected to the gas pipe.

THE WHEATSTONE BRIDGE.

A diagrammatic sketch of the Wheatstone bridge is shown in Fig. 1. This instrument is widely used for the determination of unknown resistances, and consists of such an arrangement of three circuits, *M, N, P*, of variable resistance, that

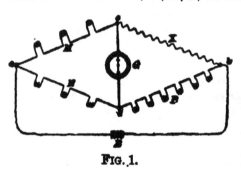

FIG. 1.

the value of a fourth may be found from their relation. This unknown resistance is connected between the points *b* and *c*, and the battery *B* between *a* and *b*. The variable resistances are then so adjusted that there shall be no difference of potential between *c* and *d*, which form the terminals of the galvanometer *G*. The drop in potential from *a* to *c* will then be the same as from *a* to *d*, and *a c* bears the same proportion to *a c b* as *a d* bears to *a d b*. From this it follows that $ac : ad = cb : db$, or the unknown resistance $X = \dfrac{MP}{N}$.

For a certain test, the ratio of the arms, $\dfrac{M}{N} = \dfrac{10}{100}$. On adjusting the resistance P, a balance is obtained when it is equal to 7,800 ohms. Then,

$$X = \frac{10 \times 7,800}{100} = 780 \text{ ohms.}$$

A commercial form of bridge is shown in Fig. 2. The same letters of reference are used as in the preceding diagram. Two keys, K and K', are added, to be used in closing the

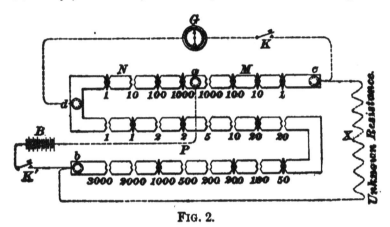

FIG. 2.

circuits. Resistances are put in by withdrawing the plugs. In the arm N there is a resistance of 10 ohms; in M, 1,000 ohms; in P, 5,838 ohms. If the galvanometer G indicates a balance, the value of the unknown resistance

$$X = \frac{1,000 \times 5,838}{10} = 583,800 \text{ ohms.}$$

CABLE TESTING.

Test for Capacity.—A condenser of known capacity k is charged by a battery and discharged through a galvanometer, producing a deflection d_1. The cable, having an unknown capacity k_2, is charged and discharged in similar manner, giving a deflection d_2. Then $k_2 = k_1 \frac{d_2}{d_1}$. The connections for the test are shown in Fig. 1. A plug commutator p may be used to make connection with the insulated line wire L or

20

with one side of the condenser *c*, by putting a plug in *1* or *2*.

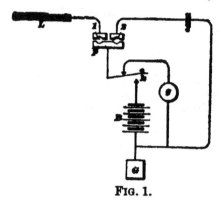

FIG. 1.

On depressing the key *k*, contact is made with one pole of the battery *B*, having about 100 cells; on releasing the key, the discharge from the line or the condenser passes through the galvanometer to the ground at *G*.

EXAMPLE.—The deflection through a condenser of 1.5 microfarads (mfds.) was 82 divisions,

and through a cable, 154 divisions. Find the capacity of cable.

SOLUTION.—From the formula given,

$$k_2 = 1.5 \times \frac{154}{82} = 2.8 \text{ microfarads.}$$

Voltmeter Method of Testing Insulation.—An ordinary Weston voltmeter with a range of 150 volts has a resistance of about 19,000 ohms. If, then, this instrument is connected across a 110-volt circuit, it will indicate the resistance of the circuit, that is, of itself, since the resistance of the armature and leads is very low. If *v* is the voltage across the mains, *r* the resistance of the voltmeter, and *x* the voltmeter reading, then the resistance to be determined, $R = \frac{v \, r}{x}$. When the voltmeter is put across the mains, $v = 110$, $r = 19,000$, and $x = 110$. The only resistance in the circuit is the voltmeter itself, for $R = \frac{110 \times 19,000}{110} = 19,000$ ohms. If we now put in series with the voltmeter a high resistance, thereby reducing the reading to 2 divisions, the total resistance $R = \frac{110 \times 19,000}{2} = 1,045,000$ ohms. From this we must subtract the voltmeter resistance in order to find the added resistance, which is $1,045,000 - 19,000 = 1,026,000$ ohms. A deflection of one division gives 2,071,000 ohms. To obtain higher readings, a special high-resistance voltmeter should be used. The connections are made as shown in Fig. 2, where *V* is the

voltmeter, F the feeder, and D the source of current. If I is the insulation resistance of a feeder, the corrected for-
mula becomes

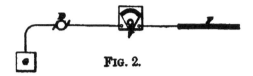

FIG. 2.

$$I = \frac{v\,r}{x} - r.$$

When a voltmeter is used having a resistance of 1 megohm (1,000,000 ohms), then a deflection of 1 division, when connected up as shown, would give an insulation resistance

$$I = \frac{110 \times 1,000,000}{1} - 1,000,000 = 109 \text{ megohms.}$$

Loss-of-Charge Method of Cable Testing.—The core of the cable must first be put to earth a sufficient length of time to be thoroughly clear from any charge due to previous electrification; then the far end is freed, and connections are

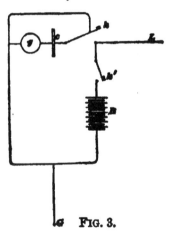

FIG. 3.

made as shown in Fig. 3. On depressing the key k, the cable is put to earth through the condenser c, which should be of very small capacity, say one-fiftieth of a microfarad. Both the cable L and the condenser c are then charged from the battery B by depressing the key k', and on releasing k, the condenser is discharged through the ballistic galvanometer g, a moment being chosen when the galvanometer is at zero, showing that the charge is steady. The deflection produced (d_1) represents the full charge held by the cable. The key k is then again depressed, and cable and condenser are charged for, say, half a minute, after which the battery is disconnected at k', and leakage of the charge is allowed to take place for perhaps 5 minutes. Selecting a moment when the charge is steady, indicating an even distribution, the key k is raised, and the condenser discharged through the galvanometer. The deflection (d_2) obtained will be less than the first one, owing to the leakage of charge during the 5 minutes, and will therefore be a measure of the conducting power of the cable covering, or its insulation resistance. The ratio of these two deflections, d_1 and d_2, will ordinarily be sufficient to indicate the condition of the cable without further calculation; the exact insulation resistance may be found by the following formula,

$$I = \frac{26.06\,t}{K \log \frac{d_1}{d_2}},$$

where I = insulation resistance of the cable in megohms;

t = time in minutes during which the charge is allowed to leak;

K = capacity of the cable in microfarads;

d_1 = initial discharge deflection;

d_2 = final deflection after t minutes.

EXAMPLE.—In a loss-of-charge insulation test, the initial deflection was 238 divisions, and the deflection after 5 minutes' leakage was 137 divisions. The capacity of the cable being 1.8 microfarads, what was the insulation resistance?

SOLUTION.— $I = \dfrac{26.06 \times 5}{1.8 \times \log \frac{238}{137}} = 301.8$ megohms.

The battery used in this test may be about 100 chloride-of-silver cells, or the same number of Leclanché cells. In the latter case it will be better to make the electrolyte of only about one-fifth the usual strength, to prevent creeping of the salts, as only very small currents are required for these tests. The battery must be very thoroughly insulated.

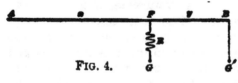

FIG. 4.

Location of Faults. A fault in a cable usually develops slowly, and there is considerable resistance at that point; therefore, in determining the location of the fault, its resistance must be taken into account. Let AB, Fig. 4, be the cable, and let a fault F connect to the ground at G through a resistance R. When the end B of the cable is insulated, the resistance is measured at the station A, and is equal to the resistance of that portion of the cable between the station and the fault *plus* the resistance of the fault, that is, $x + R$. B is then grounded at G', and the resistance is

$$x + \frac{y R}{y + R}.$$

Let
$$x + R = r.$$

Let
$$x + \frac{y r}{y + R} = r'.$$

Let
$$x + y = r''.$$

Then,
$$x = r' - \sqrt{(r - r')(r'' - r')};$$
$$y = r'' - r' + \sqrt{(r - r')(r'' - r')}.$$

If L = length of cable in feet, the distance from A to the fault is

$$\frac{L x}{x + y}.$$

EXAMPLE.—The resistance of a cable in good condition is 8 ohms. A fault develops, and, on testing, the resistance through it is 160 ohms, the far end of the cable being insulated. When the far end is grounded, the resistance is 2.95 ohms. What is the distance to the fault, the length of cable being 5,180 ft.?

SOLUTION.— $r = 160$, $r' = 2.95$, $r'' = 3$.

Then, $x = 2.95 - \sqrt{157.05 \times .05} = .15$ ohm.

$$y = 3 - 2.95 + \sqrt{157.05 \times .05} = 2.85 \text{ ohms.}$$

The distance to the fault $= \dfrac{5.180 \times .15}{3} = 259$ ft.

SURVEYING.

COMPASS SURVEYING.

The *magnetic bearing* of a line is the angle that the line makes with the magnetic needle. The length of a line, together with its bearing, is termed a *course*. To take the bearing of a line, set the compass directly over a point in it, at one extremity, if possible. This may be done by means of a plumb-bob suspended from the compass.

Bring the compass to a perfectly level position. Let a flagman hold a rod carefully plumbed at another point of the line, preferably the other extremity, if he can be distinctly seen. Direct the sights upon this rod and as near the bottom of it as possible. Always keep the same end of the compass ahead—the north end is preferable, as it is readily distinguished by some conspicuous mark, usually a *fleur-de-lis*—and always read the same end of the needle, that is, the north end of the needle if the north point of the compass is ahead, and *vice versa*. Before reading the angle, see that the eye is in the direct line of the needle, so as to avoid the error that would otherwise result from *parallax*, or apparent change of the position of the needle, due to looking at it obliquely.

The angle is read and recorded by noting, *first*, whether the N or S point of the compass is nearest the end of the needle being read; *second*, the number of degrees to which it points; and *third*, the letter E or W nearest the end of the needle being read.

Let *A B* in Fig. 1 be the direction of the magnetic needle, *B* being at the north end. Let the sights of the compass be directed along the line *C D*. The north point of the compass will be seen to be nearest the north end of the needle which is to be read. The needle, which has remained stationary while the sights were being turned to *C D*, now points to 45° between the *N* and *E* points, and the angle is read *north forty-five degrees east* (N 45° E).

A sure test of the accuracy of a bearing is to set up the compass at the other end of the line, i. e., the end first sighted

to, and sight to a rod set up at the starting point. This process is called *backsighting*. If the second
bearing is the same as the first, the reading
is correct. If it is not the same, it shows
that there is some disturbing influence at
either one or the other end of the line. To
determine which of these two bearings is the
true one, the compass must be set up at
one or more intermediate points, when
two or more similar bearings will prove
the true one.

FIG. 1.

The *magnetic meridian* is the direction of the magnetic
needle. The *true meridian* is a true north
and south line, which, if produced, would
pass through the poles of the earth. The
declination of the needle is the angle that
the magnetic meridian and the true
meridian make with each other.

**Example of the Use of the Compass in Rail-
road Work.**—Suppose CAD in Fig. 2 to be
a railroad in operation, and that it has
been decided to run a compass line from
the point A along the valley of the stream
X to the point B. The bearing of the
tangent AD cannot be determined by setting up the compass at A on account of the
attraction of the rails. The *direction* of
this tangent, however, can be obtained by
setting up at A and sighting to a flag held
at D. The point A, which is the starting
point of the line to be run, is marked 0.
Producing the line AD 440 ft., the point E
is reached, which has
been previously de-
cided on as a proper
place for changing the
direction of the line.

FIG. 2.

The compass having
been set up at E, the bearing of the line AE, which is the

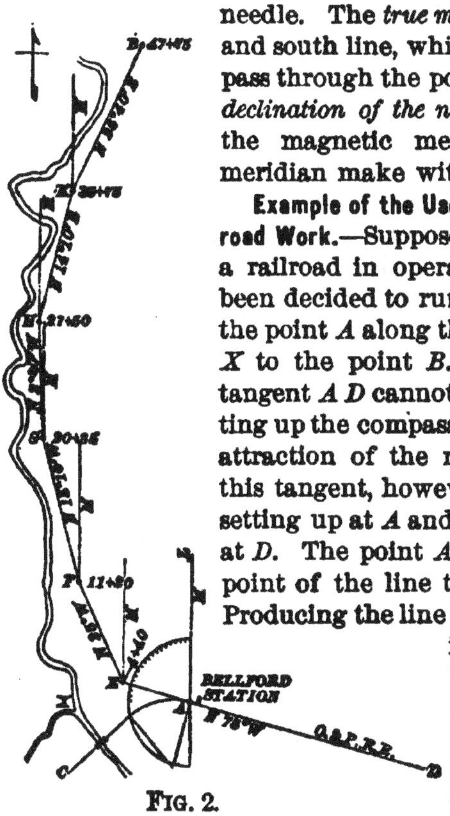

line *A D* produced, is found by sighting to *A*, or, what is still better, to the point *D*, if that point can be seen. The number of Sta. (Station) *E*, namely, 4 + 40, and the bearing of *A E* are then recorded by the compassman. By this time the chief of party has located the point *F*, and the flag is in place for sighting. The axmen, if there is work for them to do, are now put in line by the head chainman; the axmen clear only so much as would interfere with rapid chaining. The bearing of the line *E F* having been recorded, the compass is moved quickly to *F*, replacing the target left by the flagman, leveled up, and directed toward the point *G*, which is already located. The chainmen reaching *F*, its number 11 + 20 is recorded by the compassman and the instrument sighted to *G* and the work continued as before.

FORM FOR KEEPING NOTES.

A plain and convenient form for keeping compass notes is the form given on page 279, which is a record of the survey platted in Fig. 2. The first column of the table contains the station numbers, the notation running from the bottom to the top of the page. By means of this arrangement, the lengths of the courses are found by subtracting the number of the station of one compass point from the number of the station of the next succeeding compass point. Before work has commenced on the plat, the subtractions are made and the lengths of the courses are written in red ink between the station numbers.

The second column contains the bearings of the lines. The bearing recorded opposite to a station is the bearing at the course between the given station and the one next above. Thus, the bearing recorded opposite Sta. 0 is 75° 00′ W, and is the bearing of the line extending from Sta. 0 to Sta. 4 + 40 next above. The length of the course is the difference between 0 and 4 + 40 equal to 440 ft. The bearing recorded opposite to 4 + 40 is N 25° 00′ W. It is the bearing of the line extending from Sta. 4 + 40 to Sta. 11 + 20 next above. Its length is found by subtracting 4 + 40 from 11 + 20 equal to 680 ft., and so on.

In the third column, under the head of remarks, are recorded notes of reference, topography, and any information that may aid in platting or subsequent location.

Station.	Bearing.	Remarks.
47 + 75		End of line.
35 + 75	N 25° 40′ E	
27 + 50	N 14° 10′ E	
20 + 35	N 2° 30′ W	Woodland.
11 + 20	N 15° 10′ W	
4 + 40	N 25° 00′ W	
0	N 75° 00′ W	Sta. 0 is at P. C. of 14° curve to left at Bellford Sta. O. & P. R. R.

TRANSIT SURVEYING.

The Vernier.—A *vernier* is a contrivance for measuring smaller portions of space than those into which a line is actually divided. The divided circle of the transit is graduated to half degrees, or 30′. The graduations on the verniers run in both directions from its zero mark, making two distinct verniers, one for reading angles turned to the right and the other for reading those turned to the left. In reading the vernier, the observer should first note in which direction the graduations of the divided

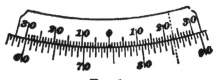

FIG. 1.

circle run. In Fig. 1 the graduations increase from left to right and extend from 57° to 91°. Next, he should note the point where the zero mark of the vernier comes on the divided circle. In Fig. 1 the zero mark comes between 74° and 74½°. Now, as the circle graduations read from left to right, we read the right-hand vernier and find that the 23d graduation on the vernier coincides with a graduation on the divided circle and the vernier reads 23', which we add to 74°, making a reading of 74° 23', an angle to the *left*. In Fig. 2 the graduations on the circle increase from r i g h t to left, and we accordingly read the left-hand vernier. The zero mark of the vernier comes between 67½° and 68°. Reading the vernier, we

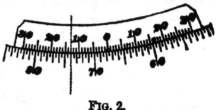

FIG. 2.

find that the 13th graduation on the vernier coincides with a graduation on the circle and the vernier reads 13'. Accordingly, we add to 67½°, the reading = 13', making a total reading of 67° 43', an angle to the *right*.

Setting Up the Instrument.—In setting up a transit, three preliminary conditions should be met as nearly as possible:

1. The tripod feet should be firmly planted.
2. The plate on which the leveling screws rest should be level.
3. The plumb-bob should be directly over the given point.

When these three conditions are met, the completion of the operation is quickly performed with the leveling screws.

How to Prolong a Straight Line.—Let *A B*, in Fig. 3, be a straight line which it is required to prolong or "produce."

FIG. 3.

The line can be prolonged in two ways: by means of *foresight* or by means of *backsight.*

1. By foresight, set up the transit at *A* and sight to *B*; measure 400 ft. from *B* in the opposite direction from *A*. Then, by means of signals, move the flag to the right or left until the vertical cross-hair shall exactly bisect the flag held at *C*. Then, the line *B C* will be the prolongation of the line *A B*.

2. By backsight, set the transit at *B* and sight to *A*. Reverse the telescope, and having measured 400 ft. from *B* in the opposite direction from *A*, set the flag at *C*; then will the line *B C* be the line *A B* produced.

Horizontal Angles and Their Measurement.—A horizontal angle is one the boundary lines of which lie in the same horizontal plane. Let A, B, and C, in Fig. 4, be three points, and let it be required to find the horizontal angle formed by the lines AB and AC joining these points. Set up the instrument precisely over the point A, and carefully level it. Set the vernier at zero, and place flags at B and C. Sight to the flag at

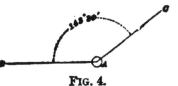

FIG. 4.

B and set the lower clamp. Then, by means of the lower tangent screw, cause the vertical cross-hair to exactly bisect the flag at B. Loosen the upper clamp. With a hand on either standard, turn the telescope in the same direction as that of the hands of a watch until the flag at C is covered or nearly covered by the vertical cross-hair. Clamp the upper plate, and with the upper tangent screw bring the line of sight exactly on the flag at C. The arc of the graduated circle traversed by the zero point of the vernier will be the measure of the angle BAC, as $143°$ $30'$. The points A, B, and C are not necessarily in the same horizontal plane, but the level plate of the instrument projects them into the horizontal plane in which it revolves.

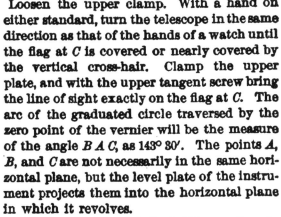

FIG. 5.

A Deflected Line.—A deflected line, or "angle line," is a consecutive series of lines and angles. The direction of each line is referred to the line immediately preceding it, the latter being, in imagination, produced, and the angle measured between it and the next line actually run. The angles are recorded R^r or L^r, according as they are turned to the right or left of the prolongation of the immediately preceding line. An example of a deflected line is shown in Fig. 5; it starts from the head block of switch at Benton Station, O. & P. R. R.

Set up the transit at A with vernier at zero. Sight to a flag

held at *F* on the center line of the track, O. & P. R. R.
Loosen the vernier clamp, the point *B* being determined, and
turn the telescope until the point *B* is distinctly seen; clamp
the vernier, and accurately sight to flag held at *B*; the angle
reads 32° 30′ and is recorded R^r 32° 30′, with a sketch showing
the connection. The bearing of the line *A B* cannot be taken
at *A* on account of the attraction of the rails. The point *A* is
in the head block of the switch (which is designated by the
abbreviation *H. B.*) at Benton Station, O. & P. R. R. The
instrument is now moved to *B*, the vernier set at zero and
backsighted to *A*; the bearing of *A B*, viz., N 75° 00′ E, is
taken, and the number of station *B*, viz., 2 + 90, together
with the bearing of *A B* recorded. The telescope is then
reversed, pointing in the direction *B B′*. The point *C* being
determined, the upper clamp is loosened and the telescope
turned to the right and sighted to *C*. The reading is found
to be 14° 30′ and recorded R^r 14° 30′. It measures the angle
B′ B C. The bearing N 89° 20′ E is then recorded. The
instrument is next set up at *C*, the vernier set at zero, backsighted to *B*, and then reversed; the deflection to *D*, viz.,
R^r 10° 00′ read and recorded, together with the number of the
station at *C*, viz., 6 + 85. This deflection measures the angle
C′ C D and gives the direction of the line *C D*. A good form
of notes for such a survey is the following:

Station.	Deflection.	Mag. Bearing.	Ded. Bearing.	Remarks.	
13+63				End of Line.	
10+31	L° 30′00′	N. 69°35′ E.	N.69°30′ E.		
6+85	R°10′00′	S. 80°30′ E.	S. 80°30′ E.		
2+90	R°14′30′	N. 89°20′ E.	N.89°30′ E.		H. B. of Switch
0		N.75°00′ E.		Sta. 0	at Benton Sta.

Checking Angles by the Needle.—In spite of the greatest care,
errors in the reading and recording of angles will occur.
The best check to such errors is the magnetic needle.

In Fig. 6, we have an example of the use of the needle in
checking angles. The bearing of the line *A B*, which corresponds to *A B* in Fig. 5, is N 75° 00′ E, and is assumed to be
correct. The bearing of the line *B C*, as read from the needle,

is N 89° 20′ E. Its *deduced bearing* is obtained as follows: To
the bearing of the line *A B*, viz., N 75° 00′ E, we add the R^r
deflection 14° 30′; the sum is 89° 30′, which is recorded in the
column headed Ded. Bearing. The deduced bearing, it will

be seen, is 10 minutes
greater than the mag-
netic bearing read from
the needle. Had the
deflection angle been
recorded L^x instead of
R^r, the deduced bearing would have been
the difference between
75° 00′ and 14° 30′, which
is 60° 30′, and would be
recorded N 60° 30′ E.

FIG. 6.

The magnetic bearing being N 89° 20′ E, would have at once
revealed the error. The confusion of the directions R^r and
L^x is the commonest source of error in recording deflections,
though sometimes a mistake of 10 degrees is made in reading
the vernier. Both angle and bearing should be read after
they are recorded, and compared with the recorded readings.

TRIANGULATION.

Triangulation is an application of the principles of trigonometry to the calculation of inaccessible lines and angles.

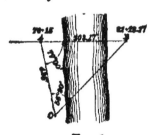

FIG. 1.

A common occasion for its use is
illustrated in Fig. 1, where the line
of survey crosses a stream too wide
and deep for actual measurement.
Set two points *A* and *B* on line,
one on each side of the stream.
Estimate roughly the distance *A B*.
Suppose the estimate is 425 ft. Set
another point *C*, making the distance *A C* equal to the estimated
distance *A B* = 425 ft. Set the transit at *A* and measure the
angle *B A C* = say, 79° 00′. Next set up at the point *C* and

measure the angle $ACB =$ say, 56° 20'. The angle ABC is then determined by subtracting the sum of the angles A and C from 180°; thus, 79° 00' + 56° 20' = 135° 20'; 180° 00' — 135° 20' = 44° 40' = the angle ABC. We now have a side and three angles of a triangle given, to find the other two sides AB and CB. In trigonometry, it is demonstrated that, *in any triangle the sines of the angles are proportional to the lengths of the sides opposite to them.* In other words, sin A : sin $B = BC : AC$; or, sin A : sin $C = BC : AB$, and sin B : sin $C = AC : AB$.

Hence, we have sin 44° 40' : sin 56° 20' = 425 : side AB;

sin 56° 20' = .83228;

.83228 × 425 = 353.719;

sin 44° 40' = .70298;

353.719 ÷ .70298 = 503.17 ft. = side AB.

Adding this distance to 76 + 15, the station of the point A, we have 81 + 18.17, the station at B.

Another case is the following: Two tangents, AB and CD (see Fig. 2), which are to be united by a curve, meet at some inaccessible point E. Tangents are the straight portions of a line of railroad. The angle CEF, which the tangents make with each other, and the distances BE and CE are required. Two points A and B of the tangent

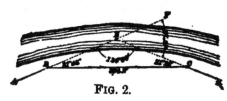

Fig. 2.

AB, and two points C and D of the tangent CD, being carefully located, set the transit at B, and backsighting to A, measure the angle EBC = 21° 45'; set up at C, and, backsighting to D, measure the angle ECB = 21° 25'. Measure the side BC = 304.2 ft.

Angle CEF being an exterior angle of triangle EBC equals sum of EBC and ECB = 21° 45' + 21° 25' = 43° 10'; angle BEC = 180° — CEF = 136° 50'. From trigonometry, we have

sin 136° 50' : sin 21° 45' = 304.2 ft. : CE;

sin 21° 45' = .37056;

.37056 × 304.2 = 112.724352;

sin 136° 50' = .68412;

side CE = 112.724352 ÷ .68412 = 164.77 ft.

Again, we find BE by the following proportion:

sin 136° 50′ : sin 21° 25′ = 304.2 : side BE;

sin 21° 25′ = .36515;

.36515 × 304.2 = 111.07868;

sin 136° 50′ = .68412;

side BE = 111.07863 ÷ .68412 = 162.36 ft.

A building H, Fig. 3, lies directly in the path of the line AB, which must be produced beyond H. Set a plug at B, and then turn an angle DBC = 60°. Set a plug at C in the line BC, at a suitable distance from B, say, 150 ft. Set up at C, and turn an angle BCD = 60°, and set a plug at D, 150 ft. from C. The point D will be in the prolongation of AB. Then, set up at D, and backsighting to C, turn the angle $CDD′$ = 120°. $DD′$ will be the line required, and the distance BD will be 150 ft., since BCD is an equilateral triangle.

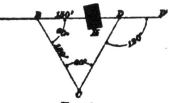

FIG. 3.

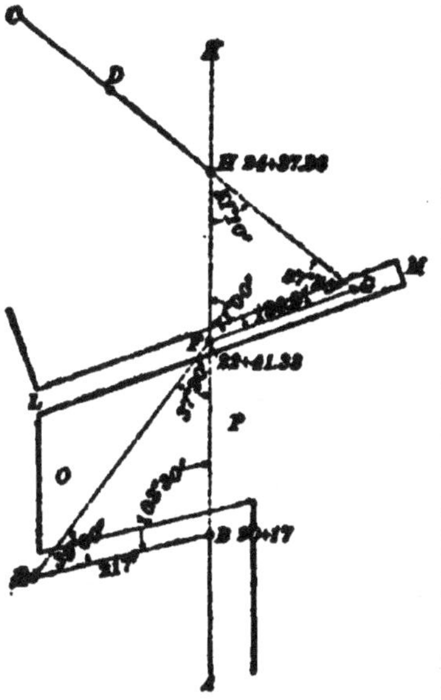

AB and CD, Fig. 4, are tangents intersecting at some inaccessible point H. The line AB crosses a dock OP, too wide for direct measurement, and the wharf LM. F is a point on the line AB at the wharf crossing. It is required to find the distance BH and the angle FHG. At B, an angle of 103° 30′ is turned to the left and the point E set 217′ from B = to the estimated distance BF. Setting up at E, the angle BEF is found to be 39° 00′.

Whence, we find the angle

FIG. 4.

BFE = 180° − (103° 30′ + 39°) = 37° 30′.

From trigonometry, we have

sin 37° 30′ : sin 39° 00′ = 217 ft. : side BF;

sin 39° 00′ = .62932;

.62932 × 217 = 136.56244;

sin 37° 30′ = .60876;

side BF = 136.56244 ÷ .60876 = 224.33 ft.

Whence, we find station F to be 20 + 17 + 224.33 = 22 + 41.33. Set up at F and turn an angle HFG = 71° 00′ and set up at a point G where the line CD prolonged intersects FG. Measure the angle FGH = 57° 50′, and the side FG = 180.3. The angle FHG = 180° − (71° + 57° 50′) = 51° 10′. From trigonometry we have

sin 51° 10′ : sin 57° 50′ = 180.3 : side FH.

Sin 57° 50′ = .84650; .84650 × 180.3 = 152.62395; sin 51° 10′ = .77897; side FH = 152.62395 ÷ .77897 = 195.93 ft.; whence we find station H to be 24 + 37.26.

CURVES.

Two lines forming an angle of 1° with each other will, at a distance of 100 ft. from the angular point, diverge by 1.745 ft.

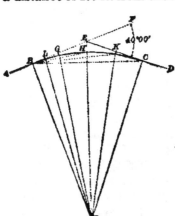

The *degree of a curve* is determined by that central angle which is subtended by a chord of 100 ft. Thus, if BOG (Fig. 1) is 10° and BG is 100 ft., $BGHKC$ is a 10° curve.

The *deflection angle* of a curve is the angle formed at any point of the curve between a tangent and a chord of 100 ft. The deflection angle is therefore *half the degree of the curve.* Thus, if the chord BG is 100 ft., the angle EBG is the deflection angle of curve $BGHKC$, and is half the angle BOG.

FIG. 1.

EXAMPLE.—Given, the deflection angle $EBG = D$ (Fig. 1), to find the radius $BO = R$.

SOLUTION.—Draw OL perpendicular to BG. In the right-angled triangle BOL, we have sin $BOL = \dfrac{BL}{BO}$; but BOL $= EBG = D$, since OL, being perpendicular to the chord BG, bisects the arc BLG. But the angle $D = \frac{1}{2}BOG$; hence, angle $BOL = D$. $BL = 50$ ft., and the radius $BO = R$. Substituting these values in the given equation, we have sin $D = \dfrac{50}{R}$; whence, $R \sin D = 50$, and $R = \dfrac{50}{\sin D}$.

For curves of from 1° to 10°, the radius may be found by dividing 5,730 ft. (the radius of a 1° curve) by the degree of the curve. The results obtained are sufficiently accurate for all practical purposes. For sharp curves, i. e., for those exceeding 10°, the above formula, viz., $R = \dfrac{50}{\sin D}$, should be used, especially if the radius is to be used as a basis for further calculation.

Tangent Distances.—When an intersection of tangents has been made and the intersection angle measured, the next question is the degree of curve that is to unite them, which being decided, the next step in order is the location of the points on the tangents where the curve begins and ends. These two points are equally distant from the point of intersection of the tangents, which is called the P. I. The point where the curve begins is called the *point of curve*, or the P. C., the point where the curve terminates is called the *point of tangent*, or the P. T. The distance of the P. C. and P. T. from the P. I. is called the *tangent distance*.

In Fig. 1, let AB and CD be tangents intersecting at the point E and forming an angle $CEF = 40°00'$ with each other. It is decided to unite these tangents by a 10° curve, whose radius is 573.7 ft. Call the angle of intersection I, the radius BO, R, and the tangent distance BE, T. From geometry we know that $BOC = CEF$, hence the angle $BOE = \frac{1}{2}CEF$. From the right triangle EBO, we have tan $BOE = \dfrac{BE}{BO}$.

Substituting the above equivalents, we have tan $\frac{1}{2}I = \dfrac{T}{R}$, or $T = R \tan \frac{1}{2}I$; $R = 573.7$; $\frac{1}{2}I = 20°$; tan $20° = .36397$;

21

578.7 × .36397 = 208.81 ft. Measure back from the point E on both tangents the distance 208.81 ft. to the points B and C. Drive plug flush with the ground at both points and set accurate center points, marked by tacks, in both. Directly opposite each of these plugs drive a stake, called a *guard stake* because it guards or rather indicates where the plug is. The stake at B, if the numbering of the stations runs from B toward C, will be marked P. C., and the stake at C will be marked P. T.

To Lay Out a Curve With a Transit.—Having set the tangent points B and C, Fig. 1, set up the transit at B, the P. C. Set the vernier at zero and sight to E, the intersection point. Suppose B to be an even or "full station," say 18, and that it has been decided to set stakes at each hundred feet. Let the central angle $B O G$, measured by the 100-ft. chord $B G$, be 10°; then, the deflection angle $E B G$, whose vertex B is in the circumference and subtended by the same chord $B G$, will be ½ $B O G$, or 5°. Turn an angle of 5° from B, which in this case will be to the right, measure a full chain 100 ft. from B and line in the flag at G; drive a stake at G, which will be marked 19. Turn off an additional 5° making 10° from zero, and at the end of another chain from G, at H, set at a stake marked 20. Continue turning deflections of 5° until 20° or one-half of the intersection angle is reached. This last deflection, if the work has been correctly done, will bring the head chainman to the point of tangent C. It is but rarely that the P. C. comes at a full station. When the P. C. comes between full stations it is called a *substation*, and the chord between it and the next full station is called a *subchord*. Had the P. C. come at a substation, say 17 + 32, the deflection for the subchord of 100 − 32, or 68 ft., the distance to the next station, is found as follows: The deflection for a full station, i. e., 100 ft., is 5° = 300', and the deflection for 1 ft. is $\frac{300'}{100}$ = 3', and for 68 ft. the deflection will be 68 × 3 = 204' = 3° 24', which is turned off from zero and a stake set on line, 68 ft. from the transit, at station 18. The length of a curve uniting two given tangents whose intersection is determined, is found as follows:

Suppose $I = 32° 40'$ and that the tangents are to be united by a 6° curve. $32° 40'$ reduced to the decimal form is 32.667°; as each central angle of 6° will subtend a 100-ft. chord or one chain, there will be as many such chords or chains as the number of times 6 is contained in 32.667, which is 5.444, that is, there will be 5.444 chains in the curve, or 544.4 ft., which is the required length of the curve. The P. C. and P. T. having been set and the station of the P. C. determined by actual measurement, say 58 + 71, the station number of the P. T. is found by adding to 58 + 71, the station number of the P. C., the calculated length of the curve 544.4 ft. 58 + 71 + 544.4 = 64 + 15.4, the station of the P. T.

Tangent and Chord Deflections.—Let $A B$ in Fig. 2 be a tangent, and $B C E H$ a curve commencing at B. Produce the tangent $A B$ to the point D. The line $C D$ is a *tangent deflection*, and is the perpendicular distance from the tangent to the curve. If the chord $B C$ is produced to the point G, making $C G = B C = C E$, the distance $G E$ is a *chord deflection* and is double the tangent deflection $D C$.

Given, the radius $B O = R$, Fig. 2, to find the chord deflection $E G$ and the tangent deflection $C D = F E$.

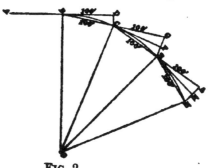

FIG. 2.

The triangles $O C E$ and $C E G$ are similar, since both are isosceles, and the angle $G C E =$ angle $C O E$. Hence, we have $O C : C E = C E : E G$. Denoting the chord $C E$ by c and the chord deflection $E G$ by d, we have, from the above proportion, $R : c = c : d$. Therefore, $d = \dfrac{c^2}{R}$. To find the tangent deflection, draw $C F$ to the middle point of $E G$. Then $F E$ is equal to the tangent deflection, or $D C$. Hence, the tangent deflection is equal to one-half the chord deflection, or the tangent deflection $= \dfrac{c^2}{2 R}$.

If the *P. C.* does not fall at a full station (and this is usually the case), compute the chord deflection by substituting for *c* in the formula for chord deflection $\frac{1}{4} c (c + c')$. Where *c'* is the length of the chord from the *P. C.* to the full station; or if the tangent deflection *f* for a chord of 100 feet has been previously found, the chord deflection for the second station beyond the *P. C.* is $d_0 = f\left(1 + \frac{c'}{c}\right)$.

Laying Out Curves Without a Transit.—During construction, the engineer is often called upon to restore center stakes on a curve when the transit is not at hand. This can be accomplished reasonably well with a tape, as follows:

In Fig. 3, *A B* is a tangent and *B*, at Sta. 8 + 25, is the *P. C.* of a 4° curve; a stake is required at each full station. The stakes at *A* and *B* are restored, determining the *P. C.* and the direction of the tangent. For a 4° curve the regular chord deflection for 100 feet is 6.98 ft., and the tangent deflection is 6.98 ÷ 2 = 3.49 ft.

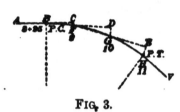

Fig. 3.

The distance from the *P. C.* to the next station *C* is 75 ft.; hence, the tangent deflection $C F = 75^2 \div (2 \times 5,730 + 4) =$ 1.96 ft. The point *F* is found by first measuring 75 feet from *B*, thus locating the point *C*, in the line with *A B*, then from *C* measuring *C F* = 1.96 feet, at right angles to *B C*; the point *F* thus determined will be Station 9. Next, the chord *B F* is prolonged 100 feet to *D*; as *B F* is only 75 feet, $D G = d_0 = 3.49 \times (1 + \frac{75}{100}) = 6.11$ feet. This distance is measured at right angles to *B D*; the point *G* thus determined will be Station 10. The position of Station 11, the *P. T.*, is determined in the same manner, except that, as the chords *F G* and *G H* are each 100 feet long, the regular chord deflection of 6.98 feet is used for *E H*. A stake is driven at each station thus located.

To Determine Degree of Curve by Measuring a Middle Ordinate. In trackwork, it is often necessary to know the degree of a curve when no transit is available for measuring it. The degree can be found by measuring the middle ordinate of any convenient chord, and multiplying its length by 8, which will give the chord deflection for that curve.

Let $A B$, in Fig. 4, be a 50-ft. chord, measured on the track, and let the middle ordinate $a b$ be .44 ft. .44 $\times$ 8 = 3.52 = chord deflection for 50 ft., which, expressed in decimal parts of a full station, is .5; .5^2 = .25. The chord deflection for 100 ft. multiplied by .25 = the chord deflection for 50 ft., which we know by calculation to be 3.52 ft. Hence, 3.52 ÷ .25 = 14.08 ft., the chord deflection for 100 ft., which, if divided by 1.745, the chord deflection for a 1° curve, gives a quotient of 8.07, nearly. The inference is that the curve is 8°.

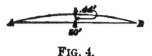

FIG. 4.

How to Keep Transit Notes.—A good form for location notes is the following:

Station.	Deflection.	Tot. Angle.	Mag. Bearing.	Cal. Bearing.		Remarks.	June 30. 1894
9							
8							
7							
6+95	6°34'P.T.	15°00'	N.35°30'E.	N.36°15'E.			
6+50	4°00'						
6	3°00'					"10-00'" Correction of Alignment	
5+50	2°00'					5+80	
5	1°00'					5+60	
4+50	0°30'	5°12'					
4	1°30'				Int.Angle=15°00'	6°Curve R^r	
3+50	0°30'				Z=153.61 ft.	Def.Angle for 50 ft.=1°30'	
3+20	P.C.0°0'				P.C.=3+20	Def.Angle for 1 ft.=1.8'	
3					Length of Curve=375 ft.		
2					P.T.=6+95		
1							
0			N.30°15'E.	N.30°15'E.			

In the first column the station numbers are recorded. In the second column are recorded the deflections with the abbreviations P. C. and P. T., together with the degree of curve and the abbreviation R^r or L^r, according as the line curves to the right or left. At each transit point on the curve, the total or central angle from the P. C. to that point is calculated and recorded in the third column. This total angle is double the deflection angle between the P. C. and the transit point. In the above notes there is but one intermediate transit point between the P. C. and P. T. The

deflection from P. C. at Sta. 3 + 20 to the intermediate transit
point at Sta. 4 + 50 is 2° 36'. The total angle is double this
deflection, or 5° 12', which is recorded on the same line in
the third column. The record of total angles at once indicates the stations at which transit points are placed. The
total angle at the P. T. will be the same as the angle of intersection, if the work is correct. When the curve is finished,
the transit is set up at the P. T., and the bearing at the forward tangent taken, which affords an additional check upon
the previous calculations. The magnetic bearing is recorded
in the fourth column, and the deduced or calculated bearing
is recorded in the fifth column.

LEVELING.

Examples in Direct Leveling.—The principles of direct leveling are illustrated in the figure.

Let A be the starting point, which has a known elevation
of 20 ft. The instrument is set at B, leveled up and sighted
to a rod held at A. The target being set, the reading, 8.42 ft.,
called a *backsight*, is the distance that the point where the
line of sight cuts the rod is above the point A, and
is to be added to the elevation of the point A. 20.00 + 8.42
= 28.42 is called the *height of instrument* and is designated
by $H.I.$ The instrument being turned in the opposite direction, a point C is chosen, which must be below the line of
sight. This point is called a *turning point*, and is designated
by the abbreviation T. P. Drive a peg at C, or take for a turning point a point of rock or some other permanent object
upon which the rod is held. The reading at this point is a
foresight, and is to be subtracted from the height of the
instrument at B to find the elevation of the point at C.

Let the rod reading be 1.20 ft. As this reading is a foresight, it must be subtracted from 28.42, the height of instrument at B; 28.42 − 1.20 = 27.22 ft., the elevation of the point
C. The leveler carries the instrument to D, which should be
of such a height above C that, when leveled up, the line of
sight will cut the rod near the top. The backsight to C gives
a reading of 11.56 ft., which, added to 27.22 ft., the elevation

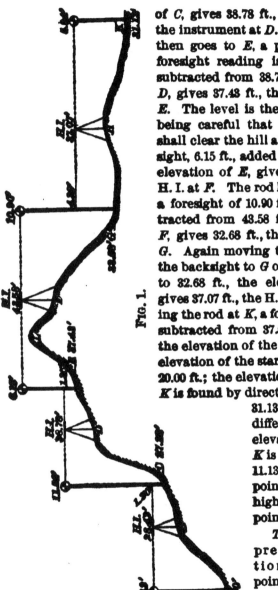

FIG. 1.

of *C*, gives 38.78 ft., the height of the instrument at *D*. The rodman then goes to *E*, a point where a foresight reading is 1.35, which, subtracted from 38.78, the H. I. at *D*, gives 37.43 ft., the elevation of *E*. The level is then set up at *F*, being careful that line of sight shall clear the hill at *L*. The backsight, 6.15 ft., added to 37.43 ft., the elevation of *E*, gives 43.58 ft., the H. I. at *F*. The rod held at *G* gives a foresight of 10.90 ft., which, subtracted from 43.58 ft., the H. I. at *F*, gives 32.68 ft., the elevation at *G*. Again moving the level to *H*, the backsight to *G* of 4.39 ft. added to 32.68 ft., the elevation of *G*, gives 37.07 ft., the H. I. at *H*. Holding the rod at *K*, a foresight of 5.94, subtracted from 37.07, gives 31.13, the elevation of the point *K*. The elevation of the starting point *A* is 20.00 ft.; the elevation of the point *K* is found by direct leveling to be 31.13 ft., and the difference in the elevations of *A* and *K* is 31.13 − 20.00 = 11.13 ft.; that is, the point *K* is 11.13 ft. higher than the point *A*.

Turning points previously mentioned are the points where backsights and foresights are taken. The backsights are plus (+) readings, and are to be added; the foresights are minus (—) readings, and are to be subtracted. A point for a foresight having been determined, the rodman drives a peg firmly in the ground and holds the rod upon it. After the instrument is moved, set up, and a backsight taken, the peg is pulled up and carried in the pocket until another turning point is called for. Turning points should be taken at about equal distances from the instrument, in order to equalize any small errors in adjustment. In smooth country an ordinary level will permit of sights of from 300 to 500 ft.

To Keep Level Notes.—Many forms are used. The distinguishing feature of one of the best (see page 295) is a single column for all rod readings. The backsights being additive and the foresights subtractive readings, they are distinguished from other rod readings by the characteristic signs + (plus) and — (minus). The turning points, whose foresight readings are —, are further abbreviated T. P.

To Check Level Notes.—A well-known method of checking level notes provides for checking the elevations of turning points and heights of instrument only, which is sufficient, as all other elevations are deduced from them. The method depends on the fact that all backsights are additive (i. e. +) quantities, and all foresights are subtractive (i.e.—) quantities. The notes given on page 295 are checked as follows: The elevation of the bench mark at station 0 is 100.00 ft., to which all backsights, or + readings, are to be added and from this sum all foresights, or — readings, are to be subtracted. The sum of the backsights, with elevation of bench mark at 0, is 122.59. Sum of foresights is 24.27, and difference is 98.82 ft., the eleva-

+	—
Thus, 100.00	10.22
5.61	2.52
5.41	11.53
11.57	24.27
122.59	
24.27	
98.32	

tion of the turning point last taken. As soon as a page of level notes is filled, the notes should be checked and a check mark ✓ placed at the last height of instrument or elevation checked. When the work of staking out or cross-sectioning is being done, the levels should be checked at each bench mark on the line. After each day's work, the leveler must check on the nearest bench mark.

Station	Rod Reading.	Ht. Instrument.	Elevation.	Grade.	Cut.	Fill.	Remarks.	Date.
B. M.	+ 5.61	105.61	100.00				On root of white oak	Stump 10′ L. Sta. 0.
0	6.1		99.5					
1	7.3		98.3					
2	8.4		97.2					
3	9.2		96.4					
T. P.	−10.22	100.80	95.39					
	+ 5.41							
4	6.3		94.5					
5	4.2		96.6					
5+50	11.5		89.3				Spring Brook.	
T. P.	− 2.52	109.85	98.28					
	+11.57							
6	6.2		103.6					
7	8.5		101.3					
8	10.1		99.7					
T. P.	−11.53		98.32					

Profiles.—A profile represents a longitudinal projection of the line of survey. In it all abrupt changes in elevation are clearly outlined. Vertical and horizontal measurements are usually represented by different scales, to render irregularities of surface more distinct through exaggeration. For railroad work, profiles are commonly made to the following scales, viz., horizontal, 400 ft. = 1 in.; vertical, 20 ft. = 1 in.

A section of profile paper is shown in the following diagram. Every fifth horizontal line and every tenth vertical line is heavy. By the aid of these heavy lines, distances and elevations are quickly and correctly estimated and the work of platting greatly facilitated. The level notes

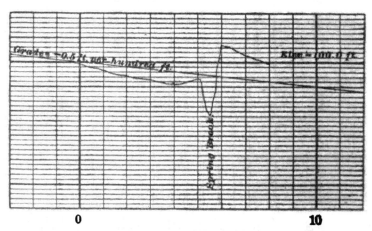

0   10

given in the preceding diagram are platted in the accompanying section. The elevation of some horizontal line is assumed. This elevation is, of course, referred to the datum plane, and is the base from which the other elevations are estimated. Every tenth station number is written at the bottom of the sheet under the heavy vertical lines. The profile is first platted in pencil and then inked in in black.

Grade Lines.—The principal use of a profile is to enable the engineer to establish a *grade line,* i. e., a line showing the slope of the road on which the amounts of excavation and embankment depend. The *rate* of a grade line is measured by the vertical rise or fall in each hundred feet of its length, and is designated by the term *per cent*. Thus, a grade line that rises or falls 1 ft. in each hundred feet of its length is called an ascending or descending 1 per cent. grade, and is written + 1.0 or − 1.0 per hundred. A rise or fall of ½ ft. in each hundred feet is called a 0.5 grade, and is written + 0.5 or − 0.5 per hundred. The grade line having been decided on, it is drawn in red ink.

EXAMPLE.—The elevation of station 20 is 140.0 ft.; between stations 20 and 100 there is an ascending grade of 75%. What is the elevation of the grade at station 71?

SOLUTION.—To obtain the elevation of the grade at station 71, we add to the elevation of the grade at station 20, or 140 ft., the total rise in grade between stations 20 and 71. Accordingly, 71 − 20 = 51; .75 ft. × 51 = 38.25 ft.; 140 ft. + 38.25 ft. = 178.25 ft., the elevation of grade at station 71.

RADII AND CHORD AND TANGENT DEFLECTIONS.

The formulas used in the computation of the following table are as follows:

For radius, $R = \dfrac{50}{\sin D}$.

For chord deflection, $d = \dfrac{c^2}{R}$.

For tangent deflection, $\tan \text{ deflection} = \dfrac{c^2}{2R}$.

In these formulas, R is the radius of the curve, D is its deflection angle (equal to one-half the degree of curve), and c is the length of chord for which the chord or tangent deflection is to be determined. The chord and tangent deflections given in the table are computed for chords of 100 feet.

Thus, for a 6° curve the deflection angle is 3°, the sine of which is .052336. Hence, for the radius and chord deflection, we have

$$R = \frac{50}{.052336} = 955.37 \text{ ft.} \qquad d = \frac{100^2}{955.37} = 10.467 \text{ ft.,}$$

as given in the table. The tangent deflection is always one-half the chord deflection.

TABLE OF RADII AND DEFLECTIONS.

Degree.	Radii.	Chord Deflection.	Tangent Deflection.	Degree.	Radii.	Chord Deflection.	Tangent Deflection.
° ′				° ′			
0 5	68,754.94	.145	.073	3 25	1,677.20	5.962	2.981
10	34,377.48	.291	.145	30	1,637.28	6.108	3.054
15	22,918.33	.436	.218	35	1,599.21	6.253	3.127
20	17,188.76	.582	.291	40	1,562.88	6.398	3.199
25	13,751.02	.727	.364	45	1,528.16	6.544	3.272
30	11,459.19	.873	.436	50	1,494.95	6.689	3.345
35	9,822.18	1.018	.509	55	1,463.16	6.835	3.417
40	8,594.41	1.164	.582				
45	7,639.49	1.309	.654	4 0	1,432.69	6.980	3.490
50	6,875.55	1.454	.727	5	1,403.46	7.125	3.563
55	6,250.51	1.600	.800	10	1,375.40	7.271	3.635
				15	1,348.45	7.416	3.708
1 0	5,729.65	1.745	.873	20	1,322.53	7.561	3.781
5	5,288.92	1.891	.945	25	1,297.58	7.707	3.853
10	4,911.15	2.036	1.018	30	1,273.57	7.852	3.926
15	4,583.75	2.182	1.091	35	1,250.42	7.997	3.999
20	4,297.28	2.327	1.164	40	1,228.11	8.143	4.071
25	4,044.51	2.472	1.236	45	1,206.57	8.288	4.144
30	3,819.83	2.618	1.309	50	1,185.78	8.433	4.217
35	3,618.80	2.763	1.382	55	1,165.70	8.579	4.289
40	3,437.87	2.909	1.454				
45	3,271.17	3.054	1.527	5 0	1,146.28	8.724	4.362
50	3,125.36	3.200	1.600	5	1,127.50	8.869	4.435
55	2,989.48	3.345	1.673	10	1,109.33	9.014	4.507
				15	1,091.73	9.160	4.580
2 0	2,864.93	3.490	1.745	20	1,074.68	9.305	4.653
5	2,750.35	3.636	1.818	25	1,058.16	9.450	4.725
10	2,644.58	3.781	1.891	30	1,042.14	9.596	4.798
15	2,546.64	3.927	1.963	35	1,026.60	9.741	4.870
20	2,455.70	4.072	2.036	40	1,011.51	9.886	4.943
25	2,371.04	4.218	2.109	45	996.87	10.031	5.016
30	2,292.01	4.363	2.181	50	982.64	10.177	5.088
35	2,218.09	4.508	2.254	55	968.81	10.322	5.161
40	2,148.79	4.654	2.327				
45	2,083.68	4.799	2.400	6 0	955.37	10.467	5.234
50	2,022.41	4.945	2.472	5	942.29	10.612	5.306
55	1,964.64	5.090	2.545	10	929.57	10.758	5.379
				15	917.19	10.903	5.451
3 0	1,910.08	5.235	2.618	20	905.13	11.048	5.524
5	1,858.47	5.381	2.690	25	893.39	11.193	5.597
10	1,809.57	5.526	2.763	30	881.95	11.339	5.669
15	1,763.18	5.672	2.836	35	870.79	11.484	5.742
20	1,719.12	5.817	2.908	40	859.92	11.629	5.814

TABLE—(*Continued*).

Degree.	Radii.	Chord Deflection.	Tangent Deflection.	Degree.	Radii.	Chord Deflection.	Tangent Deflection.
° ′				° ′			
6 45	849.32	11.774	5.887	10 0	573.69	17.431	8.716
50	838.97	11.919	5.960	10	564.31	17.721	8.860
55	828.88	12.065	6.032	20	555.23	18.011	9.005
				30	546.44	18.300	9.150
				40	537.92	18.590	9.295
7 0	819.02	12.210	6.105	50	529.67	18.880	9.440
5	809.40	12.355	6.177				
10	800.00	12.500	6.250				
15	790.81	12.645	6.323	11 0	521.67	19.169	9.585
20	781.84	12.790	6.395	10	513.91	19.459	9.729
25	773.07	12.936	6.468	20	506.38	19.748	9.874
30	764.49	13.081	6.540	30	499.06	20.038	10.019
35	756.10	13.226	6.613	40	491.96	20.327	10.164
40	747.89	13.371	6.685	50	485.05	20.616	10.308
45	739.86	13.516	6.758				
50	732.01	13.661	6.831	12 0	478.34	20.906	10.453
55	724.31	13.806	6.903	10	471.81	21.195	10.597
				20	465.46	21.484	10.742
8 0	716.78	13.951	6.976	30	459.28	21.773	10.887
5	709.40	14.096	7.048	40	453.26	22.063	11.031
10	702.18	14.241	7.121	50	447.40	22.352	11.176
15	695.09	14.387	7.193				
20	688.16	14.532	7.266	13 0	441.68	22.641	11.320
25	681.35	14.677	7.338	10	436.12	22.980	11.465
30	674.69	14.822	7.411	20	430.69	23.219	11.609
35	668.15	14.967	7.483	30	425.40	23.507	11.754
40	661.74	15.112	7.556	40	420.23	23.796	11.898
45	655.45	15.257	7.628	50	415.19	24.085	12.043
50	649.27	15.402	7.701				
55	643.22	15.547	7.773	14 0	410.28	24.374	12.187
				10	405.47	24.663	12.331
9 0	637.27	15.692	7.846	20	400.78	24.951	12.476
5	631.44	15.837	7.918	30	396.20	25.240	12.620
10	625.71	15.982	7.991	40	391.72	25.528	12.764
15	620.09	16.127	8.063	50	387.34	25.817	12.908
20	614.56	16.272	8.136				
25	609.14	16.417	8.208				
30	603.80	16.562	8.281	15 0	383.06	26.105	13.053
35	598.57	16.707	8.353	10	378.88	26.394	13.197
40	593.42	16.852	8.426	20	374.79	26.682	13.341
45	588.36	16.996	8.498	30	370.78	26.970	13.485
50	583.38	17.141	8.571	40	366.86	27.258	13.62ᵒ
55	578.49	17.286	8.643	50	363.02	27.547	13.7˙

TABLE—(*Continued*).

Degree.	Radii.	Chord Deflection.	Tangent Deflection.	Degree.	Radii.	Chord Deflection.	Tangent Deflection.
° ′				° ′			
16 0	359.26	27.885	13.917	18 10	316.71	31.574	15.787
10	355.59	28.123	14.061	20	313.86	31.861	15.931
20	351.98	28.411	14.205	30	311.06	32.149	16.074
30	348.45	28.699	14.349	40	308.30	32.436	16.218
40	344.99	28.986	14.493	50	305.60	32.723	16.361
50	341.60	29.274	14.637				
				19 0	302.94	33.010	16.505
17 0	338.27	29.562	14.781	10	300.33	33.296	16.648
10	335.01	29.850	14.925	20	297.77	33.583	16.792
20	331.82	30.137	15.069	30	295.25	33.870	16.935
30	328.68	30.425	15.212	40	292.77	34.157	17.078
40	325.60	30.712	15.356	50	290.33	34.443	17.222
50	322.59	31.000	15.500				
18 0	319.62	31.287	15.643	20 0	287.94	34.730	17.365

RETAINING WALLS.

On the Theory of Retaining Walls.—Let *a b d c*, Fig. 1, be a retaining wall with battered face and vertical back. The top *b e* of the backing is level with the top of the wall. Let *d e* represent the natural slope of the material composing the filling, viz., 1½ horizontal to 1 vertical, which is the average of materials used for back filling.

It is assumed that the wall *a b d c* is heavy enough to resist sliding along its base and that it can fail only by overturning,

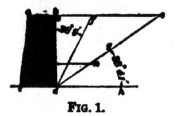

FIG. 1.

i. e., rotating about its toe *c*. Now, if the angle *o d e* (between the vertical line *o d* drawn from the inner bottom edge of the wall and the natural slope *d e*) be bisected by the line *d f*, the *angle o d f* is called the *angle,* and the *line d f* the *slope, of maximum pressure.* The triangular prism of earth *o d f* is called the *prism of maximum pressure*, because, if considered as a wedge acting against the back of the wall, it would exert a greater pressure against it than would the entire triangle *o d e* of earth considered as a single wedge. For though the latter is more than double the weight of the former, yet it receives much greater support from the underlying earth. It has been proved by experiment that, if the triangle of earth *o d e* is divided by any line *d f* into wedges, the wedge that will press most against the wall is that formed when the line *d f* divides the angle *o d e* into two equal parts.

The angle *o d h* formed by the vertical *o d* and the horizontal *d h* is 90°. The angle of natural slope *h d e* is 33° 41'; hence, the angle *o d f* of maximum pressure is equal to (90° − 33° 41') ÷ 2 = 28° 09'.

In making calculations, only *one foot* of the length of *wall* and of the *backing* is taken, so all that is necessary is to take the area of the section of the wall and backing. The material composing the backing is supposed to be perfectly dry and to possess no cohesive power, which is practically true of pure sand.

If we conceive the wall *a b d c*, Fig. 1, to be suddenly removed, the triangle *b d f* of sand included between the line of maximum pressure *d f* and the vertical back *b d* of the wall would slide downward, impelled by a force *n P*, acting in a direction *n P* at right angles to the side *b d* of the triangle, i. e., at right angles to the vertical back *b d* of the wall; the *center of pressure* being at *P* one-third of the distance between *b* and *d* measured from the bottom of the wall *d*. The amount of this force *n P* is:

$$\text{Perpendicular pressure} = \frac{\text{Wt. of triangle of earth } b\,d\,f \times o\,f}{\text{vertical depth } o\,d}.$$

This formula not only applies to walls with vertical backs, as in Fig. 1, but to those with inclined backs, as in Fig. 2, for inclinations as high as 6 in. horizontal to 1 ft. vertical, which is rarely met with and never exceeded.

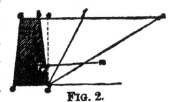

FIG. 2.

Friction Caused by Pressure of Backing.—If all the backing

material contained between the line of natural slope and the
back of the wall were unconfined, it would slide, producing

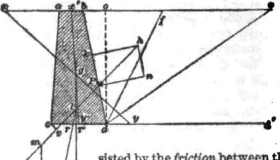

motion; but
confined by
the retaining
wall, the
force is con-
verted into
pressure of
earth against
the back of
the wall, re-

sisted by the *friction* between the compressed
earth and the wall.

　　If the wall were to begin to overturn
about its toe *c* (Figs. 1 and 2) as a fulcrum,
its back *b d* would rise, producing friction
against the backing. So long as the wall
does not move, the friction of the backing
acts constantly, and must, therefore, be one
of the forces that *prevent* overturning. We

FIG. 3.　　ascertain the amount and effect of this friction as follows: Let *a b d c*, Fig. 3, be a retaining wall, and let
n P represent to some *scale* the perpendicular pressure against
the back of the wall calculated by the preceding formula,
viz., perpendicular pressure =

$$n P = \frac{\text{weight of triangle } d\,b\,f \times o\,f}{\text{vertical depth } o\,d}.$$

　　Make the angle *n P h* equal to the angle of wall friction, viz.,
that at which a plane of masonry must be inclined to the horizontal in order that dry sand and earth may slide freely
over it, and taken at 33° 41'. Draw *n h* perpendicular to *n P*
and complete the parallelogram *n h k P*. Then will *k P* represent to the same scale the *amount of friction against the back* of
the wall. As the friction acts in the direction of the back
b d of the wall, it may be considered as acting at any point
P of the line of the back, and we will have two forces, viz.,
the perpendicular pressure *n P* and the friction *k P* acting
at *P*. By composition and resolution of forces, the diagonal

$h\,P$ measured to the same scale will give us the amount of their resultant, which is approximately the *single theoretical force* both in *amount* and *direction* that the wall has to resist. This force includes the wall friction. The force $h\,P$ is always equal to the perpendicular force $n\,P$, divided by the cosine of the angle of wall friction. The cosine of the angle of wall friction is .832 and the value of the force $h\,P$ may be expressed in the following formula:

Approximate theoretical pressure

$$= h\,P = \frac{\text{weight of triangle } b\,d\,f \times o\,f}{\text{vertical height } o\,d \times .832}.$$

When the back of the wall does not incline forward more than 6 in. horizontal to 1 ft. vertical, equal to an angle of about $26° 34'$, the following formula by Trautwine is used, viz.:

Approximate theoretical pressure

$$= h\,P = \text{weight of triangle } b\,d\,f \times .643,$$

which includes friction of earth against the back of the wall.

To Find the Overturning and Resisting Forces.—*To find the overturning tendency of the earth pressure and the resistance of the wall against being overturned about its toe c, as a fulcrum* (see Fig. 3). Find the center of gravity g of the wall, and through g draw the vertical line $g\,i$. Produce the line of pressure $h\,P$, and draw $c\,v$ at right angles to this line. To any convenient scale, lay off $l\,t$ equal to the weight of the wall and to the same scale $l\,m$ equal to the pressure $h\,P$. Complete the parallelogram $l\,m\,s\,t$. The diagonal $l\,s$ will be the resultant of the pressure and the weight of the wall. The stability of the wall will increase as the distance $c\,r$ from the toe to the point where the resultant $l\,s$ cuts the base, increases. To insure stability, $c\,r$ must be greater than $\frac{1}{3}\,c\,d$.

The pressure $h\,P$, if multiplied by its leverage $c\,v$, will give the moment of the pressure about c, and the weight of the wall $l\,t$, multiplied by its leverage $c\,r'$, will give the moment of the wall. The wall is secure against overturning in proportion as its moment exceeds that of the pressure.

For example, let the height of the wall $a\,b\,d\,c$, in Fig. 3, be 9 ft.; the thickness at the base $c\,d$, 4.5 ft., and at the top $a\,b$, 2 ft.; and the batter of $a\,c$ be 1 in. to the foot. The triangle of earth $b\,d\,f$ has a base $b\,f = 6.57$ ft. and altitude $d\,o = 9$ ft.

22

Taking the section as 1 ft. in thickness, we have the contents equal to 6.57 × 9 ÷ 2 = 29.56 cu. ft. Assuming the material to weigh 120 lb. per cu. ft., the weight of the triangle *b d f* is 29.56 × 120 = 3,547 lb.; *e f* = 4.81 ft. 3,547 × 4.81 = 17,061. 17,061 ÷ *o d* = 1,895.7 lb. = the perpendicular pressure *n P*. Lay off on a line perpendicular to the back of the wall at P, to a scale of 2,000 lb. = 1 in., *n P* = 1,895.7 ÷ 2,000 = .948 in., the perpendicular pressure. Draw *P h*, making the angle *n P h* = 33° 41'. Draw *n h* intersecting *h P* in *h*; then will *n h* to the same scale equal the friction of the earth against the back of the wall. Completing this parallelogram, *n h k P*, the diagonal *h P* = 1,139 in., which, to a scale of 2,000 lb. = 1 in., amounts to 2,278 lb., and is the resultant of the pressure and the friction.

Produce the resultant *h P* to *s*. We next find the center of gravity *g* of the wall *a b d c*. The section of the wall is a trapezoid, and the center of gravity *g* is readily found as follows: Produce the upper base of the section to *x*, and make *a x* = *c d* = 4.5 ft. Then produce the lower base in the opposite direction to *y*, and make *d y* = *a b* = 2 ft. Join *x* and *y*. Find the middle points *x'* and *y'* of the upper and lower bases of the section. Join these points. The intersection *g* of the lines *x y* and *x' y'* is the center of gravity of the trapezoid *a b d c*.

The volume of the section of wall *a b d c* is readily found. The sum of top and bottom widths = 2.0 + 4.5 = 6.5 ft. 6.5 ÷ 2 = 3.25 ft. 3.25 × 9 = 29.25 cu. ft. 29.25 × 154 = 4,504 lb. (the weight per cubic foot of good mortar rubble = 154 lb.) = the weight of the section *a b d c*. Draw through *g* a vertical line *g t*, and lay off on it, to a scale of 2,000 lb. to the inch, from the point *l*, where the line of gravity intersects the prolongation of the line of pressure *h P*, the length *l t* equal to 4,504 lb., the weight of the wall. Lay off from *l* on the prolongation of *h P*, *l m* equal to 2,278 lb. to the same scale. Complete the parallelogram *l m s t*. The diagonal *l s* represents the resultant of the pressure and of the weight of the wall. The distance *c r* from the toe *c* to the intersection of the resultant *l s* with the base *c d* is more than one-third of the width of the base, which insures ample stability.

Pressure of the Backing on Surcharged Walls.—In Fig. 4 the surcharge of backing *m b o* slopes from *b* at its natural slope,

and attains its maximum pressure where the slope of maximum pressure *d k* intersects the natural slope *b m* at *f*. Any additional height of surcharge does not increase this pressure. If the surcharge slopes from *a*, as shown by the line *a p*, or from any point between *a* and *b*, then the slope of maximum pressure must be extended, intersecting the slope from *a* in the

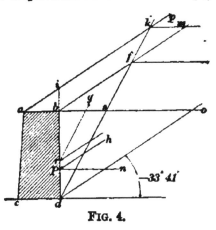

FIG. 4.

point *k*. The prism of maximum pressure will then be *d i k*. The triangle of earth *a b i* on the top of the wall exerts no pressure against the back of the wall, but adds to its stability.

Having found the weight of the triangle *b d f*, we have

approximate pressure = weight of triangle *b d f* × .643,

which includes the pressure of the backing and the friction of the earth against the back of the wall.

Draw *P n* perpendicular to the back of the wall and draw *h P* making the angle *n P h* = 33° 41, the angle of wall friction. Then, *h P* will be the direction of the pressure. The point of application of this pressure will not always be at *P*, one-third of the height of *b d* measured from *d*, but above *P*, as at *r*, where a line drawn from the center of gravity *g* of the *prism of maximum pressure d i k* (omitting any earth resting directly upon the top of the wall), and parallel to the line *d k* of maximum pressure, cuts the back *b d* of the wall. The center of pressure *P* will be at one-third the height of the wall when the sustained earth *d b s* or *d b f* forms a *complete triangle*, one of whose angles is at *b*, the *inner top edge* of the wall. For all other surcharges, the point of pressure will be *above P*.

TUNNEL SECTIONS.

Tunnel sections vary somewhat, according to the material to be excavated, but the general form and dimensions are much the same.

Section of
Double-Track Tunnel.

Section of
Single-Track Tunnel.

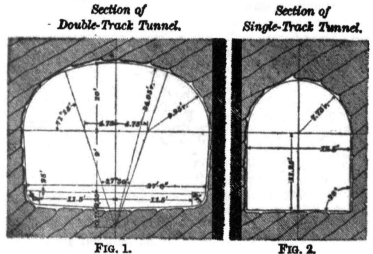

FIG. 1.   FIG. 2.

The general dimensions are as follows: For double track, from 22 to 27 ft. wide and from 21 to 24 ft. high, and for single track, from 14 to 16 ft. wide and from 17 to 20 ft. high (see Figs. 1 and 2).

In seamy or rotten rock the section is sufficiently enlarged to receive a lining of substantial rubble or brick masonry laid in good cement mortar. When the material has not sufficient consistency to sustain itself until the masonry lining is built, resort is had to timbering, which furnishes the necessary support.

CALCULATION OF EARTHWORK.

In calculating the quantity of material in excavation and embankment, two general methods are used, namely, the *end-area formula* and the *prismoidal formula*.

Calculation by the end-area method consists in multiplying the mean, or average, area in square feet of two consecutive sections by the distance in feet between them. Thus let A represent the area in square feet of one section; B, the area in square feet of the next section; C, the number of feet between the sections; and D, the total number of cubic feet in the prismoid lying between these sections. Then,

$$D = \frac{A+B}{2} \times C, \text{ approximately.}$$

The distance between sections should not be more than 100 ft., and should be less if the surface of the ground is irregular.

A more accurate result is obtained by the use of the prismoidal formula. In applying the prismoidal formula to the calculation of cubic contents, it is requisite to know the middle cross-section between each two that are measured on the ground. The dimensions of this middle section are the means of the dimensions of the end sections.

Calling one of the given sections A, the other B, the middle (not the mean) section M, the distance between the sections L, and the required contents S, we have, by the prismoidal formula,

$$S = \frac{L}{6} (A + 4 M + B).$$

EXAMPLE.—Two sections are represented by Figs. 1 and 2, and are denoted by the letters A and B. The perpendicular distance between them is 50 ft. It

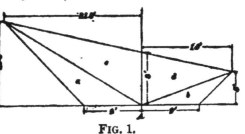

FIG. 1.

is required to find the cubical contents of the prismoid.

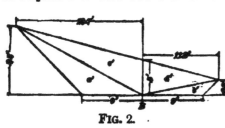

FIG. 2.

SOLUTION.—The section given in Fig. 1 is composed of the four triangles a, b, c, and d. The triangles a and b have equal bases of 9 ft., the half width of the roadway; hence, if we take half the sum of their altitudes and multiply it by the common base we shall have the sum of the areas of the triangles a and b.

The triangles c and d have a common base 8 ft., the center cut of the section, and if we take the half sum of the side distances and multiply it by 8 ft., we shall obtain the areas of the triangles c and d. Taking the dimensions of section A given in Fig. 1, we have

Areas of triangles $a + b = \dfrac{12.8 + 5}{2} \times 9 = 80.1$ sq. ft.

Areas of triangles $c + d = \dfrac{21.8 + 14}{2} \times 8 = 143.2$ sq. ft.

Total area of section $A = 223.3$ sq. ft.

Taking the dimensions of the section B given in Fig. 2, we have

Areas of triangles $a' + b' = \dfrac{9.7 + 2.2}{2} \times 9 = 53.55$ sq. ft.

Areas of triangles $c' + d' = \dfrac{18.7 + 11.2}{2} \times 5 = 74.75$ sq. ft.

Total area of section $B = 128.3$ sq. ft.

In applying the prismoidal formula we calculate the area of a section midway between the given sections, and for its

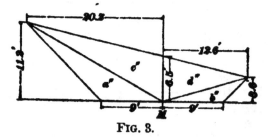

FIG. 3.

dimensions we take the mean of the dimensions of the given sections. These dimensions will be as follows:

Center cut, $\dfrac{8 + 5}{2} = 6.5$ ft.

Right-side distance, $\dfrac{14 + 11.2}{2} = 12.6$ ft.

Left-side distance, $\dfrac{21.8 + 18.7}{2} = 20.25$ ft.

With dimensions thus found, construct the section M shown in Fig. 3.

The area of section M is computed by the same method as that used with sections A and B in Figs. 1 and 2, and is as follows:

Area of triangles $a'' + b'' = \dfrac{11.2 + 3.6}{2} \times 9 = 66.6$ sq. ft.

Area of triangles $c'' + d'' = \dfrac{20.2 + 12.6}{2} \times 6.5 = 106.6$ sq. ft.

Total area of section $M = 173.2$ sq. ft.

Denoting the distance between the sections by L and the cubical contents of the prismoid by S, we have, by substituting in the prismoidal formula,

$$S = \frac{L}{6}(A + 4M + B).$$

$$S = \frac{50}{6}(223.3 + 4 \times 173.2 + 123.3) = 8,703 \text{ cu. ft.} = 322.3 \text{ cu. yd.}$$

TRACKWORK.

Curving Rails.—When laying track on curves, in order to have a smooth line, the rails themselves must conform to the curve of the center line. To accomplish this, the rails must be curved. The curving should be done with a rail bender or with a lever, preferably with the former.

To guide those in charge of this work, a table of middle and quarter ordinates for a 30-ft. rail for all degrees of curve should be prepared.

The following table of middle ordinates for curving rails is calculated by using the formula

$$m = \frac{c^2}{8R},$$

in which $m =$ middle ordinate;

$c =$ chord, assumed to be of the same length as the rail;

$R =$ radius of the curve.

The results obtained by this formula are not theoretically correct, yet the error is so small that it may be ignored in practical work.

In curving rails, the ordinate is measured by stretching a cord from end to end of the rail against the gauge side, as shown in Fig. 1. Suppose the rail *A B* is 30 ft. in length, and the curve 8°. Then, by the previous problem, the middle ordinate at *a* should be $1\frac{7}{8}$ in. To insure a uniform curve to the rails, the ordinates at the quarter *b* and *b'* should be tested. In all cases the quarter ordinates should be three-quarters of the middle ordinate. In Fig. 1, if the rail has been properly curved, the quarter ordinates at *b* and *b'* will be $\frac{3}{4} \times 1\frac{7}{8}$ in. $= 1\frac{13}{32}$, say $1\frac{3}{8}$ in.

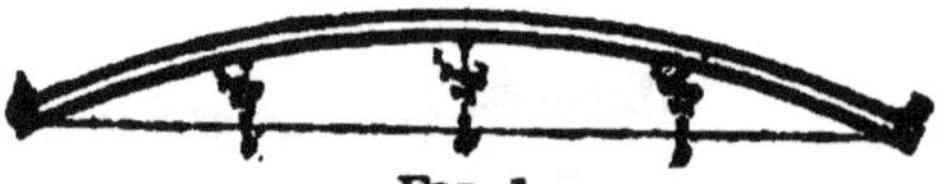

Fig. 1.

MIDDLE ORDINATES FOR CURVING RAILS.

Degree of Curve.	Length of Rail.					
	30 ft.	28 ft.	26 ft.	24 ft.	22 ft.	20 ft.
	in.	in.	in.	in.	in.	in.
1	1/4	3/16	3/16	3/16	1/8	1/8
2	1/2	7/16	3/8	5/16	1/4	3/16
3	11/16	5/8	9/16	7/16	3/8	5/16
4	15/16	13/16	11/16	5/8	1/2	7/16
5	1 3/16	1 1/16	7/8	3/4	5/8	9/16
6	1 7/16	1 1/4	1 1/16	7/8	3/4	5/8
7	1 5/8	1 7/16	1 1/4	1 1/16	7/8	3/4
8	1 7/8	1 5/8	1 7/16	1 3/16	1	7/8
9	2 1/8	1 7/8	1 5/8	1 3/8	1 1/8	15/16
10	2 3/8	2 1/16	1 3/4	1 1/2	1 1/4	1 1/16
11	2 5/8	2 1/4	1 15/16	1 11/16	1 3/8	1 1/8
12	2 13/16	2 1/2	2 1/8	1 13/16	1 9/16	1 1/4
13	3 1/16	2 11/16	2 5/16	1 15/16	1 5/8	1 3/8
14	3 5/16	2 7/8	2 1/2	2 1/8	1 3/4	1 1/2
15	3 9/16	3 1/16	2 11/16	2 1/4	1 15/16	1 9/16
16	3 3/4	3 1/4	2 13/16	2 3/8	2 1/16	1 11/16
17	4	3 1/2	3	2 9/16	2 3/16	1 3/4
18	4 3/16	3 11/16	3 3/16	2 11/16	2 5/16	1 7/8
19	4 7/16	3 7/8	3 3/8	2 7/8	2 7/16	2
20	4 11/16	4 1/8	3 9/16	3	2 9/16	2 1/8

In trackwork it is often necessary to ascertain the degree of a curve, though no transit is available for measuring it. The following table contains the middle ordinates of a 1° curve for chords of various lengths:

The lengths of the chords are varied, so that a longer or shorter chord may be used, according as the curve is regular or not.

Length of Chord. Feet.	Middle Ordinate of a 1° Curve. Inches.
20	1/8
30	1/4
44	1/2
50	5/8
62	1
100	2 5/8
120	3 3/4

The table is applied as follows: Suppose the middle ordinate of a 44-ft. chord is 3 in. We find in the table that the middle ordinate of a 44-ft. chord of a 1° curve is ½ in. Hence, the degree of the given curve is equal to the quotient of 3 + ½ = 6° curve.

Elevation of Curves.—To counteract the centrifugal force developed when a car passes around a curve, the outer rail is elevated. The amount of elevation will depend on the radius of the curve and the speed at which trains are to be run. There is, however, a limit in track elevation as there is a limit in widening gauge, beyond which it is not safe to pass.

The best authorities on this subject place the maximum elevation at one-seventh the gauge, or about 8 in. for standard gauge of 4 ft. 8½ in. The gauge on a 10° curve elevated for a speed of 40 miles an hour should be widened to 4 ft. 9¼ in.

All curves, when possible, should have an elevated approach on the straight main track, of such length that trains may pass on and off the curve without any sudden or disagreeable lurch.

A good rule for curve approaches is the following: For each half inch or fraction thereof of curve elevation, add 30 ft., for 1 rail length, to the approach; that is, if a curve has an elevation of 2 in., the approach will have as many rail lengths as the number of times ½ is contained in 2, or 4. The approach will, therefore, have a length of 4 rails of 30 ft. each, or 120 ft.

The following table for elevation of curves is a compromise between the extremes recommended by different engineers. It is a striking fact that experienced trackmen never elevate track above 6 in. and many of them place the limit at 5 in.

Degree of Curve.	Length of Approach. Feet.	Elevation. Inches.	Width of Gauge.	Speed of Train. Miles per Hour.
1	60	1	4′ 8½″	60
2	120	2	4′ 8½″	60
3	150	2½	4′ 8¾″	60
4	180	2¾	4′ 8¾″	55
5	180	3	4′ 8¾″	50
6	210	3¼	4′ 8¾″	45
7	210	3½	4′ 9″	40
8	240	3¾	4′ 9″	35
9	240	4	4′ 9″	30
10	270	4¼	4′ 9″	25
11	270	4½	4′ 9¼″	20
12	270	4¾	4′ 9¼″	15
13	240	4½	4′ 9¼″	10
14	240	4¼	4′ 9¼″	10
15	240	4	4′ 9½″	10
16	240	4	4′ 9½″	10

The Elevation of Turnout Curves.—The speed of all trains in passing over turnout curves and crossovers is greatly reduced, so that an elevation of ¼ in. per degree is amply sufficient for all curves under 16°. On curves exceeding 16°, the elevation may be held at 4 in. until 20° is reached, and on curves extending 20°, $\frac{3}{16}$ in. of elevation per degree may be allowed until the total elevation amounts to 5 in., which is sufficient for the shortest curves.

The Frog.—The frog is a device by means of which the rail at the turnout curve crosses the rail of the main track. The frog shown in Fig. 2 is made of rails having the same cross-section as those used in the track. The wedge-shaped part *A* is the *tongue*, of which the extreme end *a* is the *point*. The space *b*, between the ends *c* and *d* of the rails, is the *mouth*, and the channel that they form at its narrowest point *e* is the *throat*. The curved ends *f* and *g* are the *wings*.

That part of the frog between A and A' is called the *heel.*
The width h of the frog is called its *spread.* Holes are drilled

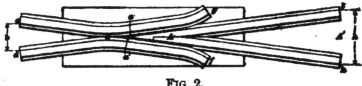

FIG. 2.

in the ends of the rails c, d, k, and l to receive the bolts used
in fastening the rail splices, so that the rails of which the
frog is composed form a part of the continuous track.

The Frog Number.—The number of a frog is the ratio of its
length to its breadth; i. e., the quotient of its length divided
by its breadth.

Thus, in Fig. 2, if the length $a'l$, from point to heel of frog
is 5 ft., or 60 in., and the breadth h of the heel is 15 in., the
number of the frog is the quotient of $60 \div 15 = 4$. Theoretically, the length of the frog is the distance from a to the
middle point of a line drawn from k to l; practically, we take
from a to l as the distance. As it is often difficult to determine the exact point a of the frog, a more accurate method
of determining the frog number is to *measure the entire length
dl of the frog from mouth to heel, and divide this length by the
sum of the mouth width b and the heel width h. The quotient
will be the exact number of the frog.*

For example, if, in Fig. 2, the total length dl of the frog is
7 ft. 4 in., or 88 in., and the width h is 15 in., and the width b
of the mouth is 7 in., then the frog number is $88 \div (15 + 7) = 4$.
Frogs are known by their *numbers.* That in Fig. 2 is a
No. 4 frog.

The Frog Angle.—The frog angle is the angle formed by the
gauge lines of the
rails, which form
its tongue. Thus,
in Fig. 2, the frog
angle is the angle
$la'k$. The amount

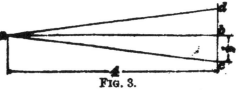

FIG. 3.

of the angle may be found as follows: The tongue and heel of the frog form an isosceles triangle (see Fig. 3). By drawing a line from the point a of the frog to the middle point b of the heel cd, we form a right-angled triangle, right-angled at b. The perpendicular line ab bisects the angle a, and, by trigonometry, we have tan $\frac{1}{2}a = \dfrac{bc}{ab}$. The dimensions of the frog point given in Fig. 3 are not the same as those given in Fig. 2, but their relative proportions are the same, viz., the length is four times the breadth. The length $ab = 4$ and the width $cd = 1$; hence, $bc = \frac{1}{2}$. Substituting these values, we have tan $\frac{1}{2}a = \dfrac{\frac{1}{2}}{4} = \frac{1}{8} = .125$. Whence, $\frac{1}{2}a = 7° 7\frac{1}{2}'$ and $a = 14° 15'$; that is, the angle of a No. 4 frog is $14° 15'$.

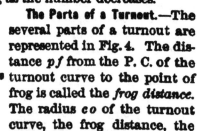

Frog numbers run from 4 to 12, including half numbers, the spread of the frog increasing as the number decreases.

The Parts of a Turnout.—The several parts of a turnout are represented in Fig. 4. The distance pf from the P. C. of the turnout curve to the point of frog is called the *frog distance*. The radius co of the turnout curve, the frog distance, the

FIG. 4.

frog angle, and the frog number bear certain relations to one another, which are expressed by the following formulas:

Tangent of half frog angle = gauge ÷ frog distance.

Frog number = $\sqrt{\text{radius } co \div \text{twice the gauge}}$.

Frog number = $1 + \frac{1}{2}$ the tangent of $\frac{1}{2}$ the frog angle.

Radius co = twice the gauge × square of the frog number.

Radius co = (frog distance pf ÷ sine of frog angle) − $\frac{1}{2}$ the gauge.

Radius co = gauge ÷ (1 − cosine of frog angle) − $\frac{1}{2}$ the gauge.

Frog distance pf = frog number × twice the gauge.

Frog distance pf = gauge pq ÷ tangent of $\frac{1}{2}$ the frog angle.

Frog distance pf = (radius co + half the gauge) × sine of frog angle.

Middle ordinate (approximate) = ¼ the gauge.

Each side ordinate (approximate) = ¼ the middle ordinate = 𝟐𝟏 (or .188) of the gauge.

Switch length (approximate) =

$$\sqrt{\frac{\text{throw in feet} \times 10,000}{\text{tan deflection for chords of 100 ft. for radius } \infty \text{ of turnout curve}}}.$$

The tangent deflection may be obtained from the table on pages 298–300.

TURNOUTS FROM A STRAIGHT TRACK.

Gauge, 4 ft. 8½ in.   Throw of switch, 5 in.

Frog Number.	Frog Angle.		Turnout Radius.	Degree of Turnout Curve.		Frog Distance.	Middle Ordinate.	Side Ordinate.	Stub Switch Length.
	°	′	Feet.	°	′	Feet.	Feet.	Feet.	Feet.
12	4	46	1,356	4	14	113.0	1.177	.883	34
11½	4	58	1,245	4	36	108.3	1.177	.883	32
11	5	12	1,139	5	02	103.6	1.177	.883	31
10½	5	28	1,038	5	31	98.9	1.177	.883	29
10	5	44	942	6	05	94.2	1.177	.883	28
9½	6	02	850	6	45	89.5	1.177	.883	27
9	6	22	763	7	31	84.7	1.177	.883	25
8½	6	44	680	8	26	80.0	1.177	.883	24
8	7	10	608	9	31	75.3	1.177	.883	22
7½	7	38	530	10	50	70.6	1.177	.883	21
7	8	10	461	12	27	65.9	1.177	.883	20
6½	8	48	398	14	26	61.2	1.177	.883	18
6	9	32	339	16	58	56.5	1.177	.883	17
5½	10	24	285	20	13	51.8	1.177	.883	15
5	11	26	235	24	32	47.1	1.177	.883	14
4½	12	40	191	30	24	42.4	1.177	.883	13
4	14	14	151	38	46	37.7	1.177	.883	11

The switch lengths in the above table merely denote the *shortest* length of *stub switch* that will at the same time form part of the turnout curve, and give 5 in. throw. *Point* or *split switches* require a throw of not more than 3½ in., though many have a throw of 5 in., with an equal space between the gauge lines at the heel. The heels of a split switch, which occupy the same position as the toes of a stub switch, should be placed at the point where the tangent deflection or offset is 5 in. The point where the tangent deflection is but 4¼ in. will answer for many rail sections, but for those above 65 lb. per yd., 5 in. should be taken.

In the table on pages 298–300, tangent deflections for chords of 100 ft. are given for all curves up to 20°; and for a curve of higher degree, the tangent deflection may be found by applying the formula $\tan \text{deflection} = \dfrac{c^3}{2R}$.

In complicated trackwork, where space is limited, curves must be chosen to meet the existing conditions, and not with reference to particular frog angles, in which case the frogs are called *special* frogs and are made to fit the particular

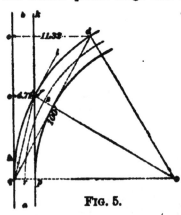

FIG. 5.

curve used. The determination of the frog distance, switch length, and frog angle may be understood by referring to Fig. 5.

Let the main track ab be a straight line; the gauge $pq =$ 4 ft. 8½ in. ($= 4.71$ ft.); the degree of the turnout curve $= 13°$; the chord $qd = 100$ ft.; $cd =$ the tangent deflection of the chord qd; and $pf =$ the frog distance. From the table on page 299, we find the

tangent deflection for a chord 100 ft. long of a 13° curve is 11.32 ft. Then, from Fig. 5, we have the proportion

$$cd : ef = \overline{qc}^2 : \overline{qe}^2.$$

Now, in curves of large radius, qc and qd are assumed to be equal. Also, $qe = pf$, the frog distance, and substituting these equivalents we have the proportion

$$cd : ef = \overline{qd}^2 : \overline{pf}^2.$$

Substituting the above given quantities in the proportion, we have $\qquad 11.32 : 4.71 = 100^2 : \overline{pf}^2;$

whence, $\qquad \overline{pf}^2 = \dfrac{100^2 \times 4.71}{11.32},$

and the frog distance, $pf = 64.5$ ft.

If the space between the gauge lines at the heels of a split switch be taken at 5 in. = .42 ft., the distance from the P. C. of the turnout curve to the heel of the switch may be found as follows:

In Fig. 5, let h, the tangent offset at the heel of the switch = .42 ft., we have the proportion

$$cd : h = \overline{qd}^2 : \overline{qh}^2,$$

and substituting known values, we have

$$11.32 : .42 = 100^2 : \overline{qh}^2,$$

whence, $\overline{qh}^2 = \dfrac{10,000 \times .42}{11.32} = 371.02,$

and $qh = 19.26$ ft.

This locates the heel of a *split switch* and the toe of a *stub switch*.

The *frog angle* is the angle kfl (see Fig. 5) formed by the gauge line of the main rail fk and the tangent to the outer rail qf of the turnout curve at the point where the two rails intersect. This angle is equal to the central angle qof. The arcs qf and rs are assumed to be of the same length. The turnout curve being 13°, the central angle for a chord of 1 ft. is $\dfrac{13 \times 60}{100} = 7.8'$, and the central angle for 64.5 ft. the *frog distance*, is $7.8' \times 64.5 = 8° 23'$, the frog angle for a 13° curve. By this process the frog distance, switch length, and frog angle may be calculated for curves of any radius.

To Lay Out a Turnout From a Curved Main Track.—There are two cases:

CASE I.—When the two curves deflect in opposite directions, illustrated in Fig. 6.

CASE II.—When the two curves deflect in the same direction, illustrated in Fig. 7.

In Fig. 6, the curve ab is 3° 30′, and it is proposed to use a No. 8 frog. By reference to the table on page 315, we find that the degree of curve corresponding to a No. 8 frog is 9° 31′. Accordingly, we use a turnout curve ae, whose degree when added to the degree of curve of the main track shall equal the degree required for a No. 8 frog; i. e., we use a 6° turnout curve, which is within 1 minute of the required degree, and close enough for practical purposes. We know that for curves of moderate radii, i. e., from 1° up to 12°, the tangent deflections or offsets increase as the degree of the curve. That is, the tangent deflection of a 2°, 4°, and 6° is two, four, and six times, respectively, that of a 1° curve. In the accompanying cuts illustrating the location of frogs and switches, each curve is represented by two lines indicating the rails, whereas only the center lines of the curves are run in on the ground. In Fig. 6, the line c d is tangent to the center lines of the curves. These center lines do not appear in the cut.

Again referring to Fig. 6, if a tangent c d be drawn at c,

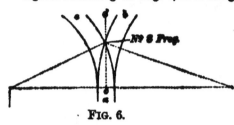

FIG. 6.

the point common to the center lines of the curves, the sum of the deflections of both curves from the common tangent will be equal, in this case, to the tangent deflection of a 9° 30′ curve from a straight line.

Accordingly, to find the frog distance for a 6° turnout curve from a 3° 30′ curve, the curves being in opposite directions, as shown in Fig. 6, we find the tangent deflection of a 9° 30′ curve for a chord of 100 ft. This deflection is 8.28 ft., as given in the table on page 299.

Assuming the gauge of track to be standard, viz., 4 ft. 8½ in. = 4.71 ft., and denoting the required frog distance by x, we have the following proportion:

$$8.28 : 4.71 = 100^2 : x^2,$$

whence, $$x^2 = \frac{10,000 \times 4.71}{8.28} = 5,688.4,$$

and the frog distance, $x = 75.42$ ft.

We use the tangent deflection for a 9° 30′ curve, which very nearly equals the tangent deflection for a 9° 31′ curve, thus saving the labor of a calculation; this will not appreciably affect the result.

We locate the heel of the switch in the same way, using for the second term of the proportion, .42 ft., the distance between the gauge lines at the heel, instead of 4.71 ft., the gauge of the track.

In Fig. 7, which comes under Case II, both curves deflect in the same direction, and the rate of their deflection from each other is equal to the rate of the deflection of a curve whose degree is equal to the difference of the degrees of the two curves from a tangent.

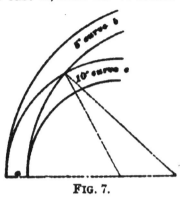

FIG. 7.

Let the main-track curve ab be 5°, and the turnout curve ac be 10°. Then, the rate of deflection or divergence of the 10° curve from the 5° curve equals the divergence of a $(10° - 5°) = 5°$ curve from a straight track or tangent.

Accordingly, we find, in the table on page 298, the tangent deflection for a 5° curve for a chord of 100 ft. $= 4.36$ ft. Denoting the required frog distance by x, we have the following proportion: $4.36 : 4.71 = 100^2 : x^2$,

whence, $x^2 = \dfrac{10,000 \times 4.71}{4.36} = 10,802.8,$

and the frog distance, $x = 103.9$ ft.

Distances are not calculated nearer than to tenths of a foot.

How to Lay Out a Switch.—In laying out a switch, locate the frog so as to cut the least possible number of rails. Where there is some latitude in the choice of location, the P. C. of the turnout curve can be located so as to bring the frog near the end of a rail.

To do this, take from the table on page 315 the frog distance corresponding to the number of the frog to be used. Locate approximately the P. C. of the turnout curve, and measure from it, along the main-track rail, the tabular frog distance. If this brings the frog point near the end of the rail, the P. C. of the turnout curve may be moved so as to require the cutting of *but one* main-track rail. Measure the total length of the frog, and deduct it from the length of the rail to be cut, marking with red chalk on the flange of the rail the point at which the rail is to be cut. Measure the width of the frog at the heel, and calculate the distance from the heel

23

to the theoretical point of frog. For example, if the width
of the frog at the heel is 8¼ in., and a No. 8 frog is to be used,
the theoretical distance from the heel to the point of frog is
8.5 × 8 = 68 in. = 5 ft. 8 in. Measure off this distance from
the point, marking the heel of the frog. This will locate the
point of the frog, which should be distinctly marked with
red chalk on the flange of the rail. It is a common practice
to make a distinct mark on the web of the main-track rail,
directly opposite the point of frog. This point being under
the head of the rail, it is protected from wear and the weather.
The P. C. of the turnout curve is then located by measuring
the frog distance from the point of frog. From the table on
page 315, we find the frog distance for a No. 8 frog is 75.3 ft.,
and the switch length, i. e., distance from P. C. of turnout
curve to heel of split switch or toe of stub switch, is 22 ft.

If a *stub switch* is to be laid, make a chalk mark on both
main-track rails on a line, marking the center of the head-
block. A more permanent mark is made with a center punch.
Stretch a cord touching these marks, and drive a stake on
each side of the track, with a tack in each. This line should
be at right angles to the center line of the track, and the
stakes should be far enough from the track not to be disturbed when putting in switch ties. Next, cut the switch
ties of proper length; draw the spikes from the track ties,
three or four at a time, and remove them from the track,
replacing them with switch ties, and tamping them securely
in place. When all the long ties are bedded, cut the main-
track rail for the frog, being careful that the amount cut off
is just equal to the length of the frog. If, by increasing or
decreasing the length of the lead 5%, it is possible to avoid
cutting a rail, do not hesitate to do so, especially for frogs
above No. 8.

Use full-length rails (30 ft.) for moving, or switch, rails,
and be careful to leave a joint of proper width at the head-
chair. Spike the head-chairs to the head-block so that the
main-track rails will be in perfect line. Spike from 8 to 11 ft.
of the switch rails to the ties, and slide the cross-rods on to
the rail flanges, spacing them at equal intervals. The cross-
rods are placed between the switch ties, which should not

be more than 15 in. from center to center of tie. The switch ties, especially those under the moving rails, should be of *sawed oak timber.* Southern pine is a good second choice. Attach the connection-rod to the head-rod and to the switch stand. With these connections made, it is an easy matter to place the switch stand so as to give the proper throw of the switch.

It is common practice to fasten the switch stand to the head-block with track spikes, but a better fastening is made with bolts. The stand is first properly placed, and the holes marked and bored, and the bolts passed through from the under side of the head-block. This obviates all danger of movement of the switch stand in fastening, which is liable to occur when spikes are used, and insures a *perfect throw.*

The use of track spikes is quite admissible when holes are bored to receive them, in which case a half-inch auger should be used for standard track spikes. The switch stand should, when possible, be placed facing the switch, so as to be seen from the engineer's side of the engine—the right-hand side.

Next stretch a cord from a, Fig. 8, a point on the outer main-track rail opposite the P. C. of the turnout curve, to b', the point of the frog. This cord will take the position of the chord of the arc of the outer rail of the turnout curve. Mark the middle point c and the quarter points d and e. Whatever the degree of the turnout curve, the distance from the middle point c of the chord to the arc $a b'$ is 1.18 ft., and the distances from the quarter points d and e are .88 ft.; hence, at c lay off the ordinate 1.18 ft., and at both d and e the ordinate .88 ft., three-quarters of the middle ordinate. These offsets will mark the gauge line of the rail $a b'$. Add to these offsets the distance from the gauge line to outside of the

FIG. 8.

rail flange, and mark the points on the switch ties. Spike a lead rail to these marks, and place the other at easy track gauge from it. Spike the rails of the turnout as far as the point of frog to exact gauge, unless the gauge has been widened owing to the sharpness of the curve. Beyond the point of frog the curve may be allowed to vary a little in gauge to prevent a kink showing opposite the frog. In case the gauge is widened at the frog, increase the guard-rail distance an equal amount. For a gauge of 4 ft. 8½ in., place the side of the guard rail that comes in contact with the car wheels at 4 ft. 6½ in. from the gauge line of the frog. This gives a space of 1½ in. between the main and guard rails.

In case the gauge is widened ¼ or ½ in., increase the guard-rail distance an equal amount.

When the turnout curve is very sharp, it will be necessary to curve the switch rails, to avoid an angle at the head-block. The lead rails should be carefully curved before being laid, and great pains should be taken to secure a perfect line.

If a *point*, or *split*, *switch* is to be laid, the order of work is nearly the same. The same precautions must be taken to avoid the unnecessary cutting of rails, with the additional precaution of keeping the switch points clear of rail joints, as the bolts and angle splices will prevent the switch points from lying close to the stock rails. As already stated, these conditions can usually be met where there is some range in the choice of the location of the switch. Where there is none, the main-track rails must be cut to fit the switch.

Having located the point of frog, the P. C. of the turnout curve, and the heel line of the switch, measure back from the heel line a distance equal to the length of the switch rails, and place on the flange of each rail a chalk mark to locate the ends of the switch points. This will also locate the head-block. Prepare switch ties of the requisite number and length, and place them in the track in proper order. As in the case of stub switches, see to it that all long switch ties are in place before cutting the rail for placing the frog; also, that the ends of the lead rails, with which the switch points connect, are exactly even; otherwise, the switch rods will be skewed, and the switch will not work or fit well. Fasten the switch rods in place, being careful to place them in their proper order, the head-rod being No. 1. Each rod is marked with a center punch, the number of the punch marks corresponding to the number of the rod.

Couple the switch points with the lead rails, and place the

sliding plates in position, securely spiking them to the ties. Connect the head-rod with the switch stand, and close the switch, giving a clear main track.

Adjust the stand for this position of the switch, and bolt it fast to the head-block. Next, crowd the stock rail against the switch-point so as to insure a close fit, and secure it in place with a rail brace at each tie; then continue the laying of the rails of the turnout.

If there is no engineer to lay out the center line of the turnout, the section foreman can put in the lead from ordinates, as explained in Fig. 8. In modern railroad practice, however, most trackwork is done under the direction of an engineer, in which case the center line of the turnout is located with a transit. This insures a correct line and expedites work. For ordinary curves, center stakes at intervals of 50 ft. are sufficient, excepting between the P. C. of the turnout and the point of frog, where there should be a center stake at each interval of 25 ft. Place a guard rail opposite the point of frog on both main track and turnout. The guard rail should be 10 ft. in length; this is an economical

FIG. 9.

length for cutting rails, as each full-length rail makes three guard rails.

Two styles of guard rails are shown in Fig. 9. That shown at B is in general use, but the style shown at A is growing in favor. The latter is curved throughout its entire length. At its middle point a, directly opposite the point of frog, the guard rail is spaced 1¼ in. from the gauge line of the turnout rail b c. From this point the guard rail diverges in both directions, giving at each end a flangeway of 4 in. This allows the wheels full play, excepting at the point of frog, where the guard rail is exactly adjusted to the track gauge, and holds the wheels in true line, preventing them from climbing, or mounting, the frog. The style of guard rail shown at B, though still much used, has two objectionable features;

viz., first, the abruptly curved ends *d* and *e* often receive an almost direct blow from the wheel flanges, which causes a car to lurch violently; and second, the flangeway of uniform width, though proper for the main track when straight, as in Fig. 9, is unsuited for sharp curves on either a main track or a turnout, as it compels the wheels to follow a curved line; whereas the normal position of the wheel base of each truck is that of a chord of, or a tangent to, the curve. These two defects alone produce what is known as a *rough-riding* frog, even though the frog is well lined and ballasted.

Location of Crotch Frog.—A *crotch,* or *middle,* frog is a frog placed at the point where the outer rails of both turnouts of

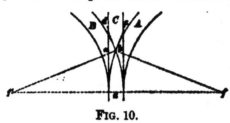

a three-throw switch cross each other. When both turnouts are of the same degree, the crotch frog comes midway between the main-track rails. Its location

FIG. 10.

and angle may be determined as follows: Let the turnout curves *A* and *B*, Fig. 10, be each 9° 30′, uniting with the main track *C* by a three-throw switch. Let *a* be the P. C. common to both curves, and *b*, the location of crotch, or middle, frog.

It is evident that the point of the crotch frog should be exactly midway between the gauge lines of the main-track rails, and if the gauge is 4 ft. 8½ in. = 4.71 ft., the point of the crotch of the frog will be $\frac{4.71}{2}$ = 2.35 ft. from each rail.

Now, the problem is to find the frog distance from *a*, the P. C., to the point *c*, where the tangent deflection will equal 2.35, or half the gauge. From the table on page 299, we find the tangent deflection of a 9° 30′ curve is 8.28 ft. Applying the principle explained in connection with Fig. 5, and letting *x* represent the required frog distance, we have the following proportion: $8.28 : 2.35 = 100^2 : x^2;$

whence, $x^2 = \dfrac{100^2 \times 2.35}{8.28} = 2,838.2 \text{ ft.,}$

and the required frog distance *x* = 53.3 feet, nearly.

Now, there are two curves starting at the common point a; the outer rails intersect at b, and the angle dbe, formed by tangents drawn at the point of intersection, is the angle of the crotch, or middle frog. The angle is equal to the sum of the angles afb and $af'b$; that is, equal to double the central angle of either curve between the P. C. and the point of intersection b. The degree of the curve is $9° 30' = 570'$, and the central angle or total deflection for each foot is $\dfrac{570'}{100}$ $= 5.7'$; and for the frog distance of 53.3 ft., the central angle is $53.3 \times 5.7' = 303.8' = 5° 03.8'$. The angle of the crotch frog is double this angle; i. e., $5° 03.8' \times 2 = 10° 07.6'$. The crotch frog should be accurately located and spiked in place before the lead rails are placed.

The one objection to the three-throw switch is the open joint at the head-block, the inevitable attendant of the stub switch, but its advantages are so great that it will continue to be used, especially in yard service.

Crossover Tracks.—A *crossover* is a track by means of which a train passes from one track to another. The tracks united are usually parallel, as are the tracks of a double-track road. Such a crossover is shown in Fig. 11. The tracks $a\,b$ and $c\,d$ are 13 ft. apart from center to center, which is the standard distance for double tracks. The crossover consists of two

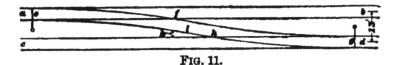

FIG. 11.

turnout curves, ef and gh. These curves are usually, though not necessarily, of the same degree. The curves terminate at the points of frog f and h, between which the track fh is a tangent. The essential point in laying out a crossover is to so place the frogs that the connecting track shall be tangent to both curves. In Fig. 11, suppose the frogs are No. 9, requiring $7° 31'$ turnout curves.

From the table on page 315, we find the required frog distance is 84.7 ft., and the switch length 25 ft. As previously noted, if there is considerable range in choice of location, the frogs can be so placed as to largely avoid the cutting of rails; but usually crossovers are required at certain precise places, and the rails must be cut as occasion demands. Having located the point of frog at f, we determine the point of the next frog at h, as follows: A No. 9 frog is one that spreads 1 in. in width to every 9 in. in length; and, as the track between the frog points is straight, the distance fh between these points will be as many times 9 in. as is the space k between the tracks at the frog point f. The main-track centers are 13 ft. apart, making the space between the gauge lines of the inside rails 8 ft. $3\frac{1}{2}$ in. As it is the rail l of the turnout that joins the second frog at h, we subtract the gauge, 4 ft. $8\frac{1}{2}$ in. from 8 ft. $3\frac{1}{2}$ in., leaving 3 ft. 7 in., the distance k, between the gauge line of the rail l, opposite the frog point f, and the gauge line of the nearest rail of the track $c\,d$. This distance multiplied by 9 in. will give the distance from the frog point f to the frog point h; 3 ft. 7 in. = 43 in.; $43 \times 9 = 387$ in. = 32 ft. 2 in. Accordingly, having located the point or frog f, we mark a corresponding point on the nearest rail of the opposite track. From this point we measure along the rail the distance 32 ft. 3 in., locating the second frog point h, and again the frog distance 84.7 ft. to the P. C. of the second turnout curve at g.

If frogs of different numbers, say 7 and 9, were to be used, the distance between the frogs is found as follows:

As the No. 7 frog spreads 1 in. in 7 in., and the No. 9 frog 1 in. in 9 in., the two will together spread 2 in. in $7 + 9 = 16$ in., or 1 in. in 8 in. Now, if the rails to be united are 3 ft. 7 in., or 43 in., apart, as in the previous problem, the distance between the frog points will be $43 \times 8 = 344$ in. = 28 ft. 8 in.

In locating crossover tracks, regard should be paid to the direction in which the bulk of the traffic moves, and the crossover tracks should be so placed that loaded cars will be backed, not pushed, from one track to the other.

At all stations on double-track roads there should be a crossover to facilitate the exchange of cars and the making up of trains.

TWO-HUNDRED-YEAR CALENDAR.

By means of the table given on the following pages, the day of the week corresponding to any date between 1752 and 1956 (new style), may be readily found. Before every leap year there is a blank space. To find the day of the week on which January 1 of any year fell, find that year in the table; glance down the column containing that year, and the day of the week at the foot of the column will be the day of the week required. Thus, to find on what day of the week January 1, 1895, fell, we find under 1895 in the table, Tuesday. For leap years, we look for day of week under the blank space before the year. Thus, January 1, 1896, fell on Wednesday, Wednesday being in the column containing the blank space before 1896. To find the day of the week for any other date, add (mentally) to the day of the month the first number under the day of the week that is contained in the column containing the year of the century; to this sum, add the number above the month at the top of the table. Find the number thus obtained in the columns of figures under the days of the week; the day of the week at the head of the column containing this number will be the day required. Thus, to find on what day of the week September 10, 1813, fell, we find 1813 in the table. The number under the day of the week in the column containing 1813 is 6, and the number above September at the top of the table is 4. Hence, $10 + 6 + 4 = 20$. The day of the week above 20 is Friday.

For dates in January and February of leap years, take one day less, or add the number beneath the day of the week under the blank space preceding the year. Thus, for February 12, 1896, we have $12 + 4 + 2 = 18$, and the day of the week above 18 is Wednesday.

Thanksgiving Day is the last Thursday in November; on what day of the month did it fall in 1897? Since the earliest day on which it can fall is the 24th, we find on what day of the week November 24 falls, and then count ahead to Thursday. Referring to the table, $24 + 6 + 2 = 32$; the day of the week above 32 is Wednesday, and since Thursday is one day later, it follows that Thanksgiving Day in 1897 fell on the 25th.

TWO-HUNDRED-YEAR CALENDAR.

3	4	5	6	0	1	2
June.	Sept. Dec.	April. July.	Jan. Oct.	May.	Aug.	Feb. Mar. Nov.
1752	1758	1754	1755		⁚ 1756	1757
1758	1759		1760	1761	1762	1763
	1764	1765	1766	1767		1768
1769	1770	1771		1772	1778	1774
1775		1776	1777	1778	1779	
1780	1781	1782	1783		1784	1785
1786	1787		1788	1789	1790	1791
	1792	1798	1794	1795		1796
1797	1798	1799	1800	1801	1802	1808
	1804	1805	1806	1807		1808
1809	1810	1811		1812	1818	1814
1815		1816	1817	1818	1819	
1820	1821	1822	1828		1824	1825
1826	1827		1828	1829	1880	1831
	1882	1888	1834	1835		1836
1837	1838	1839		1840	1841	1842
1848		1844	1845	1846	1847	
1848	1849	1850	1851		1852	1858
Sun.	Mon.	Tues.	Wed.	Thur.	Fri.	Sat.
1	2	8	4	5	6	7
8	9	10	11	12	18	14
15	16	17	18	19	20	21
22	28	24	25	26	27	28
29	80	31	32	88	84	85
86	87	88	39	40	41	42
48	44					

3	4	5	6	0	1	2
June.	Sept. Dec.	April. July.	Jan. Oct.	May.	Aug.	Feb. Mar. Nov.
1854	1855		1856	1857	1858	1859
	1860	1861	1862	1863		1864
1865	1866	1867		1868	1869	1870
1871		1872	1873	1874	1875	
1876	1877	1878	1879		1880	1881
1882	1883		1884	1885	1886	1887
	1888	1889	1890	1891		1892
1893	1894	1895		1896	1897	1898
1899	1900	1901	1902	1903		1904
1905	1906	1907		1908	1909	1910
1911		1912	1913	1914	1915	
1916	1917	1918	1919		1920	1921
1922	1923		1924	1925	1926	1927
	1928	1929	1930	1931		1932
1933	1934	1935		1936	1937	1938
1939		1940	1941	1942	1943	
1944	1945	1946	1947		1948	1949
1950	1951		1952	1953	1954	1955
Sun.	Mon.	Tues.	Wed.	Thu.	Fri.	Sat.
1	2	3	4	5	6	7
8	9	10	11	12	13	14
15	16	17	18	19	20	21
22	23	24	25	26	27	28
29	30	31	32	33	34	35
36	37	38	39	40	41	42
43	44					

In England the new-style calendar was adopted in September, 1752, by making September 3 legally September 14, in order to allow for the error in the Julian calendar, which went into use 45 B. C. According to the Julian calendar, every fourth year was made a leap year, with the result that the Julian year was a trifle longer than the true year, as measured by the time it takes the earth to make a complete circuit of its orbit. The new style, or Gregorian, calendar allows for this error by making every secular year (a secular year is one divisible by 100, as 300, 1400, 1900, etc.) a common year unless it is divisible by 400, in which case it is a leap year. Hence, the years 400, 800, 1200, 1600, and 2000 are leap years, while the other secular years preceding 2000 are common years. In 1752 the seasons had been advanced 11 days, and to correct this, 11 days were dropped by changing September 3 to September 14. The change was greatly opposed by the people and for many years afterwards, it was customary to use two dates; or when one date was used to annex the letters N. S. or O. S. to the date in order to signify whether the date was new style or old style. Thus, George Washington was born on February 22, 1732 (N. S.) or February 11, 1732 (O. S.). To find what day of the week this was, proceed as follows: $1752 - 1732 = 20$; $20 + 4 = 5$, the number of leap years between 1732 and 1752. Divide the sum of 20 and 5 by 7 and count the remainder backwards from 1752; thus $(20 + 5) + 7 = 3 + 4$ remainder, and counting backwards 4 columns from the right we stop at the column headed 1755. This operation indicates that if the table continued backwards to 1732, the year 1732 would occur in the column headed 1755. Since 1732 was a leap year, we use the preceding column, and $3 + 22 + 2 = 27$; hence, February 22, 1732 (N. S.) was Friday.

MEMORANDA

MEMORANDA

MEMORANDA

MEMORANDA

Promotion
Advancement in Salary
and
Business Success

**Secured
Through the**

Mechanical Drawing, Mechanical Engineering, Shop Practice, Engine and Dynamo Running, Steam-Electric, and Gas Engines

COURSES OF INSTRUCTION

OF THE

International Correspondence Schools

International Textbook Company, Proprietors

SCRANTON, PA., U. S. A.

SEE FOLLOWING PAGES

24

From $45 to $300 a Month

In 1906 I was working in a large steel plant, for the small salary of $45 a month. I saw opportunities of responsible positions open up before me almost daily, but was without the training to take such a position. Knowing of the I. C. S., I decided to take up a Course and push for a better position. Two months after enrolment I was rewarded with an increase in wages. In November, 1906, I was offered a position in the same plant at a salary of $90 a month. January 30, 1907, I accepted a position at $115 a month; and, in July of the same year, my salary was increased to $150. In the meantime I completed my Course and received my Diploma. In October, 1909, I again accepted a position, netting me $175 a month; and on June 1, 1910, my salary was increased to $300 a month. I can only find words of the highest praise for the training the I. C. S. affords, and will gladly lend my services to the grand cause that has done me so much good.

O. J. Gibboney, Erecting Engineer,
330 Eagle St., Mt. Pleasant, Pa.

OWES MUCH TO THE I. C. S.

J. G. BENEDICT, Waynesboro, Pa., had been a country-school teacher before enrolment for a Mechanical Course. He is now treasurer and general manager of the Landes Machine Company. He feels that he owes much to the I. C. S. and has been instrumental in persuading a number of his employes to enroll.

STUDY BROUGHT PROMOTION

When GEO. C. KIMMEL, 4654 Edgewood Ave., Winton Place, Cincinnati, Ohio, enrolled with the Schools for a Complete Mechanical Course he was working in the shops. He says that his Course kept him out of saloons and advanced him from the drafting room to the position of chief designer and then to vice-president and works manager of the Cincinnati Grinder Company. His salary has been advanced several hundred per cent.

MASTER MECHANIC AT $2,000 A YEAR

REUBEN BASTOW, Bay City, Mich., quit school at 13 years of age. He was earning $15 a week as a machinist when he enrolled with the I. C. S. for a Mechanical Course. He is proud of his I. C. S. education, which has made him master mechanic of the Michigan Chemical Company at a salary of $2,000 a year.

MANY TIMES HIS FORMER SALARY

W. R. C. MILLER, 1421 E St., Lincoln, Neb., says he was a common laborer when he enrolled with the I. C. S. for the Mechanical Engineering Course. When he had finished Mechanical Drawing he obtained a position by showing one of his plates. He is now employed by the C. B. & Q. R. R. Company, in the maintenance-of-way department, at a salary several times greater than what he received at the time of enrolment.

BECAME GENERAL MANAGER

A. R. LENTZ, Jackson, Mich., was earning $10.50 a week as a patternmaker when he took up an I. C. S. Mechanical Course. This proved of inestimable benefit to him, since he has become general manager of the Central Foundry, receiving $2,000 a year.

PRAISES THE I. C. S.

C. B. FARQUHARSON, Tulsa, Okla., praises the I. C. S. as the bridge which carried him safely to success. His Mechanical Course gave him the start in business which resulted in his becoming proprietor of the Tulsa Boiler and Sheet-Iron Works.

3

An I. C. S. Diploma a Sufficient Recommendation

At the age of 35, I enrolled with the Schools for a Complete Mechanical Course, from which I graduated. Later on I took the Gas Engines Course. When I enrolled I was working as a traveling salesman, at $1,000 a year. I did a great deal of studying on trains, but soon gave up the road to accept a position in the drafting room. I am acting as consulting engineer for the Carroll Foundry and Machine Company, at a salary of $3,000, which position does not take all my time; and I am doing some expert work for other companies. I am the inventor of the Hallett Tandem Gas Engine built by the Carroll Company. I cannot say too much in praise of the Schools, which have done everything for me. Any person applying to me for a position having a Diploma from the I. C. S. would be sufficiently recommended.

W. E. HALLETT, Bucyrus, Ohio

350 PER CENT. GREATER

A. T. ANDERSON, 14th and Cass Ave., St. Louis, Mo., enrolled for the Mechanical Drawing Course while he was working as a stationary fireman. He is now the president and manager of the Anderson Company, electric grinders and drills. He recommends the Course to any one desiring to obtain a thorough knowledge along his chosen line, since it has enabled him to handle his own business in the manufacture of electric machinery. His income has increased some 350 per cent.

NOW MANAGER—SALARY DOUBLED

LOUIS J. BIGNELL, Hill City, S. Dak., was running a steam engine at the time he enrolled for the Mechanical Drawing Course. He says he cannot thank the Schools enough, since he is considered the best draftsman in his part of the country. He is now manager of the Hill City Electric Light and Power Company, earning twice what he did at the time of enrolment.

HIS COURSE BROUGHT PROMOTION

CHARLES BENNET, 614 N. Franklin St., Saginaw, Mich., was working in a furniture factory at $1.50 a day with no hope of promotion at the time he enrolled for his Mechanical Drawing Course. He now has charge of all Babbitt work on passenger and freight engines in the Saginaw round house and shops of the Pere Marquette Railroad. His salary is about $210 a month.

SALARY INCREASED 500 PER CENT.

When HARRY BAILEY, 7113 E. 17th St., Kansas City, Mo., enrolled for the Mechanical Drawing Course, he was earning 70 cents a day as an apprentice machinist. He is now partner in a machine shop and his earnings have increased more than 500 per cent. He ascribes his success largely to his Course.

IN BUSINESS FOR HIMSELF

ARTHUR D. BAUM, 204 S. Franklin St., Kirksville, Mo., laid a good foundation for his later business life when he obtained his diploma in the Mechanical Drawing Course. After working for a time as a draftsman, he went into business for himself in the heating and plumbing line and immediately enrolled for the Complete Plumbing and Heating Course. He now carries on a successful and growing business and he attributes his success to his study done during spare moments.

RANCH HAND TO PROPRIETOR

R. L. BALLARD, Orange, Calif., worked as a ranch hand when he enrolled for the Mechanical Drawing Course. He is now proprietor of a garage and machine shop.

From $2 a Day to $3,000 a Year as Inspector

Being one of your students, it will interest you to hear that I have been appointed State Boiler Inspector.

I enrolled for the Mechanical Course while employed as a boilermaker, at $2 a day, and the Course has fitted me for the position I now hold, as it has given me the necessary education to pass successfully the very rigid Civil Service examination, without which I could not be appointed. My position is that of Inspector of Boilers and Engines, Bureau of Navigation, State of New York, at a salary of $3,000 a year and expenses.

I had a fair education in my native country, but received no school instruction in this country, except through the I. C. S., which I think speaks well for your Schools. I believe the methods of instruction are such as to lead a student easily and gradually to success.

As your Schools have been so large a factor in my success I venture to say no young man with ambition can afford to be without a Scholarship in the I. C. S.

G. C. WEHLING,
418 W. Dominick St., Rome, N. Y.

ROSE TO BE SUPERINTENDENT

F. A. DAGLES, Sheldon Springs, Vt., rose to be superintendent through our instruction, although having very little education when he enrolled for our Mechanical Course. He is superintendent of the Missisquoi Pulp Company at a salary 500 per cent. larger than when he began to study.

GRATIFYING ADVANCEMENT

HERMAN A. FREYLER, 1313 W. Benton Ave., Helena, Mont., was earning $2.50 a day working around the mines when he enrolled for the Steam-Electric Course. When he had nearly finished the steam part of his Course he was able to install and operate any kind of mining and milling machinery and his pay as an erecting engineer was $5 a day.

400 PER CENT. LARGER

E. KARL WHITENER, Edgemont, N. C., was working as a machinist at the time he enrolled for the Shop Practice Course. He is now foreman in the shops of the Carolina & Northwestern Railway Company. Mr. Whitener gives great credit for his advancement to his Course, which has also increased his salary about 400 per cent.

NOW FOREMAN

CARL AHLSTROM, 8 Leeds St., Worcester, Mass., enrolled for the Shop Practice Course while in the employ of the Norton Grinding Company. He now holds the position of foreman and he declares that his success is largely due to the excellent training he received from the I.C.S.

LABORER TO FOREMAN

When EARL BAYLESS, 1235 E. Wheeler St., Macomb, Ill., enrolled for the Shop Practice Course, he was employed as a laborer. He is now shop foreman in the employ of the Illinois Electric Porcelain Company, with a corresponding increase in salary.

HOLDS A GOOD POSITION

GEORGE HEALD, 427 Adams St., Buffalo, N. Y., had just started to learn the machinists' trade and was receiving $4.50 a week at the time he enrolled for the Shop Practice Course. This enabled him to advance more rapidly than the other boys in the shop and in the end secured for him the position of teacher in machine shop practice at the Seneca Vocational School, where he nows earns $1,500 a year.

7

Now General Manager and Secretary

When I enrolled with the I.C.S. I was working as an apprentice machinist at $5 a week. Within a month after finishing my apprenticeship I was recommended for a position as draftsman by the superintendent of the shop where I was employed. There I completed my Course and received my diploma. I cannot speak too highly of the I.C.S. and their method of instruction, and I have been glad to recommend the Schools to many who have been in my employ. After becoming head draftsman in the employ of the Deere & Mansur Company, I organized the Moline Tool Company, manufacturers of gang drilling and boring machinery. I am now secretary and general manager of this company.

WILSON P. HUNT,
Moline, Ill.

9

IN BUSINESS FOR HIMSELF

Geo. J. Spengler, Seemsville, Siegfried, R. D. No. 2, Pa., was serving as an apprentice machinist at 90 cents a day when he enrolled for the Shop Practice Course. This enabled him to become foreman for the Atlas Portland Cement Company, and later to open a machine shop of his own, having two men regularly at work for him.

BECAME FOREMAN

John A. Brewer, 1416 Osborne Court, Niagara Falls, N. Y., declares that he would not be where he is today if it had not been for the help of his Shop Practice Course. At the time of his enrolment he was just a machinist having a very meager common-school education. His Course has enabled him to become foreman in a shop employing at times as many as fifty machinists and his pay is more than twice what it was at the time of his enrolment.

NOW PROPRIETOR

H. R. Chase, Fortuna, Calif., worked as a carpenter at the time he enrolled for the Shop Practice Course. His object was to be able to go into business for himself. He is now partner in the Eel River Garage, and he ascribes his success to his Course.

NOW SUPERINTENDENT

P. Bendixen, Bettendorf, Iowa, enrolled for the Mechanical Engineering Course and also for the Shop Practice Course while employed as a machine shop foreman. He is now superintendent for the Bettendorf Company. He highly recommends the I.C.S.

TOOK IT FROM HIS WIFE

When D. R. Noonan, 220 W. Washington St., Paris, Ill., enrolled for the Shop Practice Course, he says that he took the money for his first payment from his wife. At the time he was working 14 hours a day for $1.25. He is most enthusiastic in his praise of the Schools which have enabled him to become the proprietor of his own machine shop, the best equipped in his part of the state. He has also the leading automobile business in his county. He ascribes all this to the I.C.S.

Master Mechanic at $1,800 a Year

Had it not been for the opportunity of home study, the very best of instruction, and textbooks that, to my knowledge are unequaled by those of any other school, it would have been impossible for me to have attained the mechanical prestige I now enjoy with my present employers, the Hoosac Cotton Mills, of North Adams, Mass. Through the aid of the Bound Volumes I equipped myself with a thorough knowledge of the theoretical points of steam sufficient to pass examination for first-class engineer's license before the State Board of Engineers. I have now held the position of master mechanic for nearly 4 years, and at present am receiving a salary of $1,800 a year.

FRANK R. BROWN,
199 E. Quincy St., North Adams, Mass.

PRAISES HIS COURSE

CLIFFORD HIGBY, Idaho City, Idaho, declares that his I.C.S. Dynamo Running Course is the best in the world, since it enables him to handle any position in the electrical field. He is now chief electrician for the Boston-Idaho Gold Dredging Company besides having all the city lights and meters to look after. His salary is $135 a month.

A MEMBER OF THE FIRM

W. E. STEED, Tremont, Utah, was working on the farm when he enrolled for the Dynamo Running Course. He says that the I.C.S. have enabled him to become an equal partner in the Steed Brothers garage and general machine shop. In his new position he has greatly increased his income.

SALARY DOUBLED

ROY E. FOSTER, Crown Hill, W. Va., was working as a fireman when he enrolled for the Dynamo Running Course. He recommends this Course because it has enabled him to take charge of the power plant of the National Bituminous Coal and Coke Company, increasing his salary at least 100 per cent.

PASSED A SUCCESSFUL EXAMINATION

AUGUST BEHRSIN, care Cottonwood Lumber Co., De Roche, B. C., Canada, was working as a fireman at the time he enrolled for the Engine and Dynamo Running Course. Although at the time he understood but very little English, he had no trouble in completing his course which has enabled him to pass successfully both U. S. and Canadian Civil Service examinations. His wages have been more than doubled.

SALARY $4,000

C. H. BURROWS, 234 Oneida St., Fulton, N. Y., had left school at 9 years of age and was working in a paper mill at $2 a day when he enrolled with the Schools for a Mechanical Course. He says that his Course has advanced him to the position of general superintendent of the Victor Paper Mills Company, at a salary of $4,000 a year, and has also made it possible for him to take charge of a department of the *Paper Trade Journal*.

SALARY NEARLY DOUBLED

C. C. ELLISOR, Sabine, Tex., was earning $55 a month when he first enrolled with the I.C.S. He says that his Mechanical Drawing Course has been of great help to him. He is now foreman for the Union Sulphur Company at a salary of $150 a month.

Increased Salary Over 1,000 Per Cent.

At the time I first enrolled for a Mechanical Engineering Course in your Schools, I was working in a bake shop. My present employment is that of chief engineer of the Holyoke Street Railway power house, at an increase of over 1,000 per cent. I obtained the greater part of my education after marriage. I attribute my success entirely to your instruction. Of course I do not mean to say that I have not had to do anything; I have studied hard and worked hard. Any one who will apply himself diligently to one of your Courses is assured of success from the start. I have regularly 18 men under my control, and during this summer have had 12 extra men most of the time.

RALPH F. BLANCHARD,
21 Walter St., Willimansett, Mass.

SHORTER HOURS—BETTER PAY

Wm. M. Webster, 719 S. Los Angeles St., Los Angeles, Cal., was working as a night engineer, earning $2 for 12 hours' work, when he enrolled for our Steam-Electric Course. He had attended public school only three terms, but he was able to pursue his Course without difficulty. He is now the engineer for the Anchor Laundry, earning $5 for 10 hours' work.

PRAISES THE SCHOOLS

L. H. Brown, Lynn, Mass., was earning $1.50 a day as a coal passer when he subscribed for the Steam Engineering Course. He praises the Schools for advancing him to the position of chief engineer in the Lynn Power Station of the Boston & Northern Railway Company, having 19 men under his direction. His salary has increased more than 200 per cent.

NOW SUPERINTENDENT

Chas. E. Beckwith, Sprague, Wash., had just finished the eighth grade in common school and was running a small portable engine during vacation, earning $45 a month, when he enrolled for the Steam-Electric Course. Since graduating, he has advanced through various stages to the position of superintendent and local manager at Sprague, for the Big Bend Light and Power Company, at a salary which is substantially more than 100 per cent. larger than what he received at the date of enrolment.

DOUBLED HIS FORMER WAGES

Howard J. Ande, Bitumen, Pa., was working as an oiler when he enrolled for the Mechanical Engineering Course. Since obtaining his diploma he was given charge of an electric power plant, receiving twice the wages he did at the time of enrolment.

BECAME FOREMAN—SALARY DOUBLED

M. L. Cope, 923 Hazel St., Akron, Ohio, was employed as an ordinary rubber worker when he enrolled for the Mechanical Engineering Course. Since acquiring the study habit he has helped his employers to get out several patents. He is now a foreman, earning double what he was paid at the time of enrolment.

EARNS MORE THAN $2,000 A YEAR

James Crombie, 51 Bissell Ave., E., Oil City, Pa., left school at the age of 12 to go to work. When he enrolled for the Mechanical Engineering Course, he was earning twenty-four cents an hour, as a boilermaker. His Course has enabled him to become a frequent contributor to the trade journals and he is now general foreman boilermaker in the Oil City Boiler Works, earning more than $2,000 a year.

A´Prosperous Inventor

I landed in this country some years previous to my enrolment—a green Scandinavian. For some time I worked at various trades, doing the best I could, but finding advancement slow. At all times I felt the need of a technical education. When I heard of the I. C. S. I was working as a machinist, and though advanced in years, I decided to enroll at once. The education received enabled me to become foreman, and then superintendent of that shop. While working I kept thinking, and in due time invented a level on which I secured three patents. I interested capital in the invention and resigned my position to become president of my own company. We are doing well and have sold between three and four thousand instruments, which are giving perfect satisfaction. Your system is a blessing to all working people.

E. A. BOSTROM,
Atlanta, Ga.

14

HIS COURSE DID IT

Noticing that the man with technical knowledge was given the best position, even over the expert workman, JAS. E. HANCOCK, 217 DeGrassi St., Toronto, Ont., Canada, determined to get in line by enrolment for the Mechanical Engineering Course. Fourteen months later he became machine-shop foreman for the Fletcher Manufacturing Company, a position which he could not hold without the help of his I.C.S. Course. His salary has been largely increased.

400 PER CENT. INCREASE

IRA I. HOVES, 1423 Milvia St., Berkeley, Calif., was working as a draftsman when he enrolled for the Mechanical Engineering Course. He is now chief estimator and outside salesman for the Jarvis Crude Oil Burner Company at a salary exactly 400 per cent. greater than what he received at the time of enrolment.

NOW MASTER MECHANIC

HARRY T. McCULLEY, 92 Catherine St., Lyons, N. Y., had not received even a common-school education when he enrolled for the Mechanical Engineering Course. At that time he was working as a steam engineer. He has now complete charge of the new steel hull dredge of the Crowell-Sherman-Stalter Company, general contractors, being also chief engineer and master mechanic. His salary is $125 a month.

NOW PROPRIETOR

THEO. A. RUNYAN, 913 McPherson St., Elkhart, Ind., declares that no one needs any better qualifications than can be obtained through the I.C.S. Mechanical Engineering Course. Mr. Runyan says that his Course enables him to operate his own factory, making fine tools and special machinery. It has also largely increased his income.

A MEMBER OF THE FIRM

F. M. CHEESEMAN, Freeport, Ill., was working as a bookkeeper and stenographer at the time of his enrolment for the Mechanical Drawing Course. He is now secretary and treasurer of the Northern Steel and Concrete Company, Inc.

RAPID ADVANCEMENT

FOREST L. CARPENTER, Battle Creek, Mich., enrolled for the Mechanical Drawing Course while working on a farm for $15 a month. Although he was only 15 years old at the time and had nothing but a common country-school education, he was able by the time he became 18 to take a position with the Kellogg's Toasted Corn Flake Company, at a salary 320 per cent. larger than when he enrolled.

265 PER CENT. INCREASE

J. L. BROOKS, Roanoke, Va., was working as helper in a riveting gang when he enrolled for the Mechanical Drawing Course. He says that a little push on his part together with the training gained from his Course have made him the general foreman of the assembling and riveting department for the Virginia Bridge and Iron Company, increasing his salary 265 per cent.

THE I.C.S. PLACED HIM AT THE TOP

L. T. RASMUSSEN, 704 S. 4th St., Council Bluffs, Iowa, was working as a common laborer when he enrolled for a Mechanical Drawing Course. When his employer knew that he was taking I.C.S. instruction he gave him an opportunity to advance. He is now superintendent of the machine shop and foundry of the Walker Manufacturing Company.

CLIMBED FROM THE BOTTOM RUNG

FRED W. DOLL, 1422 Liberty St., Allentown, Pa., was on the bottom rung of the ladder in the Balliet box factory at the time of his enrolment for the Mechanical Drawing Course. He says that his climb upward was swift and steady, thanks to the I.C.S. He is now foreman of the factory, with an increase in earnings of 150 per cent.

HIS BEST INVESTMENT

JOSEPH C. WILSON, 59 Newport St., Lynn, Mass., considers his I.C.S. Course in Mechanical Engineering his best investment, since it raised him from a place in the coal mines to the position of tool designer for the United Shoe Manufacturing Company, Beverly, Mass. His income has been increased from $10 to $29.50 a week.

250 PER CENT. LARGER

HENRY J. GUTH, Box 363, Denver, Colo., was earning $10.50 a week as a patternmaker when he enrolled for the Mechanical Drawing Course. He is now supervisor of the patternmaking and wood-work division in the Colorado School of Mines. His salary is 250 per cent. larger than when he enrolled.

300 PER CENT. INCREASE

A machinist's apprentice named C. C. Barber, 1070 47th St., Brooklyn, N. Y., enrolled with the I.C.S. for the Draftsmen's Course. He is now employed by the *Railroad Age* as a draftsman with an increase in salary of over 300 per cent.

NOW WORKING FOR THE GOVERNMENT

HARRY B. BOTHWELL, 1904 G St., N. W., Washington, D. C., was an apprentice draftsman at $4 a week when he enrolled with the I.C.S. His studies in the Mechanical Engineering Course have secured for him a position as skilled draftsman under Civil Service rules, at a salary of $1,000 a year.

16

CPSIA information can be obtained
at www.ICGtesting.com
Printed in the USA
LVHW101036151219
640545LV00007BA/210/P